**Current Topics in
Developmental Biology**
Volume 53

Current Topics in Developmental Biology

Volume 53

Edited by

Gerald P. Schatten

Director, PITTSBURGH DEVELOPMENT CENTER
Deputy Director, Magee-Womens Research Institute
Professor and Vice-Chair of Ob-Gyn-Reproductive
Sci. & Cell Biol.-Physiology
University of Pittsburgh School of Medicine
Pittsburgh, PA 15213

ACADEMIC PRESS

An imprint of Elsevier Science

Amsterdam   Boston   London   New York   Oxford   Paris
San Diego   San Francisco   Singapore   Sydney   Tokyo

Academic Press
An imprint of Elsevier Science.
525 B Street, Suite 1900, San Diego, California 92101-4495, USA
http://www.academicpress.com

Academic Press
84 Theobald's Road, London WC1X 8RR, UK
http://www.academicpress.com

International Standard Book Number: 0-12-153153-8

PRINTED IN THE UNITED STATES OF AMERICA
02   03   04   05   06   07   MM   9   8   7   6   5   4   3   2   1

Contents

1

Developmental Roles and Clinical Significance of Hedgehog Signaling
Andrew P. McMahon, Philip W. Ingham, and Clifford J. Tabin

2 _____

Genomic Imprinting: Could the Chromatin Structure Be the Driving Force?
Andras Paldi

3 _____

Ontogeny of Hematopoiesis: Examining the Emergence of Hematopoietic Cells in the Vertebrate Embryo
Jenna L. Galloway and Leonard I. Zon

4 _____

Patterning the Sea Urchin Embryo: Gene Regulatory Networks, Signaling Pathways, and Cellular Interactions
Lynne M. Angerer and Robert C. Angerer

Contributors

Numbers in parentheses indicate the pages on which authors' contributions begin.

Lynne M. Angerer (159), Department of Biology, University of Rochester, Rochester, New York 14627

Robert C. Angerer (159), Department of Biology, University of Rochester, Rochester, New York 14627

Jenna L. Galloway (139), Division of Hematology/Oncology, Harvard Medical School and Howard Hughes Medical Institute, Children's Hospital, Boston, Massachusetts 02115

Philip W. Ingham (1), MRC Intercellular Signalling Group, Centre for Developmental Genetics, School of Medicine and Biomedical Science, University of Sheffield, Western Bank, Sheffield S10 2TN, United Kingdom

Andrew P. McMahon (1), Department of Molecular and Cellular Biology, Harvard University, Cambridge, Massachusetts 02138

Andras Paldi (115), Institut Jacques Monod, CNRS, École Practique des Hautes Études, 75005 Paris, France

Clifford J. Tabin (1), Department of Genetics, Harvard Medical School, Boston, Massachusetts 02115

Leonard I. Zon (139), Division of Hematology/Oncology, Harvard Medical School and Howard Hughes Medical Institute, Children's Hospital, Boston, Massachusetts 02115

Preface

This volume of *Current Topics in Developmental Biology* is exceptional in many ways. Developmental biology has exploded in so many directions in past decades and we are now witnessing the emergence of the new field of developmental medicine—the fundamental, translational, and clinical convergence of developmental biology and clinical medicine, especially molecular medicine. Astonishing discoveries in developmental medicine as well as developmental biology are presented in thoughtful and forward-looking articles.

The Developmental Roles and Clinical Significance of Hedgehog Signaling by Andy McMahon of Harvard University, Phil Ingram of the University of Sheffield and Chris Tabin of Harvard Medical School is an extraordinary treatise that will soon be considered of historic importance for its comprehensive and integrated considerations of the *hedgehog* genes and their gene products in invertebrate and vertebrate development, and especially in human development and clinical diseases and disorders ranging from birth defects, through cancers, to most every organ system.

Genomic Imprinting: Could the Chromatin Structure Be the Driving Force? by Andras Paldi of the Institut Jacques Monod examines another important regulator in development biology also of direct clinical importance: genomic imprinting. While we think of DNA as composed of four nucleotides (ATCG), mammals have a fifth—methylated cytosine. This 5'-MeC in mammals remains mysterious but of crucial importance in human and mammalian development and disease.

The Ontogeny of Hematopoiesis: Examining the Emergence of Hematopoietic Cells in the Vertebrate Embryo by Jenna Galloway and Leonard Zon of the Harvard Medical School and the Howard Hughes Medical Institute considers another new topic in developmental medicine—stem cell ontogeny, especially in the hematopoeisis.

Patterning the Sea Urchin Embryo: Gene Regulatory Networks, Signaling Pathways, and Cellular Interactions by Lynne Angerer and Robert Angerer of the University of Rochester presents the exceptional investigations on body axis determination and patterning. These fundamental decisions on dorsal-ventral, left-right, and anterior-posterior (or their equivalents in other systems) are absolutely crucial and molecular determinants, especially β-catenin, are vital.

Together with other volumes in this Series, this volume provides a comprehensive survey of major issues at the forefront of modern developmental biology and developmental medicine. These chapters will be valuable to clinical and fundamental researchers in the fields of developmental biology and developmental medicine, as well as to students and other professionals who want an introduction to current topics in cellular and molecular approaches to developmental biology and clinical problems of aberrant development. This volume in particular will be essential reading for anyone interested in stem cells, signaling, medical implications of developmental determinants, hematopoeisis, axis specification, and molecular genetics of development.

This volume has benefited from the ongoing cooperation of a team of participants who are jointly responsible for the content and quality of its material. The authors deserve the full credit for their success in covering their subjects in depth, yet with clarity, and for challenging the reader to think about these topics in new ways. The members of the Editorial Board are thanked for their suggestions of topics and authors. I also thank Michelle Emme and Kimberly Wagle for their exemplary administrative support. Finally, we salute the Pittsburgh Development Center of Magee-Womens Research Institute at the University of Pittsburgh School of Medicine for providing intellectual and infrastructural support to *Current Topics in Developmental Biology*.

Jerry Schatten

1

Developmental Roles and Clinical Significance of Hedgehog Signaling

Andrew P. McMahon,[1], Philip W. Ingham,[2]
and Clifford J. Tabin[3]*
[1]Department of Molecular and Cellular Biology
Harvard University
Cambridge, Massachusetts 02138

[2]MRC Intercellular Signalling Group
Centre for Developmental Genetics
School of Medicine and Biomedical Science
University of Sheffield
Western Bank, Sheffield S10 2TN, United Kingdom

[3]Department of Genetics
Harvard Medical School, Boston, Massachusetts 02115

*To whom correspondence should be addressed. E-mail: amcmahon@mcb.harvard.edu.

Current Topics in Developmental Biology, Vol. 53

Cell signaling plays a key role in the development of all multicellular organisms. Numerous studies have established the importance of Hedgehog signaling in a wide variety of regulatory functions during the development of vertebrate and invertebrate organisms. Several reviews have discussed the signaling components in this pathway, their various interactions, and some of the general principles that govern Hedgehog signaling mechanisms. This review focuses on the developing systems themselves, providing a comprehensive survey of the role of Hedgehog signaling in each of these. We also discuss the increasing significance of Hedgehog signaling in the clinical setting.   © 2003, Elsevier Science (USA).

I. Introduction

Genetic screens in *Drosophila* have been extremely successful in uncovering the molecular pathways regulating development. As the genes involved in those pathways first began to be identified, it quickly became apparent that many had highly related sequences in the vertebrate genome, providing starting points for investigating problems in vertebrate embryology that were previously intractable. Of the vertebrate genes discovered in such a manner thus far, perhaps none have provided as many new inroads into such a plethora of developmental phenomena or led to so many fundamental insights as the vertebrate homologs of the *Drosophila* segment polarity gene *hedgehog* (*hh*).

hh was first identified in a screen for mutations disrupting the patterning of the *Drosophila* larva (Nusslein-Volhard and Weischhaus, 1980). Genetic arguments strongly suggested that the Hh protein could act as an intercellular signal

(Ingham *et al.,*1991; Mohler, 1988), an idea that was confirmed when the gene was finally cloned (Lee *et al.,*1992; Mohler and Vani, 1992; Takata *et al.,*1992; Tashiro *et al.,* 1993). Rapidly, vertebrate homologs of *hh* were reported from chick, mouse, frog, zebrafish, and human (Chang *et al.,* 1994; Echelard *et al.,* 1993; Ekker *et al.,* 1995a,b; Krauss *et al.,* 1993; Marigo *et al.,* 1995; Riddle *et al.,* 1993; Roelink *et al.,* 1994). Subsequently, *hh* homologs have also been described in crickets (Inoue *et al.,* 2002) and the cephalochordate *Amphioxus* (Shimeld, 1999).

The discovery of this new and important class of signaling molecules opened the door to studies of the biological events they regulate as well as the biochemical pathways they employ. While Hedgehog signal transduction has been extensively reviewed (most recently by Ingham and McMahon, 2001), the broad topic of Hedgehog biology has not been addressed on a comprehensive scale. The aim of this review is to attempt to cover each developmental system in which Hedgehog proteins act in sufficient depth to provide an entry point for readers interested in various developmental roles of this family. In addition, in the final two sections we review the role of Hedgehog signaling in the etiology of human congenital abnormalities and a variety of cancers.

Because the field of Hedgehog signaling has been extensively reviewed on a biochemical level, we do not discuss Hedgehog signal transduction in depth; the reader is referred instead to a review that covers this area (Ingham and McMahon, 2001). Nonetheless, we briefly describe the molecular players in the Hedgehog signaling pathway as a glossary for this review (Fig. 1; see color insert) (see Ingham and McMahon, 2001, for details and references cited therein).

There are three distinct *hedgehog* genes in mammals and birds, *Sonic hedgehog* (*Shh*), *Indian hedgehog* (*Ihh*), and *Desert hedgehog* (*Dhh*). These represent three different subfamilies, some of which have been expanded by gene duplication in various other taxa. Thus, zebrafish has two *Shh* subfamily members, *twhh* and *shh* and two *Ihh* subfamily members, *qhh* and *ehh*. In addition, *Ihh* is given the name *bhh* in *Xenopus,* and *Xenopus* has two *Dhh* subfamily members, *hh4* and *chh.*

All Hedgehog proteins undergo autocatalytic cleavage to yield a highly conserved N-terminal mature signaling peptide that is tethered to cholesterol at its C-terminal end and is also palmitoylated at its N-terminal cysteine. In *Drosophila,* release of Hh from the cells that produce it depends on the activity of Dispatched, a multipass transmembrane protein. Movement of cholesterol-modified Hh protein across multiple cells in *Drosophila* also requires the expression of *tout-velu* in cells beyond those producing Hh. *tout-velu* encodes a glycosaminoglycan transferase, implicating proteoglycans in Hedgehog movement. Although their equivalent role in transport of vertebrate Hedgehog proteins has not been demonstrated, there are several vertebrate homologs of Tout-velu that are members of the EXT family of glycosyltransferases.

The transmembrane receptor for Hh is a multipass membrane protein, Patched (Ptc), of which there are two vertebrate homologs, Ptc1 and Ptc2. Ptc1 appears to play a role equivalent to that of the *Drosophila* Ptc protein, whereas the function of Ptc2 is currently unclear. Generally speaking, Ptc plays at least two critical roles in this pathway. First, in the absence of bound Hedgehog ligand, it negatively regulates the signaling activity of a second downstream effector molecule called Smoothened (Smo). (Note that there is a single vertebrate homolog of *Smo, SMOH,* in humans.) When bound to Hedgehog protein, however, Ptc is functionally inactivated, allowing Smo to signal. Inactivation of Ptc thus results in the cell autonomous constitutive activation of the Hh response pathway. A second key role of Ptc is to limit the range of diffusion of Hedgehog proteins by binding and sequestering them. A second, unrelated protein called hedgehog-interacting protein (HIP) may also share this role in vertebrates.

Both *Ptc* and *Hip* are themselves transcriptional targets of Hedgehog signaling, providing a feedback loop limiting the range of Hedgehog activity. Downstream of *Smo* is a complex of proteins including the positively acting Fused (fu) and negatively acting Suppressor of Fused [Su(fu)], costal2 (Cos2), protein kinase A (PKA), and *slimb*. Through a complex series of interactions involving phosphorylation, proteolytic cleavage, and association with microtubules, this complex acts to regulate the processing and subcellular localization of a transcription factor encoded by the *Cubitus interruptus* (*Ci*) gene. The unprocessed Ci protein acts as a transcriptional activator whereas a proteolytically cleaved form acts as a repressor. *Gli1, Gli2,* and *Gli3* are vertebrate homologs of *Ci. Gli1* is itself transcriptionally activated by Hedgehog signaling whereas both *Gli2* and *Gli3* are repressed by Hedgehog signaling. When produced, Gli1 and Gli2 both act as Hedgehog-dependent transcriptional activators. Conversely, Gli3 acts predominantly as a Hedgehog-regulated transcriptional repressor, and Gli2 may also possess transcriptional repressor activity. These proteins have thus adopted different aspects of the regulatory functions exhibited by Ci.

The target genes regulated by these transcription factors vary between organisms and cell types. However, it is worth noting that the *TGF-β* family member *decapentaplegic* (*dpp*) and its vertebrate homologs *Bmp2* and *Bmp4* are downstream targets of Hedgehog signaling in a variety of systems. Although there are places (such as the lung buds) where *hedgehog* and *Bmp* genes are expressed adjacent to each other but are regulated independently, and other plases (such as the vertebrate neural tube) where the activities of these genes actually oppose each other, the pattern of a Hedgehog signal inducing *dpp/Bmp* gene expression occurs frequently enough that it is likely to represent a conserved regulatory cassette that has been evolutionarily co-opted to control various developmental processes.

The novel Hedgehog signaling pathway can be manipulated by several chemical compounds. The most important of these have been two steroidal alkaloids, jervine and cyclopamine. Jervine or cyclopamine treatment blocks Hedgehog signal transduction downstream of Ptc in the recipient cell. Another useful pharmacological reagent has been forskolin, a PKA activator. Because PKA is a negative modulator of Hedgehog signaling, forskolin treatment has also been an effective way of interfering with Hedgehog activity both *in vivo* and *in vitro*. The use of these reagents, along with an increased understanding of Hedgehog signaling pathways gained in the last decade, has facilitated the dissection of the biological roles played by the *hedgehog* genes, as outlined in detail in this review.

II. Cement Gland

The cement gland, an ectodermally derived organ, develops at the anterior of the frog embryo to allow the embryo to adhere to a solid surface (Sive and Bradley, 1996). Ectopic cement gland formation can be induced by the bone morphogenetic protein (BMP)/activin antagonists Noggin and Follistatin, and several *Xenopus* Hedgehog family members (Ekker *et al.*, 1995a). As ectopic *hedgehog* expression does not result in ectopic expression of *Noggin* or *Follistatin*, Hedgehog signaling appears to operate independently of these factors. The anterior expression of *Cephalic, Banded*, and *Sonic hedgehog*s suggests that one or more of these family members participate in cement gland induction during normal development (Ekker *et al.*, 1995a).

III. Central Nervous System

A. Neuronal Specification

The CNS of the vertebrate embryo, which consists of an anterior brain and posterior spinal cord, is established from neurectoderm cells at gastrulation. Species-specific morphogenetic processes generate a primitive neural tube along the length of the anterior–posterior axis of the embryo. Specific neuronal subtypes are formed at particular positions along the anterior–posterior and dorsal–ventral axes of the neural tube in response to local positional cues. Embryonic microsurgery, tissue grafting, explant culture, and mutant analyses have highlighted the critical importance of midline tissues in patterning presumptive brain and spinal cord regions (Brand *et al.*, 1996; Dale *et al.*, 1997; Pera and Kessel, 1997; Placzek, 1995).

In anteriormost regions, prechordal plate mesendoderm underlying the presumptive forebrain initiates ventral pattern. Similarly, in more posterior positions, the underlying midline notochord plays an equivalent role in presumptive midbrain, hindbrain, and spinal cord regions. Although initial ventral patterning signals originate outside of the neural tube, their inductive action induces secondary signaling centers at the ventral midline of the neural tube, notably the floor plate of the hindbrain and spinal cord. Floor plate-derived signals are thought to play a continuing role in ventral patterning.

Shh is expressed in the prechordal plate, notochord, and floor plate of the vertebrate embryo (Chang *et al.*, 1994; Echelard *et al.*, 1993; Ericson *et al.*, 1995; Krauss *et al.*, 1993; Marti *et al.*, 1995; Riddle *et al.*, 1993; Roelink *et al.*, 1994, 1995). In addition, several other members of the Hedgehog family have midline expression in the fish (Currie and Ingham, 1996; Ekker *et al.*, 1995b). Hedgehog signaling by midline tissues plays a central role in specifying ventral cell fates along the length of the neural tube. Much of this work has focused on the spinal cord of the mouse and chick (for reviews see Briscoe and Ericson, 2001; Jessell, 2000), but it is likely that the same general principles operate within the developing brain and in lower vertebrates. Further, the Hedgehog pathway has also been linked to neuroblast specification in *Drosophila* (Bhat, 1996; Matsuzaki and Saigo, 1996; McDonald and Doe, 1997), although there is some dispute as to the extent of Hedgehog action in this process (Deshpande *et al.*, 2001). For the purposes of this review, we primarily address the vertebrate CNS.

Analyses of a number of homeobox-containing transcriptional regulators in presumptive spinal cord regions of the early chick and mouse embryo demonstrate five spatially distinct, neural precursor domains that occupy specific positions above the midline floor plate within the ventral half of the neural tube, each characterized by the expression of a distinct set of genes (Briscoe *et al.*, 1999, 2000; Ericson *et al.*, 1996, 1997; Pierani *et al.*, 1999; Tanabe *et al.*, 1998). From ventral to dorsal these neural precursor populations are as follows: p3 (Nkx2.2; Nkx6.1), pMN (Nkx6.1; low Pax6), p2 (Nkx6.1; Pax6; Irx3), p1 (Dbx2; Irx3; Pax6), and p0 (Dbx1; Dbx2; Pax6; Irx3). A given combination of homeodomain-containing factors within a cell prefigures the differentiation of that cell into a particular postmitotic neural subtype that can itself be identified by the expression of other homeodomain factors. Thus, p3, p2, p1, and p0 precursors give rise to v3 (Sim1), v2 (Chx10), v1 (En1), and v0 (Evx1; Evx2) interneurons, respectively, whereas pMN cells form motor neurons (Isl1; Isl2; HB9). Several lines of evidence indicate that Shh acts directly as a concentration-dependent signal over most, if not all, of the ventral neural tube-inducing floor plate at the ventral midline and distinct neural progenitors in more dorsal positions. Thereafter, the particular combination of homeodomain factors determines the differentiation of specific neural subtypes through processes that are largely independent of Shh signaling.

Although Shh is transcribed only medioventrally in the floor plate and in the underlying notochord, Shh protein is present throughout the ventral half of the neural tube, a finding that supports a more widespread signaling role (Gritli-Linde *et al.*, 2001). Further, upregulation of *Ptc1* and *Gli1* within the same domain indicates active signaling throughout the ventral neural tube and possibly extending into dorsal regions (Goodrich *et al.*, 1996; Hui *et al.*, 1994; Lee *et al.*, 1997; Marigo and Tabin, 1996). Application of Shh to intermediate explant cultures also demonstrates that each ventral progenitor domain can be induced in response to a specific concentration of Shh. Importantly, there is an excellent correlation between the position of a progenitor population and the concentration of Shh required for its induction. Floor plate requires the highest concentration (4 nM or greater) and v3, MN, v2, v1, and v0 precursors require progressively less (3, 2, 0.5, 0.25, and 0.05 nM, respectively) (Ericson *et al.,* 1996, 1997; Pierani *et al.,* 1999; Roelink *et al.,* 1995). Morever, the antibody-mediated inhibition of Shh activity within neural explants supports the notion that Shh acts directly on neural tissue to initiate ventral patterning *in vivo* (Briscoe *et al.,* 2000; Ericson *et al.,* 1996).

Given that these experiments support a model in which Shh acts as a long-range morphogen specifying multiple ventral cell fates in the presumptive spinal cord, the phenotype of Shh mutants is surprising. Although induction of floor plate, v3, and motor neuron progenitors requires Shh, v2, v1, and v0 progenitors form in Shh mutant embryos, albeit in reduced numbers (Chiang *et al.,* 1996; Litingtung and Chiang, 2000). Thus, Shh is not essential for induction of the most dorsal populations of ventral interneurons.

Shh may also act in combination with other factors to induce specific cell types. Thus, Shh and retinoic acid may cooperate in the induction of v1 and v0 interneuron progenitors (Pierani *et al.,* 1999). In addition, vitronectin, a secreted matrix protein that binds Shh, and NT3, a neurotrophic factor, are both reported to synergize with Shh in the induction of motor neurons (Dutton *et al.,* 1999; Martinez-Morales *et al.,* 1997; Pons and Marti, 2000).

To address the cellular requirement for Hedgehog signaling, a dominant-negative form of Ptc1 that lacks the first extracellular loop of the protein has been expressed in the chick spinal cord (Briscoe *et al.,* 2001). This mutant form of the receptor fails to bind Hedgehog ligand but is still able to inhibit Smoothened activity. Consequently, Hedgehog signaling is cell autonomously inhibited, even in the presence of high concentrations of ligand (Briscoe *et al.,* 2001). Interestingly, these exper iments indicate that all ventral interneurons are Hedgehog dependent, including v0 and v1 interneuron progenitors, which can be induced by retinoic acid (Pierani *et al.,* 1999). How can these conflicting observations be reconciled?

Although it is always possible that the results reflect distinct differences in the requirement for Shh between chick and mouse, there is the possibility that some Hedgehog signaling may still occur within the neural tube of *Shh* mutants. *Ihh* is expressed in the underlying endoderm and could thus be sufficient for the activation of those homeodomain factors that depend on a low threshold of Hedgehog signal. *Ihh* and *Shh* are partially redundant; both signals induce paraxial sclerotome precursors immediately adjacent to the ventral neural tube (X. Zhang *et al.,* 2001). Examining neural patterning in *Smo* mutants may help to determine whether Ihh plays a redundant role in the induction of some ventral cell identities in the neural tube of *shh* mutants.

How, then, are individual progenitor domains specified in response to Hedgehog signaling? Briscoe *et al.* (2000) have proposed a model whereby there are two classes of response genes. Class I genes (*Dbx1, Dbx2, Pax6, Pax3,* and *Irx3*) are repressed and class II genes (*Nkx2.2* and *Nkx6.1*) are activated at distinct concentrations of Shh. Thus, the ventral boundaries of class I expression and dorsal boundaries of class II genes demarcate progenitor domains. Whether any of the class I or class II genes are direct targets of Shh signaling remains unclear.

The apparent repression and activation may be explained by several different mechanisms (reviewed in Briscoe and Ericson, 2001). For example, moderate concentrations of Shh repress *Dbx2* (Pierani *et al.,* 1999). However, an ability to respond to Shh is essential for expression of *Dbx2* in its normal domain (Briscoe *et al.,* 2001). One could explain these observations if Shh activates *Dbx2* expression over a wide range of concentrations from low levels, but Shh induces a repressor of *Dbx2* transcription at some higher level, such that *Dbx2* is expressed only dorsal to this repressor domain. Indeed, loss of *Nkx6.1* results in the expansion of *Dbx2* activity into more ventral neural progenitors (Sander *et al.,* 2000). Importantly, because the concentration gradient of Shh is not expected to alter in *Nkx6.1* mutants, ventral cells now appear to activate *Dbx2* at a concentration threshold that normally inhibits *Dbx2* expression. Similarly, in *Pax6* mutants there is a dorsal expansion of *Nkx2.2* (Ericson *et al.,* 1997). Here the absence of class II-mediated repression extends the dorsal boundary of *Nkx2.2* expression. Thus, the ventral expression of *Nkx2.2* may reflect the absence of repression rather than more direct Shh-mediated activation.

Repressive interactions among these potential homeodomain targets may also help create sharp boundaries between adjacent neural progenitor domains. Dbx2 can repress *Nkx6.1* and Nkx2.2 can repress *Pax6* (Briscoe *et al.,* 2000). Thus, mutual pairwise repression between Dbx2 and Nkx6.1, and between Nkx2.2 and Pax6, likely demarcates tight progenitor boundaries, a mechanism that may also act at other boundaries where putative repressors have not yet been identified.

Clearly, the next important step in revealing the molecular logic of the Shh-mediated patterning process is to detemine the direct targets of regulation. Although there is compelling evidence that *Hnf3β* is both a direct target of Shh signaling (Sasaki *et al.,* 1997) and a direct activator of *Shh* within the floor plate (Epstein *et al.,* 1999), the *cis*-regulatory mechanisms governing transcription of homeo-domain genes expressed within ventral progenitors have yet to be studied in any depth.

Cross-regulatory repressive interactions may do more than establish tight borders between adjacent progenitor populations; they may also obviate the requirement for continued Shh input. Such a mechanism would ensure that a specific pattern of homeodomain factors, once established in a cell, becomes independent of changes in the Shh concentration gradient, which might be expected to occur because of rapid growth of the neural tube. Although exactly when the maintenance of a particular profile of homeodomain expression becomes independent of Shh input is unclear, antibody blocking experiments point to a period of 12–15 h after signal presentation (Briscoe *et al.,* 2000). This timing would encompass two cell cycles at most.

Once a specific combination of homeodomain factors is established within a cell, this homeodomain code determines the subsequent specification of postmitotic neurons arising from a given progenitor domain. For example, Nkx6.1 induces motor neurons, but Nkx6.1 and Irx3 together specify v2 interneuron fates (Briscoe *et al.*, 2000). *Nkx6.1* is also coexpressed with *Nkx2.2*. However, *Nkx2.2* appears to be dominant in these cells specifying v3 interneuron fates (Briscoe *et al.*, 2000; Ericson *et al.*, 1997).

How does the activity of Shh relate to more general mechanisms that regulate neurogenesis that are governed by the activity of the neurogenic proteins, which include Notch, its ligands, and members of the basic helix–loop–helix (bHLH) family of transcriptional regulators (Chan and Jan, 1999)? In the *Drosophila* CNS, Hedgehog appears to act on ectodermal neuroblast precursors, upstream of proneural genes, in the formation of a specific population of segmental neuroblasts (Matsuzaki and Saigo, 1996). *Ptc* activity is also linked to the subsequent specification of these neuroblasts (Bhat, 1996; Duman-Scheel *et al.*, 1997). In *Xenopus, Gli* expression correlates with neurogenic zones, and Gli stimulates neurogenesis while inhibiting the antineurogenic activity of Zic factors (Brewster *et al.*, 1998). However, Shh signaling is reported both to induce neurogenic gene expression (Blader *et al.*, 1997) and inhibit primary neurogenesis (Franco *et al.*, 1999), apparently conflicting results. The relationship between Shh and the neurogenic genes in vertebrate CNS patterning will only be clarified by additional studies.

Analysis of Gli functions in the neural tube has shed light on how their activities are regulated by Shh signaling. Whereas both *Gli2* and *Gli3* are expressed in neural cells before Shh signaling (Ding *et al.*, 1998; Hui *et al.*, 1994), *Gli1* is induced only after Shh signal transduction begins in ventral neural tissue. *Gli1* mutants have no discernible phenotypes in the CNS. Thus, if Gli family members are the sole transcriptional effectors of Hedgehog signaling, Gli2 and/or Gli3 must be responsible for initiating the primary transcriptional response to Shh. If not, there must be Gli-independent mechanisms of Hedgehog signaling. In *Drosophila,* where this problem has been rigorously addressed, there are conflicting reports as to whether all Hedgehog signaling is mediated through the unique *Ci* gene (Methot and Basler, 2001; Suzuki and Saigo, 2000).

Gli2 mutants have no floor plate and v3 interneuron progenitors are markedly reduced and abnormally positioned at the ventral midline (Ding *et al.*, 1998; Matise *et al.*, 1998). As *Gli1* is not induced, both Gli1 and Gli2 activity appear to be absent in the *Gli2* mutant (Ding *et al.*, 1998). Indeed, *Gli1/Gli2* compound mutants closely resemble *Gli2* mutants (Matise *et al.*, 1998). These results demonstrate that the primary role of Gli2 is floor plate induction. Studies indicate that Gli1 and Gli2 are interchangeable (Bai and Joyner, 2001). As Gli1 appears to act exclusively as a Hedgehog-independent activator (reviewed in Ingham and McMahon, 2001), Shh-mediated activation of Gli2 is likely the normal mechanism for floor plate induction.

These results question how the remaining cell fates could be specified by Shh and Gli3, as all ventral progenitor domains are present in *Gli3* mutants. Interestingly, the loss of Gli3 in a Shh mutant background rescues motor neuron fates and increases the numbers of v0, v1, and v2 progenitors (Litingtung and Chiang, 2000), whereas there is no rescue of floor plate and v3 progenitors (Litingtung and Chiang, 2000). Importantly, each ventral progenitor population is reported to occupy its correct position relative to each other and the ventral midline. Together with the earlier data, these studies support the notion that Gli3 acts, at least in part, as a repressor of motor neuron fates. Shh signaling is therefore required to suppress this inhibitory activity. The failure to rescue floor plate is consistent with Shh-mediated induction of floor plate through Gli2. Presumably, the absence of v3 neurons also reflects the positive input of Gli2, and possibly Gli1, as the v3 population is markedly reduced in Gli2 mutants.

The apparent preservation of ventral polarity in *Shh/Gli3* compound mutants is most intriguing. A similar result has been reported in *Shh/opb* compound mutants. *opb* encodes rab23, a small GTP-binding protein that plays an as yet undefined role in Ptc1-mediated inhibition of Smo (Eggenschwiler and Anderson, 2000; Eggenschwiler *et al.*, 2001; Gunther *et al.*, 1994). Thus, cell-autonomous, Hedgehog-independent activation of Hedgehog signaling occurs in *opb/Shh* double mutants (Eggenschwiler and Anderson, 2000). With no reference to a localized source of Shh secreted by midline cells, how does one account for the observed polarity of ventral progenitor populations in these compound mutants?

One mechanism invokes a second ventral patterning signal that establishes polarity in the absence of Shh. Ventral gradients of BMP antagonists are an attractive possibility. Several *BMP* genes, and other members of the *TGF-β* superfamily, are expressed in the dorsal neural tube, where their signaling activities regulate dorsal cell identities (for review see Lee and Jessell, 1999). The presence of BMPs modifies the response to Shh in neural explants requiring higher concentrations of Shh to induce a particular progenitor population (Liem *et al.*, 2000). The notochord is a source of several BMP antagonists, including Noggin and Follistatin, and Noggin activity is required for normal Shh-mediated patterning (Liem *et al.*, 2000; McMahon *et al.*, 1998). Thus, a ventral gradient of BMP antagonism may provide polarity to ventral cell types (Patten and Placzek, 2002). However, as discussed earlier, Ihh signaling by ventral endoderm may provide an alternative source of Hedgehog signal (X. Zhang *et al.*, 2001). *Smo/Gli3* compound mutants should be informative in this regard.

Much of our detailed mechanistic understanding of the role of Hedgehog signaling comes from studies of chick and mouse spinal cord progenitors, but the same general mechanisms operate in lower vertebrates and are responsible for patterning the ventral brain. In contrast to mouse and chick, zebrafish has three *hedgehog* genes that are expressed in midline tissues. *Ehh* is present in the notochord (Currie and Ingham, 1996), *shh* in the notochord and floorplate (Krauss *et al.*, 1993), and *twhh* in the floorplate (Ekker *et al.*, 1995b). Whereas *Shh* mutants have a weaker phenotype in zebrafish than in the mouse (Odenthal *et al.*, 2000; Schauerte *et al.*, 1998), loss of both *Shh* and *twhh* together, or the lone fish *Smo* gene, leads to a failure of lateral—but not medial—floorplate, and primary and secondary motor neuron, induction (Beattie *et al.*, 1997; Bingham *et al.*, 2001; Chandrasekhar *et al.*, 1998; Chen *et al.*, 2001; Etheridge *et al.*, 2001; P. M. Lewis *et al.*, 2001; Varga *et al.*, 2001).

Shh induces ventral cell identities in the forebrain (Barth and Wilson, 1995; Chiang *et al.*, 1996; Dale *et al.*, 1997, 1999; Ericson *et al.*, 1995; Nakagawa *et al.*, 1996), midbrain (Agarwala *et al.*, 2001; Chiang *et al.*, 1996; Hynes *et al.*, 1995, 2000; Wang *et al.*, 1995; Watanabe and Nakamura, 2000; Ye *et al.*, 1998), and hindbrain (Chiang *et al.*, 1996; Hynes *et al.*, 1997; Ye *et al.*, 1998). Shh also regulates dorsoventral patterning of the basal telencephalon. This action both promotes basal ganglion specification and inhibits dorsal cortical development (Gaiano *et al.*, 1999; Kohtz *et al.*, 1998, 2001; Sussel *et al.*, 1999). The necessity for Hedgehog signaling at all axial levels to regulate ventral cell identities raises the question of specificity. What mechanisms ensure that the response is appropriate for a given region of the brain or spinal cord? One mechanism involves the regional interplay of additional signaling factors. In this model, neural tissue is developmentally plastic at different axial levels, but the combination of Shh and specific local signals dictates the cell types that are generated.

For example, Shh and BMP7 produced by the prechordal plate act cooperatively in the induction of *Nkx2.1* in the ventral forebrain of the chick. Only application of both factors induces *Nkx2.1* at caudal levels where *Nkx2.1* expression is normally absent (Dale *et al.*, 1997). The timing of this inductive interaction *in vivo* may depend on the activity of a BMP antagonist, chordin (Dale *et al.*, 1999). In contrast, similar experiments in the mouse indicate that the prechordal plate is unable to induce *Nkx2.1* at caudal levels of the neural plate, whereas notochord, on the other hand, can induce ectopic *Nkx2.1* expression in the forebrain (Shimamura and Rubenstein, 1997). Further, ectopic BMP7 appears to inhibit *Shh* expression in the mouse hindbrain (Arkell and Beddington, 1997). In the midbrain, ventral cell identities, including dopaminergic neurons, are induced where Shh and fibroblast growth factor 8 (FGF8) signaling intersect (Carl and Wittbrodt, 1999; Ye *et al.*, 1998). These same factors induce serotonergic neurons in the ventral hindbrain, where FGF4 is thought to control the specificity of the response (Ye *et al.*, 1998). Although signal combinations are likely significant, prior patterning events that regionalize the anteroposterior axis of the neural plate may dictate the response to an equivalent Hedgehog input. This would ensure a regional specific output. If so, neural tissue at any axial level may have more limited developmental potential than the neural plate as a whole.

B. Oligodendrocytes

Shh plays a critical role in the induction of oligodendrocyte progenitors and, hence, the production of glial cells responsible for myelinating axon fasicles in the central nervous system. Oligodendrocytes represent a relatively late-emerging cell type (Orentas *et al.,* 1999). At spinal cord levels, oligodendroctye progenitors appear to arise from a subset of cells that express *Nkx2.2* and the genes encoding two bHLH factors, Oligo1 and Oligo2 (Lu *et al.,* 2000; Soula *et al.,* 2001; Sun *et al.,* 2001; Zhou *et al.,* 2001). *Oligo1* expression is sustained in maturing oligodendrocytes (Lu *et al.,* 2000).

Induction of oligodendrocytes (Pringle *et al.,* 1996; Poncet *et al.,* 1996; Orentas *et al.,* 1999) and expression of *Oligo1/2* (Lu *et al.,* 2000) are Shh dependent. Further, *Oligo1* and *Oligo2* can induce ectopic oligodendrocyte formation, although induction requires Nkx2.2 (Sun *et al.,* 2001; Zhou *et al.,* 2001). Interestingly, the *Oligo1/2* expression domain overlaps the region where Nkx6.1 progenitors are thought to generate motor neurons, suggesting a second role for these factors in motor neuron induction (Lu *et al.,* 2000). The close linkage between Shh signaling, *Oligo2* expression, and oligodendrocyte formation extends into the brain, suggesting that Shh is a general inductive factor for oligodendrocyte progenitors (Alberta *et al.,* 2001; Davies and Miller, 2001; Lu *et al.,* 2000; Nery *et al.,* 2001; Tekki-Kessaris *et al.,* 2001). Whether a Shh-inductive pathway is the exclusive route to oligodendrocyte specification is unclear (Sussman *et al.,* 2000; Wada *et al.,* 2000). Although forebrain explants from *Shh* mutants do generate oligodendrocytes (Nery *et al.,* 2001), their induction appears to be cyclopamine sensitive, suggesting an involvement of Hedgehog signaling (Tekki-Kessaris *et al.,* 2001). *Ihh* is reported to be ectopically expressed in forebrain explants. If so, its activity could counteract a loss of Shh signaling (Tekki-Kessaris *et al.,* 2001).

C. Proliferation and Survival

In addition to the role of Shh in cell fate specification, Shh is a potent mitogen within the CNS, a function with obvious implications in the involvement of Hedgehog pathway components in CNS tumors (see Section XXVII, Cancer). The cerebellum exemplifies the mitogenic function of Shh. Here, Purkinje cell-derived Shh expands cerebellar granule cell precursors in the external granule cell layer by a mechanism that acts in part through the control of *cyclinD1* (Dahmane and Ruiz-i-Altaba, 1999; Kenney and Rowitch, 2000; Wallace, 1999; Wechsler-Reya and Scott, 1999). Shh also promotes proliferation of neural precursors in several other regions of the brain and spinal cord, although the mechanism in these regions is less clear (Agarwala *et al.,* 2001; Goodrich *et al.,* 1997; Kalyani *et al.,* 1998; Rowitch *et al.,* 1999). Shh continues to be expressed in ventricular regions of the adult mammalian brain, where neural stem cells proliferate in the adult (Traiffort *et al.,* 1999). Whether Shh plays a role in the expansion of neural precursors in the adult brain has not been determined.

Shh rescues apoptosis induced by the removal of midline tissues in the chick (Charrier *et al.,* 2001) and also promotes the survival of dopamine- and γ-aminobutyric acid (GABA)-producing neurons *in vitro* (Miao *et al.,* 1997). Thus, Shh may have an anti-apoptotic activity in the developing and mature CNS. However, ectopic application of Shh also enhances apoptosis within the ventral neural tube, although this cell death may be secondary to the misspecification of ventral cell types (Oppenheim *et al.,* 1999). No clear molecular link has been demonstrated between Hedgehog signaling and the apoptotic machinery.

D. Mature Neurons

Shh expression continues in several mature neurons after the establishment of a functional neural circuitry. Although Shh could play a role in neuron–target interactions in the adult CNS, this possibility has not been addressed. For example, Shh is present in adult retinal ganglion cells and appears to undergo axonal transport in lipid rafts (Traiffort *et al.,* 2001). Shh in the optic nerve regulates proliferation of adjacent astrocytes (Wallace and Raff, 1999). Whether axonal transport of Shh plays any role in the optic tectum, the target field innervated by retinal axons, remains to be determined. In *Drosophila,* transmission of Hedgehog in retinal axons triggers proliferation of neuronal precursors in the optic lamina of the fly brain, an essential step in the generation of lamina synaptic cartridges (Huang and Kunes, 1996, 1998). In the cerebellum of the mouse, Shh protein is observed on dendrites and axonal projections of Purkinje cells (Gritli-Linde *et al.,* 2001). In the spinal cord, *Shh* is expressed in motor neurons, although it is not clear whether protein is actually produced (Marti *et al.,* 1995; Oppenheim *et al.,* 1999). Finally, in the melatonin-producing pineal gland, *Ptc1* exhibits circadian regulation (Borjigin *et al.,* 1999). However, the functional significance of this mode of regulation and its relationship to Hedgehog production are unclear.

IV. Circulatory System

The earliest blood vessels and erythrocytes are observed as blood islands within the yolk sac of the developing mammalian conceptus, shortly after the onset of gastrulation (Palis *et al.,* 1999). The yolk sac consists of two distinct populations of cells, the endoderm and mesoderm, that have quite different lineages. The yolk sac mesoderm, or visceral mesoderm, is a population of mesoderm cells that derives from the posterior aspect of the primitive streak. The visceral ndoderm, an early-formed cell lineage, arises from the primitive or extraembryonic endoderm. Several lines of evidence indicate that the blood vessels and blood cells of the blood islands develop from a common visceral mesoderm-derived progenitor, the hemangioblast (Choi *et al.,* 1998; Lacaud *et al.,* 2001), in response to visceral endoderm-derived signals.

Ihh is expressed in the visceral endoderm and is able to recapitulate blood vessel and blood cell induction by visceral endoderm in certain *in vitro* assays (Belaoussoff *et al.,* 1998; Dyer *et al.,* 2001). Additional support for Ihh signaling in the development of yolk sac blood cells and vessels comes from the analysis of embryonic stem (ES) cell differentiation. Under appropriate conditions, ES cells differentiate *in vitro* into embryoid bodies that contain blood islands similar to those in the yolk sac. In contrast, embryoid bodies derived from *Ihh* mutant ES cells show little evidence of blood island formation, although an occasional, cell-free, endothelial cell-lined pocket is observed (Byrd *et al.*, 2002). A similar phenotype is also observed in *Smo* mutant embryoid bodies (Byrd *et al.*, 2002), although the vascular phenotype appears to be more severe. *Dhh* is expressed at low levels within the visceral mesoderm (Farrington *et al.*, 1997); consequently, the loss of both Ihh and Dhh signaling in *Smo* mutants may result in a more extreme phenotype than that of *Ihh* mutant embryoid bodies.

Phenotypic and molecular analyses indicate that these blood island deficiencies reflect a reduction in hemangioblasts. Together with the aforementioned inductive studies, the data support a model in which Ihh (and possibly Dhh) signaling promotes hemangioblast development. However, *Ihh* mutants exhibit a comparatively mild failure in large vessel remodeling within the yolk sac (Byrd *et al.*, 2002). Further, *Smoothened* mutants are able to form a rudimentary, yolk sac vasculature that contains primitive erythrocytes (Byrd *et al.*, 2002). Thus, there does not appear to be an absolute requirement for Hedgehog signaling in the development of the hemangioblast or its derivatives.

Whereas the precise role that Hedgehog signaling plays within the yolk sac awaits further clarification, studies have investigated Hedgehog function in the fetal and adult circulatory system. Shh signaling has been shown to stimulate angiogenesis in the adult through the regulation of vascular endothelial cell growth factor (VEGF) and angiopoietins (Pola *et al.*, 2001). This result might explain the hypervascularization phenotype that is observed when *Shh* is ectopically expressed in the neural tube (Rowitch *et al.*, 1999) and the hypovascular skeleton of *Ihh* mutants when growth plate-derived Ihh is absent (St-Jacques *et al.*, 1999).

Hedgehog signaling may also play a later role in blood cell development. Detmer *et al.* (2000) report a reduction of erythroid differentiation by mononuclear marrow cells in response to cyclopamine-mediated inhibition of Shh. In contrast, Shh-N (Shh with cholesterol linked to the amino-terminal domain)-mediated stimulation expands erythroid cells. Furthermore, Bhardwaj *et al.* (2001) suggest that a primitive hematopoietic progenitor, expressing *Shh*, *Ptc1*, and *Smo*, may require Hedgehog signaling for its maintenance. As this activity appears to be blocked by the BMP antagonist Noggin, Hedgehog signals may act indirectly by regulating BMP activity. BMP signaling has also been implicated in the *Ihh*-mediated activation of hematopoiesis and vasculogenesis in early epiblast cultures (Dyer *et al.*, 2001). Finally, the thymic maturation of T cells from CD4/CD8 double-negative cells to CD4/CD8 double-positive cells is inhibited by Shh signaling and promoted by Shh-blocking antibodies (Outram *et al.*, 2000). *Shh* is expressed within the thymic stroma, consistent with a role in T cell differentiation (Outram *et al.*, 2000).

V. Face and Head

The dramatic facial abnormalities observed in *Shh* mutants (Chiang *et al.,* 1996), and in embryos exposed to the small, alkaloid Hedgehog antagonists cyclopamine and gervine (Cooper *et al.,* 1998, Incardona *et al.,* 1998, 2000), suggests a role for Shh signaling in the regulation of facial morphogenesis. In mice that have no *Shh* activity, the frontal–nasal processes fuse to form a long, medial proboscis-like structure and the upper and lower jaws are absent. In contrast, ectopic activation of Hedgehog signaling achieved by implanting Shh-N-soaked beads (Hu and Helms, 1999), or in gain-of-function mutants such as *talpid* (Munoz-Sanjuan *et al.,* 2001) and *Gli3* (Hui and Joyner, 1993; Mo *et al.,* 1997), expands frontal–nasal and maxillary primordia.

Shh is expressed from early somite stages in the ectoderm of the developing frontal–nasal processes and in the endoderm and ectoderm of the first branchial arch (Bitgood and McMahon, 1995; Hu and Helms, 1999); both upper (maxillar) and lower (mandibular) jaw structures originate from the first arch. *Ptc1* is upregulated in both the ectoderm and underlying cranial mesenchyme of the first arch, suggesting that active signaling is occurring in association with early facial morphogenesis (Hu and Helms, 1999; Marigo and Tabin, 1996). Thus, expression of *Shh*, and the response within potential target tissues, appear to correlate well with the observed phenotypes. However, Shh is also required for normal patterning of the developing brain (see Section III, Central Nervous System) and for survival of cranial neural crest progenitors (see Section XVI, Neural Crest). Thus, loss of brain structures or of facial skeletal precursors may result in secondary facial anomalies.

By employing a careful surgical procedure, Hu and Helms (1999) removed the ectodermal source of Shh from the developing frontal–nasal and maxillary processes. The manipulation resulted in a truncation of facial structures, demonstrating the importance of this epithelial population for normal outgrowth of the frontal–nasal and normal maxillary primordia. That Shh activity is required for outgrowth was demonstrated by implanting beads soaked in an Shh-neutralizing antibody. When Shh signaling was blocked in this way, facial outgrowth was markedly perturbed (Hu and Helms, 1999). Thus, *Shh* appears to play a specific role within facial primordia, possibly promoting the proliferation of skeletal precursors (Hu and Helms, 1999).

Some studies have started to address the mechanisms that might regulate *Shh* itself. Retinoic acid adminstration at high doses is known to perturb facial development, a result that correlates with reduced levels of *Shh* expression (Helms *et al.,* 1997). Loss of *Shh* expression and inhibition of mid and upper facial structures also follow from inhibition of endogenous retinoic acid receptor function (Schneider *et al.,* 2001). Although the retinoic acid synthetic enzyme, aldehyde dehydrogenase 6, is present within the facial ectoderm, analysis of retinoic acid receptor expression suggests that the actual target of retinoic acid action is the cephalic neural crest (Schneider *et al.*, 2001). These results indicate that the neural crest maintains *Shh* expression in the overlying ectoderm of facial primordia via a retinoic acid-dependent mechanism. Reciprocal interactions between mandibular ectoderm and mandibular mesenchyme (in this case nonneural crest) have also been invoked in the *Prx1/Prx2*-dependent maintenance of *Shh* expression (Lu *et al.*, 1999; ten Berge *et al.*, 2001).

VI. Gonads

The generation of mature eggs and sperm depends on complex interactions between somatic and germ line cells within the gonads of all organisms. In *Drosophila*, the development of the gonad begins during embryogenesis with the migration of the primordial germ cells, known as pole cells, toward the clusters of somatic gonadal precursor (SGP) cells within the mesoderm. The pole cells traverse the posterior midgut endoderm before migrating toward and along the mesoderm. Relatively little is known about this process, although genes mediating both attractive and repulsive effects on the pole cells have been isolated. Loss of Hh function results in the disruption of germ cell migration (Moore *et al.*, 1998). Whereas Deshpande *et al.* (2001) have presented evidence that Hh acts as an attractant for the migrating pole cells, other studies point to defective migration being an indirect consequence of the requirement for Hh in the specification of the SGP cells themselves (Azpiazu *et al.*, 1996; Riechmann *et al.*, 1998).

Each *Drosophila* egg develops from an egg chamber, consisting of a clone of 16 germ line cells ensheathed by somatically derived follicular cells. Interactions between the follicular and germ line cells single out one of the latter to become the oocyte and also serve to establish the axes of the oocyte, and consequently those of the future embryo. The remaining 15 germ line cells function as nurse cells that supply the oocyte with maternally expressed gene products. Assembly of germ line and somatic cells into egg chambers takes place in the germarium, a complex structure at the end of each ovariole, the long strings of maturing egg chambers that comprise the ovary. The germarium contains the stem cells for both the germ line (GSC) and somatic (SSC) lineages. The GSCs lie directly adjacent to a specialized structure, the terminal filament, at the anterior tip of the germarium whereas the SSCs are located some 10 cell diameters away, closer to the center of the germarium. Cystoblasts, the daughters of the GSCs, undergo four divisions to generate the 16-cell cyst, which moves along the germarium and past the SSCs before associating with about 16 SSC-derived prefollicle cells. Most of these prefollicle cells continue to proliferate as they move along the germarium in association with the cyst, eventually encapsulating it to form a nascent egg chamber. As the latter buds off, a few specialized follicle cells form a stalk that links the egg chamber to the germarium and that later separates it from its neighboring chamber.

Strikingly, Hh is produced by the terminal filament cells and another closely associated specialized somatic cell type, the cap cells, whereas Ci protein is present throughout the rest of the germarium (Forbes *et al.*, 1996a,b). High-level *ptc* expression occurs in somatic cells in the region of the germarium extending from the terminal filament up to and including the SSCs, consistent with the latter receiving and responding to the Hh signal (Forbes *et al.*, 1996b). In line with this interpretation, the inactivation of Hh in adult females results in a cessation of oogenesis; egg chambers become fused, reflecting a reduction in follicle cell number, and subsequently all follicle cell production is lost (Forbes *et al.*, 1996a). Conversely, overexpression of Hh or removal of Ptc activity results in an overproliferation of follicle cells, the stalks that separate adjacent egg chambers becoming massively enlarged (Forbes *et al.*, 1996b; Y. Zhang and Kalderon, 2000). Moreover, the enlarged stalks induced by Hh overactivity are composed of prestalk rather than mature stalk cells, suggesting that persistent Hh signaling blocks the terminal differentiation of these cells and maintains them in a proliferative state (Y. Zhang and Kalderon, 2000). These effects strongly suggest a role for Hh in stimulating proliferation of SSCs or their progeny. Removal of Smo activity from follicle cell precursors does not, however, prevent them from dividing or differentiating (Y. Zhang and Kalderon, 2000). However, when Smo activity is removed from individual SSCs, such cells give rise to relatively few progeny, suggesting that they have lost their self-renewal capacity and become prefollicle cells (Y. Zhang and Kalderon, 2001). These findings have been taken to imply that the normal role of Hh is to maintain the stem cell properties of SSCs; consistent with this interpretation, removal of Ptc activity from SSCs leads to an increase in SSC numbers within the germarium (Y. Zhang and Kalderon, 2001). Thus, Hh signaling may establish a niche in the germarium within which stem cell proliferation is maintained.

The proximity of the GSCs to the terminal filament implies that Hedgehog signaling could play a similar role in maintaining the division of these cells; however, evidence for such a role is less clear cut. Loss of function of the *Yb* gene—which normally regulates *hh* expression in the terrminal filament and cap cells—eliminates GSCs and reduces SSC division. Significantly, both of these defects can be reversed by restoring high-level Hedgehog signaling activity in the germ line, either by removing Ptc activity or inducing *hh* expression throughout the germarium, implying that Hh can stimulate GSC as well as SSC division (King *et al.*, 2001). The relevance of this effect for the normal regulation of GSC division is, however, unclear. King *et al.* (2001) report defects in about 20% of germ line cells when Hh is inactivated in adult ovaries, implying that any such role is partially redundant.

Another line of evidence points to a role for Hh in regulating the proliferation of cystoblasts, through its effect on the stability and subcellular distribution of the Sxl protein (Vied and Horabin, 2001). Germ cells that lack Sxl develop as tumorous cysts of many small undifferentiated cells, reflecting a role for Sxl in controlling germ cell mitosis. Sxl protein undergoes a characteristic shift in distribution in the germarium, being exclusively cytoplasmic in the GSCs but entering the nucleus in cyst cells as they progress down the germarium. This transition is blocked in the absence of Hedgehog activity and also in *fused* mutant ovaries, consistent with nuclear import of Sxl being mediated by transduction of the Hedgehog signal by the cyst cells. Surprisingly, however, this translocation is independent of *smo* and *ptc* activity within the germ line, implying a distinct mechanism for Hh signal transduction in these cells (Vied and Horabin, 2001).

In the mammalian gonad, *Dhh* is expressed in Sertoli precursors, the chief support cell of the testis, shortly after their Y chromosome-dependent differentiation. Sertoli cell signaling is postulated to regulate spermatogenesis directly within the germ line and indirectly through the control of androgen-secreting Leydig cells. *Dhh* mutants display a number of defects that are influenced by the genetic background, including a loss of the germ line, abnormal seminiferous tubule formation correlating with a loss of peritubular myoid cells, and a failure of Leydig cell differentiation (Bitgood *et al.,* 1996; Clark *et al.,* 2000; Pierucci-Alves *et al.,* 2001). As a consequence, gonad size is dramatically reduced, all male mice are infertile, and a significant portion of males display a feminized phenotype that most likely reflects the absence of Leydig cell-derived androgens that drive the masculanization process (Bitgood *et al.,* 1996; Clark *et al.,* 2000). Although the upregulation of *Ptc1* in peritubular cells, pre-Leydig cells, and within the germline itself points to direct roles for Hedgehog signaling in the regulation of each of these cell types, the precise mechanisms by which *Dhh* controls testis development and function have not been explored in depth.

VII. Gut Development

All metazoa are characterized by a specialized gut for the absorption of nutrients. In the primitive state, this consisted of a simple closed sac formed by invagination from one side of the body during gastrulation. The inner lining of this primitive gut was formed by an endodermal epithelium. Stomodeal and proctodeal openings evolved secondarily to facilitate ingestion and egestion (Wolpert, 1994). The resulting linear gut tube is found in all modern metazoans with the exceptions of Cnidaria, Plathelminthes, and Porifera. The secondary connections of the endoderm to the exterior form by fusion with ectodermally derived epithelium, which in modern forms can be rather minor components of the gut (e.g., the oral and anal epithelia of vertebrates) or quite extensive (e.g., the entire foregut and hindgut of insects). These distinct tissue origins of different segments of the gut are underscored by the utilization of quite different signaling networks in different regions.

The linear epithelial gut tube is surrounded by mesenchymal cells that assist in the movement of material through the gut. In higher animals, the linear tube is further regionalized to form organs with specialized functions for processing and absorbing nutrients. Moreover, in higher forms, additional organs form by branching off from the gut tube to take on accessory functions in digestion as well as functions in gas exchange and metabolic homeostasis; these include, for example, the lungs and pancreas in vertebrates, and the malpighian tubules in *Drosophila* (functionally analogous to the vertebrate kidneys but more similar to the vertebrate ceca or appendix in their morphogenesis and in their position at the hindgut–midgut boundary).

The development of the gastrointestinal tract and the gut-derived organs can be viewed as encompassing several distinct steps. These include the invagination of the endoderm and recruitment of associated mesoderm to form a primitive gut tube, regionalization of the gut tube along the rostral–caudal axis, organogenesis, and finally, establishment of a stem cell population within the adult morphology to allow self-renewal of the organs in postnatal life. *hedgehog* genes have been shown to play critical roles in many of these processes at different stages of gut development in both vertebrates and *Drosophila*.

In all vertebrates, the formation of the gut begins with the ventral infolding of the definitive endoderm to form the anterior intestinal portal (AIP), and a second infolding to form the caudal intestinal portal (CIP). These elongate toward each other, forming two open-ended tubes that eventually meet and fuse at the yolk stalk. As these two tubes form and grow, there are endodermal signals that recruit the splanchnic mesoderm to the gut tube and induce it to undergo gut-specific differentiation to form visceral mesoderm (Kedinger *et al.*, 1986). In both mouse and chick embryos, *Shh* and *Ihh* are expressed in broad overlapping patterns that are initiated as the AIP and CIP form (at stage 8 in the chick, 8.5 days postcoitum [dpc] in the mouse) and are maintained as the AIP and CIP extend to form a linear tube (Bitgood and McMahon, 1995; Echelard *et al.*, 1993; Marigo *et al.*, 1995; Narita *et al.*, 1998; Roberts *et al.*, 1995). The possibility that hedgehog protein might act as a signal for mesodermal recruitment at this stage is raised by the observation that several Hedgehog targets, including *Ptc1* (Marigo *et al.*, 1995) and *BMP4* (Roberts *et al.*, 1995), are expressed adjacent to the regions of endoderm expressing *hedgehog* genes. Moreover, these genes can be ectopically induced in the chick splanchnic mesoderm by Shh at this stage (Roberts *et al.*, 1995). Whether this early *hedgehog* expression really plays a role in mesodermal recruitment, or indeed has any functional significance, remains unproven as neither *Ihh* nor *Shh* mouse mutants show early gut defects. This could be explained by redundancy between these signals, but as the double mutant dies at the same time as gut induction (8.5 dpc) (Ramalho-Santos *et al.*, 2000), this possibility has not yet been examined.

At least some aspects of the regional specification of the gut actually precede the formation of the CIP and AIP and the expression of *hedgehog* genes. This is demonstrated by regional differences in response to ectopic Shh in the chick splanchic mesoderm, which can activate *BMP4* expression in the midgut and hindgut, but not in the future stomach region (Roberts *et al.,* 1998). As the gut tube forms, *Hox* genes are activated in expression domains that correlate with morphological borders of different future gut organs (Roberts *et al.,* 1995; Yokouchi *et al.,* 1995). The timing of *hedgehog* gene expression and the ability of Shh to induce ectopic expression of *Hoxd-13* in the chick (Roberts *et al.,* 1995) suggest that Hedgehog signaling may play a role in establishing these *Hox* domains, although, again, the lack of an effect on *Hox* expression in either the *Shh* or *Ihh* mutant and the possibility of redundancy prevent a definitive conclusion.

In the next phase of gut development, the gut tube expands in thickness and length, with differential proliferation in different domains. During this time, *Shh* and *Ihh* continue to be expressed in broad domains in the endoderm of the gut (Ramalho-Santos *et al.,* 2000; Roberts *et al.,* 1995), with the exception of the pancreatic buds (Apelqvist *et al.,* 1997). Viral misexpression of *Shh* in the chick suggests that one role of these signals is to increase proliferation of the adjacent mesoderm, resulting in a thicker muscle layer after differentiation (Roberts *et al.,* 1998; Smith *et al.,* 2000). Interestingly, one way that regional differences in proliferation appear to be achieved is by the induction of *BMP4* by Shh in the hind and midguts, but not the stomach (Roberts *et al.,* 1998). In particular, because ectopic *BMP4* expression results in a decrease in mesodermal proliferation and because the presumptive stomach is one region of the gut where Shh does not induce *BMP4* expression, these findings are consistent with the occurrence of a thicker layer of mesoderm surrounding the stomach. In addition, there is an increase in cell proliferation in the midgut of *Xenopus* in response to an activated form of Smo (J. Zhang *et al.,* 2001). Consistent with the increase in gut muscle thickness seen after *Shh* overexpression, both *Shh* and *Ihh* mutant mice show a reduction in smooth muscle (Ramalho-Santos *et al.,* 2000).

As gut organogenesis begins between 11.5 and 14.5 dpc in mice, *Shh* expression is downregulated in the posterior stomach, jejunum, and ileum, whereas *Ihh* is expressed from the posterior stomach to the anus (Bitgood and McMahon, 1995). The downregulation of Hedgehog activity at this stage appears to be significant because maintaining Hedgehog signaling via an activated form of Smo results in a number of gut defects in *Xenopus,* including a failure of the intestine to lengthen and coil. In addition, there is a loss of cytodifferentiation in the midgut epithelium (J. Zhang *et al.,* 2001). This latter effect likely reflects a secondary signal downstream of Hedgehog that signals from the mesoderm to the endoderm, as *Ptc1* (Marigo *et al.,* 1996a) and downstream effectors of Hedgehog signaling (Grindley *et al.,* 1997; Platt *et al.,* 1997) are expressed exclusively in the mesodermal compartment of the gut. Significantly, classic transplantation experiments have demonstrated that mesodermally derived signals instruct the morphological differentiation of the overlying endoderm (Haffen *et al.,* 1983; Kedinger *et al.,* 1986, 1998; Yasugi, 1993). The effects of activated Smo thus suggest that at least some of these mesodermal signals are dependent on downregulation of Hedgehog signals. However, other regions of the gut epithelium continue to express *hedgehog* genes at this time, and in these regions signals from the mesoderm responsible for patterning the epithelium may be downstream of Hedgehog activity, as the stomach epithelium of *Shh* mutants is partially transformed to an intestinal character (Ramalho-Santos *et al.*, 2000). This effect is in any event indirect, because as noted above, the *hedgehog* signal transduction machinery does not exist in the gut endoderm.

At 18.5 dpc in the mouse, *Shh* and *Ihh* continue to be expressed in the epithelium of the glandular stomach, the small intestine, and the colon. Both *Shh* and *Ihh* are expressed at the base of the villi in the small intestine, a region where epithelial stem cells are believed to be located (Ramalho-Santos *et al.*, 2000). Their expression appears to be critical for the maintenance of a stem cell population, as the *Ihh* mouse mutant shows reduced epithelial stem cell proliferation (Ramalho-Santos *et al.*, 2000). Conversely, *Shh* appears to play an antiproliferative role, affecting the gastric epithelial gland cell but not the lumenal pit compartment in the mature mouse stomach (van Den Brink *et al.*, 2001).

Another late function of Hedgehog signaling in the prenatal gut is in radial patterning of the mesoderm. In particular, application of endodermal explants, Shh protein, or the Hedgehog antagonist cyclopamine to chick mesodermal explants has established that Shh inhibits expression of *cFKBP/SMAP* (a key regulator of smooth muscle differentiation), in mesoderm near the endodermal source of Hedgehog signaling. Thus, ectopic Shh blocks smooth muscle differentiation, whereas cyclopamine induces it (Sukegawa *et al.*, 2000). Similarly, although activated Smo stimulates proliferation of the visceral mesenchyme in *Xenopus,* and hence leads to a thickening of the muscle layer (as discussed above), it also attenuates terminal differentiation of smooth muscle and the expression of smooth muscle actin (J. Zhang *et al.*, 2001). The result is a specific arrangement of cell types with inner nonmuscle lamina propria and submucosa layers and an outer smooth muscle layer. In addition, *BMP4* induced by Shh adjacent to the epithelium inhibits enteric neuron differentiation; as a result neural plexi are restricted to the outer edge of the mesodermal gut (Sukegawa *et al.*, 2000). This indirect inhibitory role of Shh on neural differentiation is substantiated by abnormal and decreased patterning of the enteric nervous system in the *Shh* mouse mutant (Ramalho-Santos *et al.*, 2000). The *Ihh* mutant also displays defects in the enteric nervous system. In contrast to the *Shh* mutant, however, the neural crest cells of the *Ihh* mutant seem to initially form a proper pattern and differentiate appropriately but then fail to survive and/or proliferate, indicating an ongoing role for Hedgehog signaling in radial patterning of the gut (Ramalho-Santos *et al.*, 2000).

Not surprisingly, given the importance of Hedgehog signaling at multiple stages of gut development, perturbation of Hedgehog signaling through either misexpression, targeted gene inactivation, or the manipulation of Hedgehog signal transduction components leads to defects in gut morphogenesis that parallel various common human congenital malformations. These include phenocopies of

Hirschsprung's disease (aganglionic colon), failure of gut rotation, intestinal transformation of the stomach, annular pancreas, hyposplenism, imperforate anus, duodenal stenosis, tracheal–esophageal fistula, and tracheal–esophageal atresia (Apelqvist *et al.,* 1997; Hebrok *et al.,* 1998; Kim *et al.,* 2001; Kim and Melton, 1998; Litingtung *et al.,* 1998; Motoyama *et al.,* 1998b; Park *et al.,* 2000; Pepicelli *et al.,* 1998; Ramalho-Santos *et al.,* 2000). These latter effects on esophageal atresia and tracheal–esophageal fistula may be due to a requirement for the forkhead gene *Foxf1*, a target of Shh signaling in the gut mesoderm that is essential for normal tracheal–esophageal development (Mahlapuu *et al.,* 2001).

Although less fully explored, Hedgehog signaling also acts at multiple stages during gut development in *Drosophila*. The visceral mesoderm of the gut is derived from cells of the anterior portion of each mesodermal parasegment. Once the visceral mesodermal layer is formed, it itself becomes segmented into anterior–posterior subdivisions, much like the better studied ectoderm but utilizing distinct genetic circuitry. These subdivisions, visualized by metameric expression of *connectin,* are established in response to Hh and Wg secreted from the ectoderm. Hh activates *connectin* expression in patches adjacent to Hh-expressing posterior ectodermal compartments, while Wg represses *connectin* in cells adjacent to anterior compartments (Azpiazu *et al.,* 1996; Bilder and Scott, 1998). Induction of these visceral mesodermal subdivisions is important in refining the *Hox* expression domains in the gut, which establish later boundaries of gut morphogenesis (Bilder and Scott, 1998).

At later stages of *Drosophila* gut development, Hh is not found in the endodermally derived midgut epithelium; however, it is expressed in the ectodermally derived fore and hindgut epithelia. In the foregut, this expression has been shown to be important for morphogenesis. In particular, the anterior portion of the invaginating foregut forms the esophagus while the posterior region gives rise to the proventriculus; in addition, the anterior portion expresses *dpp* while the posterior domain expresses both *wg* and *hh* (Pankratz and Hoch, 1995). Significantly, the maintenance of *wg* expression depends on Hh signaling whereas *dpp* appears to be independently regulated in the adjacent anterior tissue. Subsequently, the morphogenesis of the foregut at late stages involves a migration of the foregut epithelium outward, followed by a folding over itself and a movement back inward to produce a multiply folded proventricular organ. Notably, *hh* and *wg* are both expressed in the portion of the foregut epithelium that moves inward, and in both the *hh* and *wg* mutants this final inward movement fails to occur (Pankratz and Hoch, 1995).

In the hindgut, *hh* is expressed in the prospective rectum in response to Wg signaling from the adjacent prospective anal pads. Hh, in turn, is required for expression of *dpp* at the posterior of the adjacent large intestine. Thus, this cascade acts to subdivide the developing hindgut. Subsequently, Hh is required for the morphogenesis of the rectum, where it is also expressed (Takashima and Murakami, 2001).

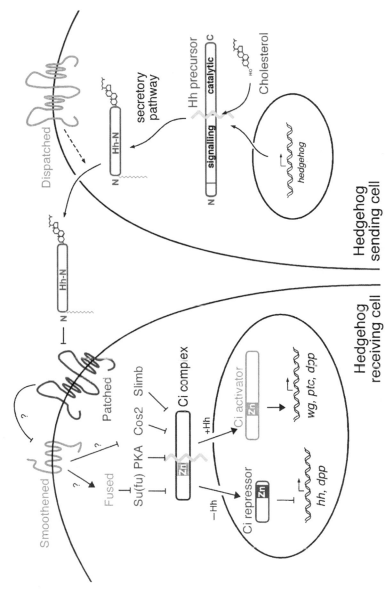

Figure 1.1 Schematic representation of the *Drosophila* Hh signaling pathway. See text for details. Components shown in red act negatively, whereas those in green act positively, in Hedgehog signal transduction. Reproduced with permission from McMahon, A. P. (2000). More surprises in the Hedgehog signaling pathway. *Cell* **100**, 185–188. The authors are grateful to Dr. Konrad Basler for producing this figure.

VIII. Hair and Feather Morphogenesis

In the developing skin, the interaction between focal cellular aggregates in the epithelia (called placodes) with adjacent mesenchymal condensations is responsible for the formation of various specialized structures such as scales, feathers, hair, nail, and claws. These are collectively referred to as epithelial appendages, epithelial organs, or adnexa. Significantly, there are both morphological and molecular similarities in the early stages of formation of the different epithelial appendages (reviewed in Chuong *et al.*, 2000; Thesleff *et al.*, 1995).

Both hair and feather development begin with changes in the presumptive bud ectoderm, which undergoes a transition from cuboidal to columnar to stratified epithelium while growing down into the adjacent dermis (Hardy, 1992; Oro and Scott, 1998; Wessells, 1965). Transplantation experiments have shown that this epithelial transition is dependent on signaling from the underlying dermis. As the placode forms, it signals back to the mesenchyme, inducing a local cellular condensation immediately below the placode. Concurrently, inhibitory lateral signals from the epidermal placode act to prevent additional placodes from forming in the immediate vicinity, contributing to the spacing of the epithelial appendages. Subsequent reciprocal signaling between cells of the placode and the condensation leads to growth and morphogenesis of the bud.

In the case of the hair follicle, the dermal condensation is responsible for secreting proliferative signals that cause the nascent follicle keratinocytes to grow down into the developing dermis. The dermal condensation itself gives rise to the dermal papilla of the resultant hair follicle. The derivatives of the epithelial placode differentiate into radially organized tissues composed of the hair itself, the inner root sheath, and the outer root sheath continuous with the basal keratinocytes of the skin. Stem cells reside in this outermost layer (Cotsarelis *et al.*, 1990; Rochat *et al.*, 1994) and are important in cycling hair growth.

Shh is expressed at the tip of the ingrowing epidermal placode (Bitgood and McMahon, 1995; Iseki *et al.*, 1996). Significantly, mice deficient in *Shh* have normally spaced epidermal placodes and associated dermal condensations, showing that the inductive signals involved in the initiation of hair follicles are intact in this mutant. However, subsequent morphogenesis of the hair follicle is strongly disrupted (Chiang *et al.*, 1999; St-Jacques *et al.*, 1998). Specifically, in the absence of Shh, there is a dramatic decrease in proliferation in the early follicle and a disruption of epithelial–mesenchymal interactions, as demonstrated by a lack of normal *Msx2* expression. In wild-type animals, the general Hedgehog target genes *Ptc1* and *Gli1* are expressed at high levels in the dermal papilla, showing that Shh acts from the epithelia on the mesenchymal component of the hair follicle. As expected, in the absence of Shh, these target genes are no longer upregulated. Taken together, these data suggest that at least one of the roles of Shh is to promote growth and expansion of the dermal papilla once it has formed. Finally, it is worth noting that the genes encoding several other secreted signals typical of the early hair bud epidermis, including *Bmp2*, *Bmp4*, and *Wnt10b*, continue to be expressed in the absence of Shh, suggesting that although these factors may be involved in initiating the formation of the hair bud, they are not sufficient to support its early morphogenesis without Shh signaling.

Another signal required for proper hair follicle morphogenesis is platelet-derived growth factor (PDGF). The gene encoding the PDGF receptor is normally expressed in aggregates of densely packed cells in the early mesenchymal condensation of the hair bud. In *Shh* mutants, genes encoding PDGF and its receptor are both expressed but the clusters of postmitotic receptor-expressing cells fail to form, suggesting that their aggregation is one aspect of dermal morphogenesis controlled by Shh (Karlsson *et al.*, 1999).

Because the *Shh* mutation is lethal, investigators have also grafted mutant skin to host animals to observe later development of the follicles in the absence of Shh (Chung *et al.*, 1999; St-Jacques *et al.*, 1998). Interestingly, late-stage differentiation markers are evident in those grafts, as are histologically distinct postmitotic cells with characteristics of hair components such as cuticle and hair. There is also evidence of proliferation at the proximal limit of the follicles that is comparable to normal levels. However, the resultant large, abnormal follicles contain small dermal papilla, fail to produce any hairlike material, and, most importantly, remain superficial without the extensive ingrowth of the epithelial component of the follicle that normally results in deep placement of the hair bulb in the dermis.

The regulation of *Shh* expression during hair follicle development has been related to the actions of members of the Wnt family. In particular, the ectopic expression of a stabilized form of β-catenin in the epidermis, using an epidermal promoter, induces *de novo* hair follicle morphogenesis, including *Shh* expression (Gat *et al.*, 1998). Conversely, in the absence of epidermal β-catenin, *Shh* is never expressed (Huelsken *et al.*, 2001). Together, these results suggest that *Shh* is positively regulated by canonical Wnt signaling during hair follicle morphogenesis. Candidates for this activity include *Wnt10a* and *Wnt10b*, which are expressed in the epithelium from the earliest stages of placode formation, preceding the expression of *Shh* (Reddy *et al.*, 2001). In addition, a noncanonical *Wnt* gene, *Wnt5a*, is expressed in the dermal condensation, which is the target tissue for Shh. Significantly, *Wnt5a* is not expressed in *Shh* mutant follicles (Reddy *et al.*, 2001).

In spite of its importance in hair development, Shh does not appear to play a role in the growth or differentiation of the stratified epithelium of the nonfollicular skin. In particular, none of the vertebrate *hedgehog* genes or their targets are expressed in the skin outside of the hair follicles (Bitgood and McMahon, 1995; Gat *et al.*, 1998; Wang *et al.*, 2000). Moreover, *Shh* mutants and mice injected with anti-Shh antibodies display normal development of a typical differentiated stratified epithelium (Chiang *et al.*, 1999; St-Jacques *et al.*, 1998; Wang *et al.*, 2000). Finally, although this initially normal epithelium in *Shh* mutant mice does undergo hyperplasia and displays abnormal keratin expression after grafting to wild-type mice (Chiang *et al.,* 1999; St-Jacques *et al.,* 1998), this is likely a secondary consequence of the superficial, abnormally juxtaposed hair follicles acting on the adjacent skin after their defective morphogenesis.

Once established during development, the number and location of the hair follicles do not change; however, the follicle itself undergoes cycles of growth (anagen), regression (catagen), and rest (telogen) (reviewed in Stenn and Paus, 2001). The cells of the postnatal follicle derive from a stem cell population localized to a specialized epithelial compartment located in the middle region of the mature follicular epithelium, referred to as the "bulge" (Fuchs and Segre, 2000; Oshima *et al.,* 2001; Taylor *et al.,* 2000; Watt, 1998). It is the shaft of the follicle below the bulge that periodically regresses and extends through the hair cycle. Cells within the bulge give rise to two populations: one set of progenitors that migrates upward to populate the basal layer of the epidermis and hence gives rise to the stratified epithelium of the skin, and a second population that responds to the dermal papilla and forms the growing hair epithelium (Oshima *et al.,* 2001; Taylor *et al.,* 2000).

Shh is expressed during the anagen growth phase in the distal cells of the growing hair follicle epithelium, while target genes such as *Ptc1* and *Gli1* are expressed in the hair stem cells of the bulge and in the adjacent dermal papilla (Gat *et al.,* 1998). The proper expression of *Shh* in the mature hair follicle is in part regulated by CDP (CCAAT displacement protein), a transcriptional repressor related to the *Drosophila Cut* gene. In particular, *CDP*-deficient mice have disrupted hair follicle morphogenesis, and *Shh* expression is seen to be shifted from its normal location at the tip of the epithelium to the inner root sheath (Ellis *et al.,* 2001).

The function of Shh during anagen of the adult hair cycle seems to be similar to its role during hair development, as mice injected with blocking anti-Shh antibodies have impaired follicle growth (Wang *et al.,* 2000). Defects are seen in both the epithelium and the dermal papilla. As in *Shh* mutant animals, the epithelium that does form is capable of some follicular-specific differentiation. Other experiments show that ectopically provided Shh is capable of driving resting hair follicles into anagen (Sato *et al.,* 1999). Specifically, adenovirus-mediated infection of mouse skin with *Shh* led to acceleration of anagen, increased follicular size, and accelerated onset of growth of new hair. This could imply that Shh normally functions as a trigger to initiate anagen, although the results are also consistent with Shh being downstream of the actual biological trigger.

Shh is expressed in the early feather bud in a manner similar to that seen during hair follicle morphogenesis. In particular, *Shh* is expressed in the feather epithelial placode, initially in the center and then shifting to the distal end (Morgan *et al.,* 1998; Nohno *et al.,* 1995; Ting-Berreth and Chuong, 1996). Like the formation of the placode itself, the expression of *Shh* is dependent on the underlying mesenchyme, since separation of the epithelium and reapposition in a different orientation results in loss of *Shh* in the original locations and re-expression above the dermal condensation (Chuong *et al.,* 1996). In addition, *Ptc1* and other target genes are expressed in the dermal condensations (Chuong *et al.*, 2000) indicating that, as in hair follicle development, the mesoderm is the target of Shh signaling.

In functional assays, forced expression of *Shh* results in alterations in dermal condensations, suggesting that Shh serves to initiate and organize the outgrowth of the bud (Ting-Berreth and Chuong, 1996). However, this interpretation is unlikely as heterospecific recombination between avian and mammalian ectoderm and dermis has shown that the signals mediating feather bud and hair follicle development are conserved (Dhouailly, 1973, 1975; Garber, 1964; Garber *et al.*, 1968) and, as discussed above, hair follicle induction occurs normally in the absence of Shh. Moreover, expression analysis using early markers for the epithelial placode and dermal condensation demonstrate that these form during normal feather development before the onset of *Shh* expression (Morgan *et al.*, 1998).

Some of the differences in these experimental studies of the role of Shh in the feather buds appear related to the timing of Shh application. In particular, careful examination of the consequence of ectopic Shh at various times reveals a shifting competence in the skin (Morgan *et al.*, 1998). Before feather bud formation, Shh causes a disorganized ectodermal proliferation, presumably an indirect effect mediated by the Hedgehog-responsive dermis. However, as the skin begins to differentiate, the forced expression of *Shh* causes precocious feather bud formation in the normal feather tracts as well as large, abnormal feather buds in the region between the feather tracts. The difference between the feather-forming and nonforming tracts is determined by the dermis (Sengel, 1975), which presumably regulates the difference in Shh responsiveness. These ectopic results, which as discussed above cannot reflect the normal events of feather bud initiation, are interpreted as reflecting instead the existence of an inhibitory signal produced by the nascent feather bud and blocking Shh responsiveness in surrounding regions before the onset of *Shh* expression. Premature *Shh* expression thus triggers an inappropriate activation of reciprocal signaling between the dermis and ectoderm, leading to ectopic buds. However, once the field of feather buds is established, the entire skin is rendered nonresponsive to Shh and, hence, forced expression of *Shh* in the interfollicular epidermis has no morphological effect after this time.

An additional late role for *Shh* in feather morphogenesis is indicated by its expression at mature filament stages. At this time, *Shh* is expressed in 9–11 longitudinal stripes running to the apex of the feather filament (Nohno *et al.*, 1995; Ting-Berreth and Chuong, 1996). These expression domains coincide with the outer marginal zones of each barb ridge, although the functional significance of this expression remains to be explored.

IX. Implantation

Implantation of the mammalian embryo within the receptive uterus requires interactions between embryo and maternal tissues that lead to attachment of the blastocyst and subsequently trigger the decidual response. Ihh is one of a number of signaling factors that are expressed in association with these events (Paria *et al.*, 2001). *Ihh* is expressed along the length of the uterine epithelium including the endometrial glands before implantation, whereas *Ptc1* expression is upregulated in the underlying stroma, suggesting an epithelial–mesenchymal signaling axis. *Ihh* expression is not altered immediately on implantation, but declines thereafter, except within uterine epithelium overlying the embryo (Paria *et al.*, 2001). *Ihh* function within the uterus has not been addressed.

X. Kidney and Adrenal Gland

Shh and *Ihh* are expressed in the mammalian metanephric kidney, *Shh* in the collecting duct epithelium and *Ihh* in proximal convoluted and straight tubules (Bitgood and McMahon, 1995; Marigo *et al.*, 1995; Valentini *et al.*, 1997). *Shh* mutants have a single fused midline kidney and $Gli2^{-/-}/Gli3^{+/-}$ mutants have paired, horseshoe-shaped kidneys (Kim *et al.*, 2001). Both phenotypes are most likely secondary defects resulting from disruptions of earlier Hedgehog-dependent processes. *Shh* is also expressed in the adrenal cortex (Bitgood and McMahon, 1995). As with the kidney, the putative role of Hedgehog signaling in this organ awaits a serious study.

XI. Lateral Asymmetry

When viewed superficially, vertebrates appear bilaterally symmetric; however, the vertebrate body plan encompasses numerous internal morphological asymmetries. Developmentally, these asymmetries do not emerge until fairly late in embryogenesis—the left and right sides of the vertebrate embryo remain mirror images of each other until late neurulation when the linear heart tube, formed from paired cardiac progenitors in the left and right lateral plate mesoderm, undergoes right-sided looping. This morphogenic process is conserved among all vertebrates. Heart looping is followed by directional rotation of the embryo (180° to a ventrally flexed position in mammals, 90° to a left-facing position in avian species), and as development proceeds, other internal organs likewise adopt left–right asymmetric shapes and positions (such as the stomach), undergo left–right asymmetric morphogenesis (such as gut coiling), or form specifically on the left or right sides of the body (such as the spleen). Even paired organs can have asymmetric features (such as differential numbers of lobes in the lung on the left and right sides).

Recently, a number of genes have been described that are expressed and/or act asymmetrically on the left and right sides during early embryogenesis and later during organogenesis. The first two such genes to be described were identified in the chick and encode two extracellular signals, *Shh* and the *TGFβ* family member *Nodal* (Levin *et al.*, 1995). Both are expressed well before any morphological signs of asymmetry and act in a cascade that directs future left–right asymmetries. Interestingly, while the downstream member of this cascade, *Nodal,* is conserved in all species examined, the early role of *Shh* in this process seems to be specific to chick and other avian species. However, in spite of its lack of conservation, the left–right-asymmetric Shh signaling cascade in chick embryos is better understood than the genetic networks regulating laterality in other vertebrate species, and as such, it can serve as a general paradigm for understanding the regulation of left–right-specific morphogenesis.

The initial breaking of bilateral symmetry in mammals appears to be accomplished through a directional current set up at Hensen's node by the action of specialized cilia that turn in a consistent counterclockwise direction in the extraembryonic fluid (Nonaka *et al.,* 1998). Similar nodal cilia have been observed in other species, including chick embryos (Essner, 2002). Moreover, the earliest left–right asymmetric molecular events in chick do seem to occur in the region of Hensen's node, where asymmetric expression, including that of *Shh,* appears to be instructed by adjacent tissues. This latter is deduced from the fact that 180° rotation of the node itself at early developmental stages has no effect on the side of subsequent gene expression within the node (Pagan-Westphal and Tabin, 1998). The regulation of gene expression at the node may also involve gap junctions acting to partition left–right determinants. In particular, treatment with lindane, which decreases gap junction communication, as well as manipulations that block activity of the gap junction protein connexin 43, both lead to bilateral *Shh* expression (Levin and Mercola, 1999). These results are consistent with a model based on earlier studies in *Xenopus* implicating gap junctions in early left–right establishment (Levin and Mercola, 1998).

Regardless of the precise nature of the early environmental signals responsible for establishing asymmetric gene expression within the node, it is clear that unpolarized *Shh* expression in the node at stage 4 (Levin *et al.,* 1995; Pagan-Westphal and Tabin, 1998) gives way to a left side-specific expression of *Shh* in the node during subsequent stages (Levin *et al.,* 1995). This occurs simultaneously with the expression of *Activin βB* on the right side of the node (Levin *et al.,* 1997) and upregulation of the gene encoding Activin receptor (*ActRIIa*) on the right (Levin *et al.,* 1995). Significantly, ectopic Activin applied on the right side induces *ActRIIa* and represses *Shh* (Levin *et al.,* 1995). *Bmp4* is also induced on the right by Activin and, because it too is able to repress *Shh* when applied to the left side of the node, BMP4 is believed to be both a secondary signal downstream of Activin βB and the proximal upstream signal repressing *Shh* on the right (Monsoro-Burq and Le Douarin, 2000). From stage 5 to stage 6, when the asymmetry of the node is no longer influenced by surrounding tissues (Pagan-Westphal and Tabin, 1998), BMP4 continuously represses *Shh* on the right side of the node while Shh simultaneously represses *Bmp4* on the left side (Monsoro-Burq and Le Douarin, 2001). This sets up a robust regulatory loop, assuring that small amounts of either signal diffusing to the opposite side of the node do not destroy the asymmetric pattern.

Slightly later, the reinforced BMP4 expression on the right results in activation of *Fgf8* on the right side (Boettger *et al.*, 1999). This acts as a further buffer to the system, as *Fgf8* functions on the right side to repress *Nodal*, the same downstream target that is activated by Shh on the left (Boettger *et al.*, 1999).

The function of left-sided *Shh* expression in the chick is to initiate a cascade of left-sided gene expression patterns in the left lateral plate, thereby providing left–right-asymmetric positional information throughout the chick embryo. This positional information is used to bias the direction of asymmetric morphogenesis. It is important to note, however, that this signaling is not required for all of the morphogenetic events to take place, but only to establish the direction in which they occur. Thus, bilateral expression of Shh (or, conversely, the absence of Shh signaling) in the chick does not lead to a failure of heart looping, but rather to a randomization of the side to which the heart loops (Levin *et al.*, 1995). It is also significant that the responses of different organs to the signals downstream of Shh appear to be independent of one another; thus, in the absence of the bias provided by Shh and its downstream effectors, the direction of heart looping is completely decoupled from the direction of body rotation and gut coiling (Levin *et al.*, 1997). This phenomenon is also seen in various perturbations of laterality in other species, including heterotaxia syndromes in humans (Afzelius, 1976).

Unlike Shh, which seems to have adopted a specialized role in avian species, many of the downstream cues in the lateral plate mesoderm responsible for directing an asymmetric left–right morphogenesis appear to be conserved in all vertebrates. These include the secreted signal Nodal and the downstream transcription factor Pitx2 (reviewed in Capdevila *et al.*, 2000). Asymmetric *Nodal* expression is first seen in the chick in a small mesodermal domain directly adjacent to *Shh*-expressing cells on the left side of the node at stage 7 (Levin *et al.*, 1995). Significantly, this medial expression domain is discontinuous with the second broad domain in the lateral plate mesoderm, which subsequently expresses *Nodal*. Because Hedgehog-responsive genes such as *Ptc1* (Pagan-Westphal and Tabin, 1998) and *Ptc2* (Pearse *et al.*, 2001) are not induced beyond the region directly adjacent to the node, the lateral plate expression of *Nodal* appears to be an indirect response to Shh. Moreover, Shh is unable to induce *Nodal* in lateral plate mesoderm explants unless paraxial tissue is included as well, implicating the paraxial mesoderm as a source of a *Nodal*-inducing secondary signal downstream of Shh (Pagan-Westphal and Tabin, 1998). A candidate for this secondary signal is the putative BMP antagonist Caronte, a member of the Cerberus/Dan family (Rodriguez Esteban *et al.*, 1999; Yokouchi *et al.*, 1999; Zhu *et al.*, 1999). In particular, *Caronte* is expressed in the paraxial mesoderm at the right time to play such a role; it is induced in response to Shh and repressed by FGF8, and it can induce *Nodal* and *Pitx2* expression when applied ectopically to the right side. Furthermore, because Caronte has been shown to bind BMPs *in vitro* (Rodriguez Esteban *et al.*, 1999; Yokouchi *et al.*, 1999) and because ectopic BMPs can under some experimental conditions repress *Nodal* in the lateral plate, it has been proposed that BMPs act bilaterally to repress *Nodal* in the lateral plate mesoderm and that Caronte acts to relieve this inhibition on the left side, leading to derepression of *Nodal* locally.

Shh-induced Caronte expression also appears to play a role in derepressing expression of another *TGF-β* family member, *Lefty-1*, in the left side of the notochord (Rodriguez Esteban *et al.*, 1999; Yokouchi *et al.*, 1999). The midline expression of *Lefty-1* is known to be important as part of a midline barrier that keeps signals specific to the left side from acting inappropriately on the right (Meno *et al.*, 1998). Because Lefty-1 has been shown to function as an antagonist of Nodal activity, it is believed to act primarily as a barrier against Nodal activity. However, the midline serves as a barrier against additional asymmetric signals as well, an effect presumably mediated by additional genes. For example, studies of conjoined twin chick embryos showed that very early asymmetric signals can diffuse widely in the absence of a midline barrier, such that negative-acting factors from the right side of one embryo (perhaps Activin βB or BMP4) will repress *Shh* on the left side of a twin positioned to its right, but not vice versa (Levin *et al.*, 1996). There is also a need for a buffering function in the midline to prevent asymmetric signals such as BMP4 from repressing *Shh* inappropriately in the notochord and floorplate. Such a role seems to be mediated by midline expression of the BMP antagonists Noggin and Chordin (Monsoro-Burq and Le Douarin, 2001).

The necessity for a midline barrier was first indicated by surgical experiments with *Xenopus* embryos (Danos and Yost, 1996) and analysis of twin embryos in chicks (Levin *et al.*, 1996). This midline requirement for achieving proper left–right morphogenesis was verified by examination of the morphological effects of genetic alterations that produce midline defects in other species. Thus, there is a randomization of *Nodal* expression and left–right orientation in *notail* and *floating head* mutants in zebrafish, which lack a differentiated notochord (Danos and Yost, 1995, 1996; Halpern *et al.*, 1993; Lohr *et al.*, 1997; Talbot *et al.*, 1995). Similarly, defects in midline development and resultant randomization of left–right asymmetry are seen in a number of mouse mutations including double-heterozygous *HNF3β*[+/−]/*Nodal*[+/−] animals (Collignon *et al.*, 1996), animals deficient in the gene encoding type *IIB* Activin receptor (Oh and Li, 1997) or in *brachyury* (King *et al.*, 1998), and the SIL and *no turning* mutations (Izraeli *et al.*, 1999; Melloy *et al.*, 1998). Laterality malformations are also correlated with human syndromes displaying midline defects (Goldstein *et al.*, 1998; Roessler and Muenke, 2001).

The importance of the midline in establishing left–right differences in the embryo provides a context for understanding the consequence of loss of Shh activity in mammalian embryos. Importantly, the key role of Shh as a trigger for left-sided signaling adjacent to Hensen's node seems to be specific to avian species, as neither *Shh* nor any other *hedgehog* family member or any general *Shh* target such as *Ptc1* has been found to be left–right asymmetric in its expression in any mammalian, fish, or frog species. Nonetheless, mouse embryos lacking *Shh* display multiple laterality defects including alterations in heart looping, axial turning, and possibly, pulmonary isomerisms (Meyers and Martin, 1999; Tsukui *et al.*, 1999). These phenotypes may best be explained by the importance of Shh in establishing midline structures such as the floorplate (see Section III, Central Nervous System), which is the source of the midline barrier Lefty-1 in the mouse (Meno *et al.*, 1996).

Although asymmetric *hedgehog* gene expression is not part of the asymmetric gene cascade in mammals, there is nevertheless evidence that Hedgehog proteins are required for establishing this cascade. In particular, *Shh* (Echelard *et al.*, 1993) and, more weakly, *Ihh* (X. Zhang *et al.*, 2001) are both expressed in a uniform pattern in the node, and both *Ihh/Shh* compound mutant and *Smo* mutant embryos fail to express *Nodal* asymmetrically in the left lateral plate mesoderm and do not undergo asymmetric morphogenesis (J. Zhang *et al.*, 2001; X. Zhang *et al.*, 2001). Two other genes encoding secreted factors expressed symmetrically in the node, *Fgf8* and *Gdf1*, are also required for asymmetric *Nodal* expression (Meyers and Martin, 1999; Rankin *et al.*, 2000), and the expression of *Gdf1* in the node (but not that of *Fgf8*) is dependent on Hedgehog signaling (X. Zhang *et al.*, 2001). One way of reconciling these observations with the chick data is to assume that Hedgehog signaling is required along with a second, unknown signal to initiate the left-side signaling cascade. In chick, the asymmetric expression of *Shh* provides the left–right directional information to the system, whereas in mice the other, hypothetical pathway would provide the initial asymmetric information; nevertheless, both genes may be required to activate downstream targets such as *Nodal* in both systems. This would imply that *Gdf1* might be a downstream target of Hedgehog signaling in the chick as well, something that has not yet been examined.

There is also some evidence that Hedgehog signaling may play a role in initiating left–right patterning in the zebrafish. Specifically, ectopic right-sided production of either Shh or a dominant-negative form of PKA, which phenocopies Hedgehog signaling, results in randomization of left–right asymmetry, whereas similar misexpression on the left has no effect (Schilling *et al.*, 1999). This is completely consistent with the role of Shh in the chick in biasing the situs of the heart and visceral organs; however, as in the mouse, no asymmetric expression has been observed in the known zebrafish *hedgehog* genes (*shh*, *twhh*, *ehh*, and *hhc*). Moreover, neither the *shh* mutation, *sonic you* (Chen *et al.*, 1997; Schauerte *et al.*, 1998), nor the zebrafish *smo* mutant, which inactivates signaling downstream of all the hedgehog genes, has laterality defects (Chen *et al.*, 2001). Thus, Hedgehog signaling does not appear to be required to establish left–right asymmetry in the zebrafish, although a maternal source of Hedgehog signaling components cannot be ruled out. The most likely explanation for the misexpression data is that the experimental introduction of Hedgehog signaling on the right has the effect of ectopically activating some other signaling system that is normally involved in specifying asymmetric morphogenesis on the left side.

In addition to its role in establishing left–right asymmetry, Hedgehog signaling appears to be required for the morphogenesis of the developing heart. As discussed above, mutations that direct the left-sided signaling cascade, including those that lead to a loss of *Nodal* and *Gdf1* expression, typically result in a randomization of heart looping. However, in *Smo* mutants, the initial formation of the heart tube is delayed and, once formed, it remains a linear tube rather than undergoing the looping that characterizes the first step of heart morphogenesis (X. Zhang *et al.*, 2001). This defect is specific to morphogenesis, as ventricular differentiation proceeds despite the failure of heart looping.

Much less is known about left–right-asymmetric development in *Drosophila;* however, lateral asymmetries in the development of the gut have been described (Hayashi and Murakami, 2001). Although the genetic regulation of this process has not been extensively studied, the handedness of morphogenesis of both the foregut and midgut is randomized in *ptc* mutants, whereas the hindgut is unaffected (Hayashi and Murakami, 2001). This implicates Hh signaling as necessary for establishing left–right asymmetry in the gut, although *hh* and *ptc* themselves are not known to be asymmetric in their expression during *Drosophila* gut morphogenesis.

Unlike the vertebrate heart, the *Drosophila* dorsal vessel does not display any morphological left–right asymmetries. Nonetheless, *hh* does play a role in *Drosophila* heart formation. In particular, the homeodomain protein Ladybird is involved in the specification of a subset of heart precursors in each segment as a component of the cardiogenic Tinman cascade (Jagla *et al.*, 1997), and restriction of *ladybird* expression to particular cardioblasts requires negative regulation by *hh* (Jagla *et al.*, 1997).

XII. Limb and Fin of Vertebrates

High levels of *Shh* transcripts are localized along the posterior margins of the developing limb and fin buds (Echelard *et al.*, 1993; Krauss *et al.*, 1993; Riddle *et al.*, 1993). Both spatially and temporally (Riddle *et al.*, 1993), this striking pattern coincides with the zone of polarizing activity or ZPA, a region of limb bud mesenchyme previously identified as a key signaling center during development.

Before the advent of modern molecular tools, experimental embryologists were able to identify a few so-called signaling centers responsible for patterning regions of the vertebrate embryo, but the tools for identifying the inductive molecules responsible for their activities were lacking. An important example is the ZPA, which has served as a paradigm for theoretical models of pattern formation. The ZPA was discovered accidentally when tissue from the posterior margin of the chick limb bud was transplanted to the anterior margin of a host limb bud as an experimental test of an unrelated issue. The dramatic result, however, was a mirror image duplication of the digits (Saunders and Gasseling, 1968). This result was interpreted as suggesting that the normal role of the ZPA was to produce a signal that both induced proliferation of the limb bud mesenchyme (because the transplant caused growth of an additional set of digits) and instructed naive tissue to adopt positional information along the anterior–posterior axis (because the resultant digits were polarized such that they formed a mirror image of the normal set) (Tickle *et al.,* 1975). Furthermore, the nature of the signal produced by the ZPA was proposed to be a morphogen (i.e., a molecule that acts in a concentration-dependent manner), because transplanting the ZPA to different locations within the limb bud (Tickle, 1981a) and transplanting different numbers of ZPA cells (Tickle, 1981b) resulted in duplications of varying severity, thus providing evidence that the response to the ZPA was concentration dependent.

Shh can produce a phenotypic response indistinguishable from a ZPA transplant when delivered either by cells expressing the protein or by beads loaded with purified recombinant protein (Chang *et al.,* 1994; Lopez-Martin *et al.,* 1995; Riddle *et al.,* 1993). Moreover, like the ZPA, the number of digits duplicated is dependent on the concentration of Shh delivered to the limb bud and as concentration is increased, the duplicated digits display progressively more posterior characteristics (Yang *et al.,* 1997).

Although other members of the Hedgehog family can also produce ZPA-like limb duplications (Vortkamp *et al.,* 1996), they are the only factors identified to date that can have such an effect. Moreover, almost all other tissues that have the capacity to induce ZPA-like duplications when transplanted into the limb bud, including Hensen's node (Hornbruch and Wolpert, 1986; Saunders and Gasseling, 1983; Stocker and Carlson 1990), the notochord (Wagner *et al.,* 1990), and the floor plate of the neural tube (Wagner *et al.,* 1990), likewise express *Shh* (Echelard *et al.,* 1993; Krauss *et al.,* 1993; Riddle *et al.,* 1993). In addition, the few exceptions to this, such as the interlimb flank mesoderm, have been shown to initiate *Shh* expression on transplantation into the distal limb bud (Lu *et al ,* 1997; Tanaka *et al.,* 2000; Yonei *et al.,* 1995).

Taken together, these data provide compelling evidence that the classic experiments that identified the ZPA as a signaling center in the limb were assaying for cells that produced Shh. These experiments left open numerous questions. For example, is Shh required for normal limb patterning (and if so, is signaling acting directly or indirectly), by what mechanism does Shh affect limb pattern, and how is Shh expression polarized within the limb bud?

Evidence that the ZPA region is in fact required for normal limb development was obtained by careful surgical deletion (Pagan *et al.,* 1996), but definitive proof of the essential role of *Shh* itself came from genetic experiments in which a targeted deletion was produced in the *Shh* gene (Chiang *et al.,* 1996). In particular, *Shh* mutant embryos have recognizable humerus and femur bones (the stylopod) with readily apparent anterior–posterior polarity; however, distal to the elbow/knee joint, anterior–posterior polarity is lost (Chiang *et al.,* 2001; P. M. Lewis *et al.,* 2001). In the forelimb, the humerus extends beyond the elbow to form a continuous element with a single forearm bone; the hindlimb, by contrast, has both a short malformed tibia/fibula element (the zeugopod) and a single digit. On the basis of the number of bones in the hindlimb digit, it would appear to correspond to the most anterior digit (digit 1), and molecular analysis (e.g., of *Hox* genes) confirms these interpretations (Chiang *et al.*, 2001; Kraus *et al.*, 2001; P. M. Lewis *et al.*, 2001). As expected, mutant limb bud tissue lacks ZPA activity when transplanted into chick limb buds.

Besides confirming the necessity of *Shh* for limb skeletal patterning, the two most important conclusions to emerge from these studies are, first, that sufficient anterior–posterior polarity exists in the early limb bud to pattern the proximal stylopod elements independently of *Shh*, and, second, that the specification of the most anterior digit does not require *Shh*. The first of these conclusions was not entirely unexpected. For example, *Shh* is not detected in the limb buds of the chick *limbless* mutant, and although these limb buds degenerate shortly after budding, molecular analysis shows that the initial expression of *Hox* genes is in fact polarized along the anterior–posterior axis (Grieshammer *et al.*, 1996; Normaly *et al.*, 1996; Ros *et al.*, 1996). Moreover, it has been shown in the chick that the dynamic expression patterns of *Hox* genes progress through three phases in the developing limb bud, correlating with the formation of the stylopod, zeugopod, and autopod, and that the first of these phases is *Shh*-independent whereas the latter two appear to be induced by Shh (Nelson, 1996). However, the second conclusion (that digit 1 does not require Shh) was more surprising, as ectopic Shh can induce duplication of a full set of digits in the chick, including the most anterior digit. Although it should be noted that the anteriormost digit in the chick wing corresponds to digit 2, *Shh* also appears capable of inducing an ectopic digit 1 in some mouse mutants (Theil *et al.*, 1999).

Making sense of the *Shh* null mutant phenotype and understanding how the different digits are specified also requires an appreciation of the interdependence between the ZPA and two other signaling centers in the limb bud: the apical ectodermal ridge (AER), which is involved in limb outgrowth and proximodistal patterning, and the dorsal ectoderm, which is involved in dorsal–ventral specification. The AER is a ridge of epithelium running along the anterior–posterior axis of the distal tip of the developing limb bud. If the AER is removed, outgrowth is affected, resulting in distal truncations (Rowe and Fallon, 1982; Saunders and Fallon, 1948; Summerbell, 1974). In addition, the AER is required for maintaining ZPA activity in the posterior mesenchyme (Vogel and Tickle, 1993). Several members of the fibroblast growth factor (*Fgf*) family are expressed in the AER and seem to be responsible for both of these properties. In particular, exogenous FGF application can restore the outgrowth of a limb bud after AER extirpation (Fallon *et al.*, 1994; Niswander *et al.*, 1993), and FGFs can replace the AER in terms of maintaining ZPA activity in the posterior mesenchyme (Niswander and Martin, 1992; Suzuki *et al.*, 1992; Vogel and Tickle, 1993).

On a molecular level, FGFs act on the posterior mesenchyme to maintain the expression of *Shh* (Laufer *et al.*, 1994; Niswander *et al.*, 1994). In a reciprocal interaction, Shh maintains the expression of *Fgf4* (Chiang *et al.*, 2001; Laufer *et al.*, 1994; Niswander *et al.*, 1994) and other *Fgf* family members (Sun *et al.*, 2000) in the posterior AER. The maintenance of *Fgfs* in the AER by Shh is indirect and appears to be mediated through the action of the BMP antagonist Gremlin. Notably, blocking BMP activity with ectopic Gremlin or other functionally equivalent antagonists induces ectopic *Fgf4* expression in the anterior AER. Maintenance of *Gremlin* expression requires in turn both the inductive influence of Shh and the permissive expression of the *limb deformity* gene. In *limb deformity* mutants, *Fgf4* expression is lost but can be rescued with ectopic BMP antagonists (Capdevila *et al.,* 1999; Merino *et al.,* 1999; Zuniga *et al.,* 1999).

In addition to the positive feedback loop between Shh from the ZPA and FGFs from the AER, a further integration of FGF and Shh signaling occurs at the level of target gene induction. Although Shh is able to directly induce some target genes in the posterior limb bud mesenchyme, such as *Ptc1* and *Gli1* (Marigo *et al.,* 1996a,b), simultaneous exposure to both FGFs and Shh is required to induce *Bmp2* and *Hoxd* cluster genes (Laufer *et al.,* 1994).

In addition to the AER and ZPA, the dorsal ectoderm is a third signaling center defining the three-dimensional pattern of the limb bud. Classically, the dorsal ectoderm was known to be important for establishing the dorsal–ventral axis of the developing limb bud (MacCabe *et al.,* 1974; Pautou, 1977). More recently, it has been shown that the dorsal ectoderm produces Wnt7a, which, through its induction of *Lmx1* in the underlying mesoderm, is responsible for establishing dorsal fate in the distal limb bud (Parr and McMahon, 1995; Riddle *et al.,* 1995; Vogel *et al.,* 1995). *Wnt7a* mutant limbs, in addition to displaying dorsal–ventral patterning defects, are shorter and often lack posteriormost skeletal elements. These aspects of the phenotype are readily explained by a significant reduction in *Shh* expression in *Wnt7a* mutant animals (Parr and McMahon, 1995). As a result of the decrease in polarizing activity, the posterior-most digit is not specified and there is a concomitant decrease in *Fgf* gene expression in the AER, producing distal truncations. Supporting this interpretation, removal of the dorsal ectoderm leads to a reduced level of *Shh* expression, which can be rescued with ectopic Wnt7a (Yang and Niswander, 1995).

Many aspects of the regulatory circuitry involving *Shh* in the tetrapod limb bud are also seen in the developing fin bud, the homologous structure in the zebrafish (reviewed in Grandel and Schulte-Merker, 1998). The pectoral fin bud is first apparent at 26 h postfertilization. The distal epidermis covering this bud thickens and by 36 h has formed an apical fold in a position analogous to the limb bud AER. For approximately 12 h, the growth of the fin bud appears similar to that of the limb bud. However, at 48 h, the apical fold begins to radically expand and is invaded by mesenchyme, forming the fin fold. The proximal bud gives rise only to a few basal endochondral bones homologous to the tetrapod limb skeleton, whereas the fin fold gives rise to the dermal fin rays, the lepidotrichia, visible in the external mature fin.

During the early phase of fin bud development, gene expression patterns are similar to those seen in the limb buds. In particular, *Shh* is expressed in the posterior mesenchyme (Krauss *et al.,* 1993), while *Fgf4* (Grandel *et al.,* 2000) and other *Fgf* family members (Furthauer *et al.,* 1997; Grandel *et al.,* 2000; Reifers *et al.,* 1998) are expressed in the apical fold. Moreover, the fin bud mesenchyme expresses putative Shh target genes including members of the *Hoxd* cluster (Sordino *et al.,* 1995) and *Bmp2,* as well as *Ptc1* and *Ptc2* (Akimenko and Ekker, 1995; Neumann *et al.,* 1999). Analysis of the zebrafish *sonic you* (*syu*) mutant, a null allele of *shh,* shows that, as in the tetrapod limb, anterior–posterior polarized expression of *Hoxd* genes is initially present in the absence of Shh, but continued and elaborated *Hox* expression requires Shh (Neumann *et al.,* 1999). In addition, the *Shh–Fgf4* feedback loop seems to be operating, as *Fgf4* expression is lost in the *syu* mutant (Grandel *et al.,* 2000).

One important difference between these interactions in tetrapods and zebrafish is the regulation of *Fgf8.* In land vertebrates, *Fgf8* is expressed independently of Shh both in the early pre-AER ectoderm and in the AER after its formation (Crossley *et al.,* 1996; Vogel *et al.,* 1996). Because *Fgf8* does not rely on Shh signaling to maintain its expression (Chiang *et al.,* 2001), there may be sufficient FGF signaling in the AER even in the absence of Shh to allow survival of the limb bud and formation of some distal skeletal elements (Chiang *et al.,* 2001). In zebrafish, by contrast, *Fgf8* expression initiates significantly later and is under the control of *Shh.* Hence, in the *syu* mutant, which lacks Shh, *Fgf8* expression is not observed and the ectodermal fold fails to form, resulting in a loss of the entire fin endoskeleton.

A second major difference seen in the fin, as discussed above, is the extensive dermal skeleton, which forms from the fin fold subsequent to the period of development analogous to the formation of tetrapod limb bud. During this later phase, *Shh* is expressed in a subset of cells in the basal layer of the ectoderm, adjacent to the newly formed lepidotrichia. The expression of *Ptc1* (the homolog of the gene called *Ptc2* in other vertebrates) can be used as an indication of the tissue responding to the Shh signal. *Ptc1* is expressed in an overlapping domain encompassing the Shh-producing cells, in other cells in close proximity with the ectoderm, and also in adjacent mesodermal cells that deposit the bone matrix. *Bmp2,* encoding a potential secondary signal downstream of Shh (as discussed below), is expressed in a pattern similar to that of *Ptc1* (Laforest *et al.,* 1998). These expression patterns are consistent with a role for Shh in inducing the differentiation of the cells that produce the skeletal matrix (the scleroblasts) or in regulating the synthesis of the matrix by these cells.

Because the chick system is easier to manipulate experimentally, the developmental role of *Shh* in the early limb bud is more clearly defined than in the zebrafish fin bud. Like ZPA transplants, ectopic Shh administered to the anterior limb bud results in two effects: reorganization of anterior–posterior limb pattern and induction of proliferation in the anterior limb bud mesenchyme (Riddle *et al.,* 1993; Saunders and Gasseling, 1968; Tickle *et al.,* 1975; Wolpert, 1969). Although Shh may act directly as a mitogen in a number of developing systems (see, e.g., Section III, Central Nervous System), its effect on the proliferation of limb bud mesenchyme is indirect and can be attributed to the regulation of *Fgf* in the AER. Thus, if the AER is surgically removed from the anterior half of the limb bud, anterior growth ceases. Moreover, implantation of a bead soaked in FGF protein not only rescues normal growth, but also drives significant ectopic proliferation of the anterior mesenchyme, although there is no induction of patterning genes such as *Bmp2* or *Hoxd* genes. In parallel experiments, introducing ectopic Shh into the AER-denuded anterior limb bud has no effect on either proliferation or outgrowth. Furthermore, although coadministration of FGF and Shh in this system does result in ectopic expression of patterning genes, it produces no more proliferation or outgrowth than application of FGF alone (Laufer *et al.,* 1994). Thus, at least in the context of the developing limb bud mesenchyme, Shh does not act as a mitogen and affects proliferation only indirectly through secondary signals.

The question of whether Shh exerts a direct effect on mesenchymal patterning has been more controversial. As discussed above, it is clear that there is a dose-dependent patterning response by the limb bud mesenchyme to ZPA signaling. It has also been shown that Shh itself mimicks this effect in a concentration-dependent manner. This does not necessarily imply, however, that Shh acts as a morphogen in the limb bud (as it apparently does in the neural tube or as Hh does in the *Drosophila* ectoderm). For example, Shh could act on immediately adjacent cells to induce, at a level proportional to its concentration, the expression of a second long-range signal that is a true morphogen (similar to the role of induction of *dpp* by Hh in the *Drosophila* wing disc), or the situation could be more complex still.

A first step toward sorting out the mode of action of Shh in the limb bud is to determine the range over which it acts. Experiments on chick embryos have indicated that the ZPA itself can act over a considerable distance, on the order of 150–200 μm (Honig, 1981). This distance, approximately 10–20 cell diameters or close to half the width of an early limb bud, is close to the size of the digit field in the early limb bud, suggesting that the activity of the ZPA could be responsible for patterning the entire anterior–posterior axis. A first indication that Shh itself could act over this range came from examination of expression of direct transcriptional targets such as *Ptc1*. Significantly, in the chick, *Ptc1* is expressed in a gradient across the posterior one-third to one-half of the early limb bud, suggesting that Shh directly acts over that distance (Marigo *et al.,* 1996a). Moreover, Shh protein can be detected in a gradient over a distance of approximately 300 μm in the early mouse limb bud, a range closely correlated with the expression of Shh target genes, including *Ptc1* (P. M. Lewis *et al.,* 2001; Zeng *et al.,* 2001). Together, these results strongly indicate that Shh actively signals across the entire digit field; however, it does not provide evidence that this broad signaling is necessarily relevant for digit patterning.

If the long-range effects of Shh in the limb bud mesenchyme are mediated by a secondary signal, the best candidate at the moment is *Bmp2*. *Bmp2* is expressed in the posterior limb bud (Francis, 1994) in a much broader domain than *Shh*, in a region similar to that of *Ptc1*. Furthermore, like its homolog *dpp* in the *Drosophila* wing disc, *Bmp2* is a target of Shh signaling in the vertebrate limb bud (Laufer *et al.,* 1994). However, unlike the case in *Drosophila,* BMP2 itself is incapable of inducing the mirror image duplications seen with ZPA duplications. Indeed, in some assays, BMPs inhibit limb mesenchyme proliferation and induce cell death (Dahn and Fallon, 1999; Zou *et al.,* 1997). Nonetheless, under certain experimental conditions, BMPs can induce an ectopic digit 2, the anteriormost digit, in the chick wing (Duprez *et al.,* 1996). However, induction of digit 3 in the chick can be achieved by BMP2 only after prior exposure to Shh (Drossopoulou *et al.,* 2000), and only Shh itself can induce the posteriormost wing digit 4 (Drossopoulou *et al.,* 2000; Lopez-Martin *et al.,* 1995; Yang *et al.,* 1997).

These data have led to a two-step model in which Shh acts directly to "prime" the limb bud, making it highly responsive to BMP patterning, and then indirectly via the induction of *Bmp2* to specify different digit identities (Drossopoulou *et al.,* 2000). However, the existing data also do not exclude a direct patterning role for Shh across the entire digit field. Moreover, the putative role of BMP2 as a secondary signal is called into question by experiments in which the BMP antagonist *Noggin* is misexpressed in the posterior of developing chick limb buds, a treatment that has no effect on anterior–posterior patterning (Capdevila and Johnson, 1998). Definitive demonstration that *Bmp2* plays a key role in digit specification will require the generation of conditional alleles of *Bmp2* that remove its activity specifically in the limb bud.

Although not actually addressing whether Shh acts as a direct morphogen in the limb bud, genetic experiments in mice have provided strong support for a concentration-dependent response in establishing the anterior–posterior axis in the mouse limb bud. Although in *Drosophila,* the cholesterol modification of Hh (reviewed in Ingham and McMahon, 2001) is critical for limiting the range of Hh diffusion (due to Ptc-mediated sequestration) (Burke *et al.,* 1999; Chen and Struhl, 1996), the cholesterol-unmodified form of Shh in mice is severely restricted in its distribution (P. M. Lewis *et al.,* 2001). Thus, replacing the endogenous gene with such a cholesterol-unmodified form, which nonetheless retains a biological activity similar to wild type, results in a restriction of Shh protein to the posteriormost limb bud. This produces a limb skeleton that includes the posteriormost digits 5 and 4, but that has lost the more anterior digits 3 and 2 (P. M. Lewis *et al.,* 2001). This could be interpreted as indicating that low concentrations of Shh, such as those normally generated by Shh diffusion, are required to specify digits 3 and 2, as low concentrations of diffusible secondary signals induced by the posterior Shh are clearly not sufficient for this. However, the results can also be interpreted in terms of the Shh priming model developed on the basis of manipulations in the chick embryo (Drossopoulou *et al.,* 2000). According to this view, the primordium of digits 3 and 4 would not see the priming effect of Shh in the cholesterol-free mutant with restricted Shh distribution, and, hence, the secondary BMP2 signal would not be able to effect their specification.

In complementary experiments, a partially penetrant conditional allele of *Shh*, which produces a greatly reduced level of the wild-type, cholesterol-modified form of Shh protein, resulted in a digit pattern missing the posteriormost digits 4 and 5 but maintaining digits 2 and 3 (P. M. Lewis *et al.*, 2001). This shows that low concentrations of Shh are sufficient to pattern more anterior digits, consistent with the results in the chick (Yang *et al.*, 1997). Importantly, in both of these mutants (P. M. Lewis *et al.*, 2001), as in the *Shh* null mutant (Chiang *et al.*, 2001; P. M. Lewis *et al.*, 2001), the anteriormost digit is unaffected. This suggests that digit 1 arises from cells that are not directly exposed to Shh, perhaps in response to the secondary signal BMP2. Although *Bmp2* expression is mostly lost in a limb bud that does not produce any Shh, a small region of *Bmp2*-expressing tissue can be detected in the posteriormost limb mesenchyme (P. M. Lewis *et al.*, 2001) and could be responsible for inducing the formation of digit 1 in the *Shh* null mutant. Alternatively, formation of digit 1 could be a default pattern that occurs with sufficient proximodistal outgrowth and does not require any anterior–posterior signaling.

Regardless of whether Shh acts to pattern the anterior–posterior limb axis directly, via a secondary signal, or in combination with a secondary signal, an additional feature that must be considered in interpreting experimental studies of Shh activity is the expansion of the limb bud over time. As the width of the limb bud expands, cells initially within the range of Shh and/or its secondary signals may be displaced to a position anterior to the range of Shh signaling. Interestingly, it has been shown that anterior–posterior digit specification is progressive over time, such that cells initially seeing a low dose of Shh and therefore specified as digit 2 will, with longer exposure, be "promoted" to digit 3 and then digit 4 identity (Yang *et al.*, 1997). Such a promotion mechanism may therefore play an important role in the dynamic process of anterior–posterior limb pattering.

In addition to its well-established role in anterior–posterior patterning, Shh also plays a role in the formation of skeletal elements along the proximal–distal axis. In this instance, the role of Shh is not to establish the basic stylopod–zeugopod–autopod pattern (which still forms in the absence of any Shh signaling) (Chiang *et al.*, 1996, 2001; Kraus *et al.*, 2001; P. M. Lewis *et al.*, 2001), but, rather, to control the differential growth of skeletal elements within each limb segment via regulation of the *Hox* genes. *Hox* gene expression can be manipulated in the limb bud by ZPA transplants (Izpisua-Belmonte *et al.*, 1991; Nohno *et al.*, 1991) or by Shh application (Riddle *et al.*, 1993), and are also under the combined regulatory influence of Shh and FGF signaling (Laufer *et al.*, 1994). Although all the phases of *Hox* gene expression have not been explored in mice deficient for *Shh*, expression of *Hoxd-11*, *Hoxd-12*, and *Hoxd-13* is lost in the forelimb autopod, and *Hoxd-11* is lost and *Hoxd-12* is severely reduced in the hindlimb autopod (Chiang *et al.*, 2001; P. M. Lewis *et al.*, 2001).

This dysregulation of *Hox* gene expression in the absence of Shh is likely to be a contributing factor to the shortened and misshapen bones seen in *Shh* mutant mice. For example, whereas the individual *Hoxd-11* and *Hoxa-11* mutations result in relatively minor defects in the ulna and radius (Davis and Capecchi, 1994; Favier *et al.,* 1995; Small and Potter, 1993), the compound *Hoxd-11/Hoxa-11* mutant exhibits almost a complete loss of growth of the zeugopod of the forelimb (Davis *et al.,* 1995). This phenotype in the absence of any eleventh paralog gene expression (there is no *Hoxb-11* gene and *Hoxc-11* is exclusively expressed in the hindlimb) indicates that *Hox* genes regulate differential growth of the limb segments. However, because the zeugopod does form small but distinct radius and ulna elements, the *Hox* genes are not involved in specifying the limb segments. Thus, the alterations in *Hox* gene expression in the absence of Shh signaling is undoubtedly a major contribution to the phenotype of the skeletal elements that do form.

Given the importance of *Shh* for various aspects of limb patterning, its correct spatial and temporal regulation is critical and, indeed, *Shh* is under both negative and positive regulation in the limb bud. In particular, a number of genes are important for preventing inappropriate *Shh* expression in the anterior limb bud. Consequently, when these genes are mutated, the resulting mice display anterior expression of *Shh* and develop preaxial polydactyly (Buscher *et al.,* 1997; Masuya *et al.,* 1995, 1997; Qu *et al.,* 1997, 1998; Takahashi *et al.,* 1998). The most interesting of these is *Gli3*. Not only is Gli3 activity required to restrict *Shh* expression in the anterior limb, but *Gli3* is also repressed in the posterior limb bud by Shh (Marigo *et al.,* 1996b; Theil *et al.,* 1999). Thus, there is a negative feedback loop between these genes that helps to restrict *Shh* expression. In addition, Shh antagonizes the generation of the repressor form of Gli3 in the limb bud (Wang *et al.,* 2000), further assuring that the posterior margin is kept free of repressive Gli3 activity, thus allowing continued *Shh* expression.

As discussed above, FGF and Wnt-7a signaling contribute to the positive regulation of *Shh* expression. However, although these signals extend across the entire distal/dorsal limb bud, *Shh* expression is confined exclusively to the posterior margin. This appears to be due to a restricted competence to respond to these signals in the posterior limb bud. This competence extends into the interlimb flank mesoderm, which, as discussed above, has the ability to express *Shh* when transplanted into an appropriate environment in the distal limb bud. Two transcription factors have been identified as potential mediators of the competence to induce *Shh* expression and whose own expression in the embryo encompasses both the posterior of the limb buds and the interlimb flank mesoderm: *Hoxb-8* (Charite *et al.,* 1994; Lu *et al.,* 1997) and *dHand* (Charite *et al.,* 2000; Fernandez-Teran *et al.,* 2000). Ectopic expression of either of these genes in the anterior limb bud results in ectopic *Shh* expression in the region of misexpression that is simultaneously exposed to FGF signaling, resulting in digit duplications. Thus, either Hoxb-8 or dHand can convey competence to express *Shh* (Charite *et al.,* 1994, 2000; Fernandez-Teran *et al.,* 2000). However, because mice deficient in *Hoxb-8* have normal limbs (van den Akker *et al.,* 1999), this could imply that *Hoxb-8* and *dHand* have redundant functions in providing competence to express Shh, or that *Hoxb-8* works in concert with other redundant *Hox* genes in this role. In contrast, mice that do not express *dHand* fail to induce *Shh* expression (Charite *et al.*, 2000).

At least one key upstream determinant of *Hoxb-8* and *dHand* expression is retinoic acid. When retinoic acid is introduced into the anterior of a limb bud, it produces mirror image duplications by inducing the formation of an ectopic ZPA (Noji *et al.*, 1991; Summerbell, 1983; Wanek *et al.*, 1991) and inducing *Shh* expression (Riddle *et al.*, 1993). Retinoic acid appears to achieve this effect by first inducing the expression of *Hoxb-8* directly as an immediate-early response (Lu *et al.*, 1997), and then subsequently *dHand* (Fernandez-Teran *et al.*, 2000), before the onset of *Shh* expression. This likely recapitulates the process by which the endogenous competence to establish a ZPA is regulated, as inhibitors of retinoic acid activity (Helms *et al.*, 1996) or synthesis (Stratford *et al.*, 1996), when applied to the limb field before the emergence of the limb buds, block endogenous expression of *Hoxb-8* and prevent the induction of *Shh* (Lu *et al.*, 1997).

A positive feedback loop also exists between *Shh* and *dHand* expression. Specifically, ectopic Shh can induce *dHand* in the limb bud, and mice deficient in *Shh* fail to maintain high levels of *dHand* expression (Charite *et al.*, 2000; Fernandez-Teran *et al.*, 2000). Such a feedback loop could serve to maintain *Shh* competence within the ZPA once it is established.

The upstream regulation of *Shh* also appears to be conserved in the posterior fin bud of zebrafish. In particular, loss of zebrafish *dHand* function in the *Hands Off* mutant results, as in the mouse, in a failure to activate *Shh* expression in the fin buds (Yelon *et al.*, 2000). Moreover, as in chick embryos, application of retinoic acid to the anterior of developing fin buds causes ectopic *Shh* expression (Akimenko and Ekker, 1995).

A final important level of regulation of *Shh* in the developing limb bud is the induction of cell death by extremely high levels of Shh signaling (Sanz-Ezquerro and Tickle, 2000). Specifically, addition of extra Shh to the posterior limb bud results in cell death within the ZPA and a reduction of endogenous *Shh* expression, whereas a reduction in the number of endogenous *Shh*-expressing cells leads to a decrease in cell death and thus to an expansion of the *Shh*-expressing domain. This mechanism serves to maintain an appropriate level of Shh signaling, buffering the limb bud against inappropriately high or low levels of expression of this potent molecule that could otherwise have severe consequences for limb patterning. Buffering is also achieved at the protein level by the induction of *Ptc1* and *Hip*, both of which are antagonists of Shh produced in response to Shh signaling in the limb (see Ingham and McMahon, 2001).

Given its key role in limb patterning during development, it is not surprising that *Shh* appears to play a similar role during regeneration of amphibian limbs. *Shh* is expressed in a posterior domain during the regeneration of newt (Imokawa and Yoshizato, 1997), axolotl (Torok *et al.*, 1999), and *Xenopus* (Endo *et al.*, 1997) limbs, similar to their expression patterns during the development of limb buds in those species. The regenerative expression domain has been described as being small and brief (Roy *et al.,* 2000), but this is true only on a relative scale; the regenerative blastema is 50 times larger than the developing limb bud and the process of regeneration takes significantly longer than embryonic limb bud development. Thus, in absolute terms, the expression of *Shh* in development and regeneration is spatially and temporally similar to that seen during limb development. The expectation that Shh should have similar activity during regeneration was verified by misexpressing Shh in regenerating axolotl blastemas, which resulted in the expected duplication of skeletal elements (Roy *et al.,* 2000).

Interestingly, the newt *Ihh* homolog, named *N-bhh*, is also expressed in the early developing and regenerating limbs. In contrast to *Shh*, however, *N-bhh* expression is uniform in the mesenchyme of the regenerative blastema and of the developing newt limb bud. Moreover, whereas *N-bhh* expression is similar to amniote *Ihh* expression elsewhere in the embryo, it is not found in the expected prehypertrophic domain of the developing skeletal elements (Stark *et al.,* 1998). The functional significance of *N-bhh* expression in the limb is unclear, as such uniform Hedgehog signaling would be expected to have extreme consequences for limb patterning. It is possible, however, that these *N-bhh* transcripts do not produce functional N-bhh protein in the newt limb. In this regard, it will be critical to examine the expression domains of *hedgehog* targets such as *Patched* homologs and to test the competence of N-bhh to induce digit duplication in the chick system.

Zebrafish do not regenerate the endochondral portion of their skeleton. They do, however, regenerate their dermal lepidotrichia, although these regenerates can be imperfect. For example, amputation of caudal fins immediately after the first branch points of the lepidotrichia results in fusion of adjacent rays. This has been correlated with a transient decrease in the expression of *Shh* and its target genes at the start of the regeneration process (Laforest *et al.,* 1998). During regeneration, *Shh* expression is not centered over the fin ray as it is during normal development, but spreads, rather, over the entire surface of the bone. If an amputation is produced just distal to a branch point, the adjacent rays are close enough that the two broad *Shh* domains form a single diffuse domain, leading to fusion in the regenerate.

Shh also plays a role in the development of other vertebrate appendages, including the external genitalia, although the mechanisms by which it acts are distinct. In particular, *Shh* is expressed in the epithelium of the genital tubercle, and misexpression studies in organ culture have shown that it can regulate the mesenchymal expression of *Ptc1*, *Bmp4*, *Hoxd-13*, and *Fgf10* in the tubercle. The functional relevance of this activity is shown in *Shh*-deficient mice, which exhibit reduced proliferation and increased apoptosis in the genital tubercle, leading to a complete failure of development of the external genitalia (Haraguchi *et al.,* 2001). Defects in the external genitalia are similarly seen in mice with induced cholesterol deficiency, which compromises Hedgehog signaling (Lanoue *et al.,* 1997). Molecular analysis of the developing genital tubercle shows that *Bmp4* expression shifts in the *Shh* mutant from the mesenchyme to the epithelium, indicating that Shh is important for inducing *Bmp4* in the mesenchyme while repressing expression of the same gene in the epithelium (Haraguchi *et al.*, 2001).

XIII. Lungs and Trachea

Lung development (reviewed in Cardoso, 2001; Hogan, 1999; Warburton *et al.*, 2000) initiates in the foregut with an outpocketing of a pair of ventrally located endodermally derived lung buds into the surrounding splanchnic mesoderm (9.5 dpc in the mouse). A number of stereotypical lateral branches then establishes the lobular organization of the lungs (in the mouse, four right, one left; in the human, three right, two left). Subsequent dichotomous branching in which the tip of each branch repeatedly splits into two new branches generates much of the branched epithelium of the lung. Finally, progressive dilation of the end buds of branches generates a thin alveolar epithelium that closely juxtaposes an underlying capillary network, the site of gaseous exchange.

Elegant recombination experiments that combine distal mesenchyme with proximal nonbranching epithelium indicate that branching of the endodermal epithelium of the developing lung is regulated by the underlying mesenchyme (Alescio and Cassini, 1962; Wessels, 1970). Not surprisingly, there has been considerable interest in elucidating the molecular interactions between mesenchyme and epithelium at the branch points. Further lung development is characterized by a progressive proximal–distal differentiation of the airways. The proximal columnar, ciliated mucus-secreting epithelia of the trachea and bronchi are distinct from the thin cuboidal surfactant-secreting epithelium that gives rise to the distally located alveolar air sacs.

Although *Shh* is expressed throughout the epithelium of the early branching lung, the highest levels occur at the tips of the branching network. In later stages, *Shh* is downregulated proximally, but its expression is sustained within the distal branching epithelium (Bitgood and McMahon, 1995; Bellusci *et al.*, 1996, 1997a,b). As *Ptc1*, *Hip1*, and all three mammalian *Gli* members are expressed in the underlying mesenchyme (Bellusci *et al.*, 1997a,b; Grindley *et al.*, 1997; Chuang and McMahon, 1999), the mesenchyme is the likely target of a Shh signal, particularly at the branch points where *Ptc1* is upregulated. Ectopic expression of *Shh* within the lung epithelium leads to a relatively modest phenotype—an enhanced proliferation of mesenchyme and decreased epithelial branching. Both phenotypes are consistent with *Shh* playing a role in regulating mesenchymal proliferation (Bellusci *et al.*, 1997a,b). The late onset of the phenotype (15.5 dpc) probably reflects the kinetics of transgene expression from the lung surfactant regulatory element used in this analysis. No difference is observed in proximal–distal differentiation of the airways or in expression of several genes, including *Bmp4*, that like *Shh* are upregulated at the distal tips of the branching network (Bellusci *et al.*, 1996, 1997a,b).

An analysis of *Shh* mutants provides compelling evidence that Shh signaling plays an important role in regulating branching morphogenesis of the mammalian lung. In *Shh* mutants, initiation of the outgrowth of the lung buds is normal, but the paired buds subsequently fail to undergo lateral branching. Consequently, the lungs form two rudimentary saclike structures (Litingtung *et al.*, 1998; Pepicelli *et al.*, 1998). Despite the absence of normal branching growth, proximal–distal differentiation of the airway epithelium is relatively normal. However, an absence of smooth muscle and a failure of blood vessel invasion indicate that stromal development is disrupted to some extent (Pepicelli *et al.*, 1998). As with other mice that have branching defects in the lung, such as those that carry mutation in the transcriptional regulator *Nkx2.1* (Minoo *et al.*, 1999), the trachea and esophagus fail to separate, generating a treacheal–esophageal fistula (Litingtung *et al.*, 1998; Pepicelli *et al.*, 1998).

Analysis of *Gli* mutants provides additional evidence of a role for the Hedgehog pathway in branching morphogenesis. However, studies of the role of Gli proteins are complicated by the potential for two *Gli* members, *Gli2* and *Gli3*, to act as both activators and repressors of Hedgehog target genes (for review, see Ingham and McMahon, 2001). Whereas *Gli1* mutants undergo normal lung development, epithelial growth is defective in *Gli2* mutant lungs, leading to a fusion of the right lung lobes, this phenotype is significantly enhanced in *Gli1/Gli2* compound mutants (Park *et al.*, 2000). Although lateral branching is not as severely effected as it is in *Shh* mutants, results are generally consistent with *Gli1/Gli2* mediation of Shh signaling in the branching process. As *Gli1* appears to encode only an activating form, and the loss of *Gli1* activity enhances a *Gli2* mutant phenotype, the results suggest that together *Gli1* and *Gli2* play a positive role in the regulation of Hedgehog targets in the mesenchyme. Combining a *Gli2* loss-of-function mutant with a partial loss of *Gli3* activity ($Gli3^{+/-}$) also results in a reduction in branching, compared with the *Gli2* phenotype and a tracheal–esophageal fistula (Motoyama *et al.*, 1998b). However, *Gli2/Gli3* compound mutants actually exhibit a complete failure of tracheal, esophageal, and lung development, a more severe phenotype than that of *Shh* mutants (Motoyama *et al.*, 1998b). If Gli3 acts predominantly as a repressor, as is generally the case (see Ingham and McMahon, 2001), then Gli2 may also play a negative role in silencing certain targets, a regulatory process that is critical for early development of the foregut (Motoyama *et al.*, 1998b). Together, these results point to a balance of Gli-mediated activation and repression in the control of gut development and lung branching. Interestingly, studies also highlight the involvement of a member of the forkhead gene family of transcriptional regulators, *Foxf1*, in regulating the transcriptional response to Shh within the lung mesenchyme (Mahlapuu *et al.*, 2001).

Several *Fgf*s, including *Fgf1*, *-7*, *-9* and *-10*, are expressed in epithelial or mesenchymal aspects of the lung, and application of these ligands to mesenchyme-free lung explants triggers either cystic (FGF7) or branching (FGF1 and -10) growth

(Bellusci *et al.*, 1997b; Cardoso *et al.*, 1997; Lebeche *et al.*, 1999). Furthermore, the presence of a dominant-negative form of FGFR2b in the epithelium of the lung buds leads to a complete failure of all branching distal to the main bronchi (Peters *et al.*, 1994). FGF10, a ligand for this receptor, has been identified as a key regulator of lung branching. *Fgf10* is expressed in the ductal lung mesenchyme in a highly dynamic expression that prefigures the branching pattern (Bellusci *et al.*, 1997b). *Fgf10* mutants show a complete failure of lung development, although tracheal and esophageal development is normal (Min *et al.*, 1998; Sekine *et al.*, 1999).

Interestingly, *Shh* negatively regulates *Fgf10* (Bellusci *et al.*, 1997a; Lebeche *et al.*, 1999; Litingtung *et al.*, 1998; Pepicelli *et al.*, 1998). Wild-type lungs demonstrate a zone of non-*Fgf10* expressing mesenchyme close to the branch tip (Bellusci *et al.*, 1997a). In *Shh* mutants the upregulation of *Fgf10* expression in the distal mesenchyme includes cells immediately beneath the branch tip (Pepicelli *et al.*, 1998). Thus, Shh-mediated repression of *Fgf10* is one mechanism by which *Fgf10* expression is localized in the lung mesenchyme. By maintaining discrete domains of *Fgf10* expression at some distance from the lung epithelium, the lung epithelium receives discrete positional cues that then trigger directional epithelial branching.

Whether Shh acts directly to inhibit *Fgf10* expression is unclear. However, BMP signaling is unlikely to be involved. *Bmp4* expression overlaps with that of *Shh* at the branch tip, but *Bmp4* expression is still present in *Shh* mutants (Litingtung *et al.*, 1998; Pepicelli *et al.*, 1998). Further, interfering with BMP4 activity produces alterations in proximal–distal patterning unlike loss of Shh function (Bellusci *et al.*, 1996; Weaver *et al.*, 1999). Thus, BMP4 primarily regulates differentiation rather than branching of the lung epithelium.

Finally, studies of *Drosophila* have uncovered a mechanistic connection between the role of FGF signaling in the control of lung branching in vertebrates and FGF signaling in branching morphogenesis of the *Drosophila* tracheal airways (Metzger and Krasnow, 1999). Tracheal progenitors migrate from a tracheal pit in response to an FGF-mediated (Bnl) chemoattractant cue provided in a highly reproducible, dynamic pattern by cells along the prospective migratory pathway. In striking contrast to the mammalian airways, the tracheal network forms in the absence of any cell proliferation. Although primary branches are multicellular, secondary airways are formed by the folding over and elongation of a single cell and tertiary branches form from highly elongated, subcellular projections. All stages of branching are FGF/Bnl dependent.

Interestingly, *Drosophila* Hedgehog signaling is implicated in tracheal development (Glazer and Shilo, 2001). In *hh* mutant embryos, migration of all tracheal branches is defective, many are absent, while others terminate prematurely. Although *ptc* upregulation is observed only in the anterior half of the tracheal pit, Hedgehog signaling appears to be required directly within all tracheal progenitors. The relationship between Hh and Bnl in regulating tracheal outgrowth will be an interesting topic for future studies.

XIV. Mammary Gland

Reverse transcriptase-polymerase chain reaction (RT-PCR) analysis of the mouse mammary glands indicates that *Dhh*, *Ihh*, and *Shh* are expressed during the development of this organ (M. T. Lewis *et al.*, 1999; M. T. Lewis, 2001). At least one of these, *Ihh*, localizes to the epithelial component of the gland (M. T. Lewis *et al.*, 1999). Some evidence for a mammary gland function comes from analysis of *Ptc1* heterozygotes (*Ptc1$^{+/-}$*) and *Gli2* mutants. In the former, ductal hyplasia and dysplasia are observed, suggesting that Hedgehog signaling modulates ductal growth and branching (M. T. Lewis *et al.*, 1999). *Gli2* is expressed exclusively within the stroma and appears to be required within this tissue for normal ductal morphogenesis (M. T. Lewis *et al.*, 1999).

XV. Myogenesis

Most skeletal muscle in vertebrates derives from the somites, paired segmental structures flanking the neural tube that arise as epithelial condensations of paraxial mesodermal cells. Shortly after their formation, the somites of higher vertebrate embryos become compartmentalized along their dorsoventral axis: ventrally, cells undergo an epitheliomesenchymal transformation to form the sclerotome that will give rise to ribs and vertebrae. Dorsally, cells retain their epithelial character and form the dermomyotome that will give rise to dermis and muscle, cells at the medial lip contribute to the epaxial musculature while cells at the lateral lip generate the hypaxial muscle.

Experimental manipulations of the developing somites, mostly performed in avian embryos, have established that specification of somitic cell fate occurs in response to signals emanating from neighboring tissues. For instance, grafting of a notochord—a potent source of Shh—dorsally between the paraxial mesoderm and neural tube induces sclerotome in dorsal somitic cells at the expense of the dermomyotome (Pourquie *et al.*, 1993). Conversely, juxtaposing dorsal neural tube—a potent source of Wnt signaling proteins—with ventral somitic tissue suppresses sclerotomal specification and induces ectopic dermomyotome (Dietrich *et al.*, 1997). Such findings have been interpreted in terms of a simple model in which opposing gradients of ventralizing (Shh) or dorsalizing (Wnt) signals emanating from the notochord or dorsal neural tube and ectoderm, respectively, specify cell fate along the dorsoventral axis of the somite (C.-M. Fan *et al.*, 1997). One way in which these gradients might be mutually reinforced is by inhibitory activities induced by one or another signal: candidates for such activities include the secreted Wnt inhibitor SFRP2 that is induced by Shh in the ventral somite (Lee *et al.*, 2000), and the Shh-binding protein GAS1 that is expressed in dorsal somites in response to Wnt activity (Lee *et al.*, 2001). Support for such a model has come from *in vitro* experiments showing that Shh can induce Pax1, a marker of sclerotome, in explanted presomitic mesoderm (C.-M. Fan *et al.*, 1995) whereas Wnt proteins are capable of inducing the expression of the myogenic genes *Myf5* and *MyoD* in the same tissue (C.-M. Fan *et al.*, 1997).

Several other lines of evidence, however, have suggested that Shh also plays a positive role in muscle development, acting variously as a proliferation, survival, or inducing factor for the myotome. *In vivo* and *in vitro* manipulations of chick somites have both pointed to a requirement for signals derived from axial tissues (the notochord and neural tube) to promote the proliferation and survival of the epaxial, but not the hypaxial, cells. Surgical ablation of axial structures results in a loss of proliferation (Marcelle *et al.*, 1999) and extensive cell death (Teillet *et al.*, 1998a) of *MyoD*-expressing cells within the medial dermomyotome. Similarly, explanted newly formed somites—which when cultured *in vitro* with axial structures can proliferate and differentiate into muscle fibers—fail to proliferate or differentiate and undergo extensive apoptosis when cultured in the absence of axial structures (Buffinger and Stockdale, 1995; Cann *et al.*, 1999). In both experimental paradigms, these effects of loss of axial structures on myogenic cells can be reversed by Shh activity, added either in purified form to the culture medium of tissue explants (Cann *et al.*, 1999) or delivered by transfected cells implanted into operated embryos (Marcelle *et al.*, 1999; Teillet *et al.*, 1998a).

Whereas these findings indicate that Shh can act both as a mitogen and survival factor for previously committed epaxial cells, other studies have pointed to a role for Shh in the induction of epaxial cell fate. Indirect evidence for such a role came from ablation and grafting experiments in the quail embryo (Pownall *et al.*, 1996), which, in contrast to the earlier report of Pourquie *et al.* (1993), suggested that notochord-derived signals are both necessary and sufficient for the induction and maintenance of epaxial marker gene expression. The difference in outcomes between the different notochord-grafting experiments most likely reflects variation in the location and timing of the grafts, those of Pownall *et al.* (1996) leaving the dorsal tissues, which provide a source of myogenic Wnt signals, relatively undisturbed (see Dietrich *et al.*, 1997). Consistent with a myogenic role for Shh, explanted presomitic mesoderm that is unresponsive to Wnt activity alone can be induced to express myogenic markers when exposed to a combination of both Wnt and Shh proteins (Maroto *et al.*, 1997; Munsterberg *et al.*, 1995; Stern *et al.*, 1995; Tajbakhsh *et al.*, 1998).

As in other contexts, discriminating between the potential activities of Shh and its actual role in epaxial muscle development during embryogenesis requires the analysis of embryos in which the *Shh* gene has been inactivated. Several studies have addressed this issue by using mice carrying an ES cell-derived targeted loss-of-function mutation (Borycki *et al.*, 1999; Chiang *et al.*, 1996) or an X-ray-induced transcript null mutation (Kruger *et al.*, 2001) of the *Shh* gene. In the mouse embryo, cells of the epaxial lineage are the first to express the *Myf5* gene, activating transcription as early as embryonic day 8.5 (E8.5), whereas expression in hypaxial cells begins only 1 day later at E9.5, along with that of *MyoD*. Initial analysis of *Shh* mutant embryos revealed that both *Myf5* and *MyoD* are expressed in the dermomyotome, which was taken to imply that *Shh* is dispensible for the induction of both the epaxial and hypaxial lineages (Chiang *et al.*, 1996). More detailed studies, however, revealed a specific deficit in *Myf5* in the presumptive epaxial domain of the dermomyotome whereas expression in hypaxial cells occurred more or less on schedule (Borycki *et al.*, 1999). One caveat to this analysis is that the *Myf5* expression was analyzed in the E9.75 embryo, but activation of *Myf5* occurs more than 1 day earlier, at E8.5. Thus it remains possible that *Myf5* is induced in the absence of *Shh* but fails to be maintained. Indeed, Kruger *et al.* (2001) have reported that *Myf5* expression is detectable in the dorsal medial lip of the the dermomyotome of newly formed somites in *Shh* null embryos at E9.5, although it subsequently disappears at E10.5. Moreover, studies of mouse embryos homozygous for a null allele of *Smo* show that *Myf5* induction can still occur at E8.5, albeit at reduced levels, in the absence of all Hedgehog signaling (X. Zhang *et al.*, 2001) and similar aberrant low-level expression of *Myf5* has also been reported in E8.75 in *Shh* mutant embryos (Gustafsson *et al.*, 2002). One shortcoming of these analyses is that their interpretation rests on the assumption that early *Myf5* expression is synonomous with epaxial identity. Confirmation that this expression corresponds to the epaxial domain will require analysis of the Myf5 epaxial enhancer (Gustafsson *et al.*, 2002) in *Smo* and *Shh* mutant embryos. Thus it remains unclear whether ventrally derived Shh signal is necessary to induce the epaxial component of the myotome. Analysis of proliferation and cell death in *Shh* mutant mice, on the other hand, suggests that in normal development, Shh acts neither as a proliferative or survival factor for epaxial muscle cells (Borycki *et al.*, 1999).

By contrast, there is general agreement that the initial specification of the hypaxial lineage proceeds independently of Shh activity; however, several lines of evidence have implicated Shh in promoting the survival and/or proliferation of hypaxial cells that form the limb musculature. In *Shh* mutant embryos, *Pax3*-expressing myoblasts migrate into the limbs normally and activate *Myf5* and *MyoD* on schedule at E10.5. By E12.5, however, the limb bud lacks the primary myotubes that have formed in wild-type limbs by this stage and few if any secondary myotubes are present 2 days later (Kruger *et al.*, 2001). This loss of myogenic derivatives coincides with a loss of *Bmp4* expression in the *Shh* mutant limbs, suggesting that Shh may promote the survival/proliferation of myogenic cells via this factor (Kruger *et al.*, 2001). Consistent with these findings, misexpression of *Shh* in the developing chick limb can cause an increase in the numbers of *Pax3* and *MyoD* positive cells, whereas *in vitro* cultures of chick limb mesenchyme show an increase in proliferation and myotube differentiation when exposed to Shh protein (Duprez *et al.*, 1998).

In teleosts such as zebrafish, the organization of the somite is somewhat different from that in amniotes, the majority of cells both dorsally and ventrally giving rise to muscle. Somites of mutant embryos lacking *Shh*-expressing axial structures (the notochord and floor plate) differentiate into muscle but lack a specific population of superficially located fibers as well as the medially located "muscle pioneer cells" (Blagden *et al.,* 1997; Halpern *et al.,* 1993). All of the missing cells derive from the slow muscle lineage that arises from the adaxial cells, a specialized group of cells that lie adjacent to the notochord. Significantly, these cells are the first to activate expression of the *MyoD* gene, doing so as soon as they appear within the presomitic mesoderm (Concordet *et al.,* 1996; Weinberg *et al.,* 1996). By contrast, expression of *MyoD* in other, more lateral paraxial mesodermal cells occurs only once they have been incorporated into somites. Several lines of evidence suggest that Shh is responsible for the precocious expression of *MyoD* by adaxial cells. Misexpression of Shh is sufficient to induce MyoD throughout the presmoitic mesoderm (Concordet *et al.,* 1996; Hammerschmidt *et al.,* 1996; Weinberg *et al.,* 1996); conversely, blocking Hh signaling, through mutation of the *smu* gene that encodes zebrafish Smo (Varga *et al.,* 2001), results in a loss of MyoD in adaxial cells (Barresi *et al.,* 2000; Chen *et al.,* 2001). Thus, as in higher vertebrates, Shh activity seems to be essential for the specification of a subset of myogenic cells. Whether Hedgehog signaling activity is required for initiation or maintenance of *MyoD* expression in this subpopulation of cells remains unclear. Coutelle *et al.* (2001) have pointed to the persistent expression of *MyoD* in the most posterior region of the tailbud of *syu* mutant fish; however, the activity of other *hh* genes expressed in the midline of the zebrafish—*twhh* (Ekker *et al.,* 1995a) and *ehh* (Currie and Ingham, 1996)—could account for this activation. *MyoD* expression is certainly reduced, if not eliminated, in *smu* (*smoothened*) mutant embryos (Barresi *et al.,* 2000).

Unlike the Shh-dependent epaxial cells in amniote embryos, the zebrafish adaxial cells give rise exclusively to slow-twitch muscle fibers. Loss of Hh activity eliminates all slow muscle fibers from the embryo (Barresi *et al.,* 2000; Coutelle *et al.,* 2001; K. E. Lewis *et al.,* 1999) whereas misexpression of Shh is sufficient to transform the entire myotome to the slow-twitch fiber type (Blagden *et al.,* 1997; Du *et al.,* 1997). Evidence of a similar role for Shh in specifying fiber type in amniote embryos has come from *in vitro* experiments in which chick somitic explants are treated with recombinant Shh protein. Such treatment resulted in a significant increase in the numbers of primary fibers expressing the slow myosin heavy chain isoform (Cann *et al.,* 1999), implying a role for Shh in specifying fiber type in amniotes as well as teleosts.

XVI. Neural Crest

The neural crest is a highly migratory population of cells within the vertebrate embryo. Crest cells arise from dorsal neural tube precursors in response to BMP signaling initiated by the dorsal epidermis (Dickinson *et al.,* 1994; Liem *et al.,* 1995). Subsequently, neural crest cells generate a wide range of distinct cell types, the formation of which is governed in part both by the relative position of the neuraxis from which the crest cells originate as well as their final postmigratory locations. Among their many contributions to the vertebrate organism, crest-derived cells are responsible for the formation of much of the head skeleton, cranial nerves, pigment, Schwann cells, peripheral nerves, connective tissue of head musculature and meninges, and portions of the cardiac outflow tract. Some evidence implicates *Shh* signaling in the regulation of neural crest cell types.

Emigration of neural crest cells from the dorsal neural tube *in vivo* and *in vitro* is inhibited by Shh (Selleck *et al.*, 1998; Testaz *et al.*, 2001). This inhibition may reflect a ventralization of dorsal neural plate tissue in the presence of Shh (see Section III, Central Nervous System); however, Shh treatment does not affect dorsal *Bmp* expression. These results point to a possible role for ventral Shh signaling in regulating neural crest cell migration. Studies of the chick head show that Shh also plays an important role in supporting survival of cranial mesenchyme, including migrating cranial neural cells (Ahlgren and Bronner-Fraser, 1999). How is not clear, but the observation that Shh signaling modulates adhesiveness of neural crest cells to a fibronectin substrate *in vitro* (Testaz *et al.*, 2001) may relate to the triggering of apoptosis when Shh is inhibited *in vivo*. Currently, there is no strong evidence that cranial neural crest cells are direct targets of Shh signaling, in fact crest cells are not reported to upregulate *Ptc1* as is generally true for Hedgehog target tissues (Ahlgren and Bronner-Fraser, 1999). Thus, either signaling is indirect or Shh acts directly on neural crest cells by an alternative pathway (Ahlgren and Bronner-Fraser, 1999; Testaz *et al.*, 2001).

XVII. Pancreas Development

The pancreas forms from buds that emerge from the dorsal and ventral sides of the duodenal gut endoderm just caudal to the stomach and adjacent to the location of another gut derivative, the liver. Both the dorsal and ventral buds give rise to specialized endocrine and exocrine cells. These buds ultimately fuse and, along with associated visceral mesoderm, form the pancreatic organ (Spooner, 1970; Slack, 1995). In spite of their similar origins and common ultimate fate, the dorsal and ventral pancreatic buds are regulated differently; for example, the dorsal pancreas depends on the activity of the transcription factors Isl1 and Hlxb9, whereas the ventral pancreas does not (Ahlgren, 1997; Harrison, 1999; Li, 1999). Another important difference in the regulation of the dorsal and ventral pancreatic buds concerns the manner in which Shh activity is modulated (discussed below).

As in the rest of the gut, the embryonic stomach and duodenal endoderm express high levels of *Ihh* and *Shh*. Strikingly, however, the endoderm fated to give rise to the pancreatic buds does not express any *hedgehog* genes (Apelqvist *et al.*, 1997). Similarly, the Hedgehog target *Ptc1* is expressed throughout the gut mesoderm but is absent from the pancreatic mesoderm (Hebrok, 1998). The region lacking Hedgehog expression correlates with the domain expressing *Pdx1*, a transcription factor critical for pancreatic development (Ahlgren, 1996; Offield, 1996). The importance of this field devoid of Hedgehog expression is seen when *Shh* is misexpressed in this domain via the *Pdx1* promoter in transgenic mice (Apelqvist *et al.*, 1997); in such animals, the mesoderm is completely transformed to intestinal mesenchyme. Importantly, this effect on the mesoderm is likely to be direct because the mesoderm expresses both the Hedgehog receptor *Ptc*1 (Marigo *et al.*, 1996a) and other downstream Hedgehog effectors (Grindley, 1997; Platt, 1997). The endoderm, by contrast, does not express Hh signal transduction components. Nevertheless, an indirect effect on the endoderm (presumably mediated via an unknown secondary signal produced by the mesoderm) can be inferred from the observation that islet architecture is also disturbed in these mice (Apelqvist *et al.*, 1997). In converse experiments, application of cyclopamine, a steroidal alkaloid that antagonizes Hedgehog signaling, to the developing chick gut results in ectopic pancreas formation (Kim and Melton, 1998). Similarly, treatment of chick gut primordia with an antibody blocking Hedgehog activity leads to ectopic pancreatic gene expression (Hebrok, 1998), whereas mice deficient in either *Shh* or *Ihh* likewise develop an expanded annular pancreas (Ramalho-Santos, 2000; Hebrok, 2000). In the case of the *Ihh* mutant, this appears to be due to ectopic branching of the ventral pancreatic bud (Hebrok, 2000), whereas *Shh* mutant animals have a three-fold increase in pancreatic mass and a four-fold increase in endocrine cell number. In contrast, loss of *Ptc1*, which leads to an increase in Hedgehog signaling, results in a concomitant decrease in pancreatic gene expression (Hebrok, 2000).

From these studies, it is clear that for normal pancreatic development to occur, *hedgehog* activity must be excluded from the pancreatic endodermal anlage. Interestingly, as the endodermal gut tube forms, expression of the *hedgehog* genes is initially absent from the dorsalmost endoderm along the entire length of the gut tube. However, both *Ihh* and *Shh* quickly spread dorsally in the regions adjacent to the pancreatic bud, for example, but not within the pancreatic bud itself (Apelqvist *et al.*, 1997), raising a question as to what prevents this spread from occurring within the pancreatic anlage. Although the notochord lies dorsal to the entire gut tube, it is striking that direct contact between the two structures is limited to the pancreatic region. Moreover, experiments in the chick demonstrate that ectopic notochord grafts can repress endodermal *Shh* expression, whereas notochord removal early in gut development results in dorsal *Shh* expression throughout the gut tube, including the pancreatic anlage (Hebrok, 1998). This role of the notochord in inhibiting *hedgehog* gene expression is somewhat surprising given the fact that the notochord itself is a potent and important source of Shh for other aspects of embryonic patterning. In the case of pancreatic specification, two different signals secreted by the notochord are of relevance.

Activin-βB and *Fgf2* are transcribed in the notochord during pancreas specification, and when these factors are applied to the anterior gut, both repress endodermal *Shh* and induce ectopic expression of pancreatic genes in chick endoderm (Hebrok, 1998). Consistent with a critical role for *Activin-βB* in this process, mutations of the activin receptor-encoding genes *ActRIIA* and *ActRIIB* in mice result in a spread of *Shh* expression and a resultant hypoplastic dorsal pancreas (Kim, 2000). As noted earlier, the ventral pancreas is regulated differently from the dorsal pancreas in several respects, and there is currently no information about the regulatory factors that repress *hedgehog* gene expression in the ventral domain.

The role of Hedgehog signaling appears to be different in the specification of the pancreas in zebrafish. The two *hedgehog* genes expressed in the zebrafish gut, *shh* and *twhh*, are expressed only in the endoderm after the onset of *Pdx1* and *Insulin* expression, well after the pancreas has been specified (Roy, 2001). Therefore, repression of endodermal *hedgehog* genes by the notochord does not seem to be involved in pancreas specification in zebrafish. Surprisingly, Hedgehog signaling from the notochord specifies the pancreatic anlage. In particular, zebrafish with defects in Hedgehog signaling display a disruption of pancreatic formation, including loss of *Pdx1* expression. Reciprocally, ectopic activation of the Hedgehog pathway results in *Pdx1* induction and ectopic pancreatic endocrine cells (Roy, 2001). Thus, whereas signaling from the notochord appears to specify the pancreatic anlage in both zebrafish and amniotes, this activity is actually mediated in the fish by *hedgehog* genes themselves, whereas in amniotes other factors from the notochord appear to be involved in suppressing *Shh* and *Ihh* expression in the endoderm.

Despite these differences between fish and amniotes, there is an interesting parallel between the early positive role of Hedgehog signaling in zebrafish pancreatic induction and a later role of Hedgehog signaling in the mature amniote pancreas. In particular, *Ihh* and *Dhh* are both expressed postnatally in the endocrine islet cells (Thomas, 2000), where they appear to positively regulate pancreas-specific gene activity. Thus, blocking Ihh and Dhh signaling with cyclopamine results in a loss of *Pdx1* expression and a concomitant drop in insulin production (Thomas, 2000, 2001). However, whereas *hedgehog* genes are indeed positive regulators of pancreatic cell fate in both the embryonic zebrafish pancreas and in the mature amniote pancreas, it is unclear whether this really reflects a common regulatory pathway. Specifically, in the mature amniote pancreas, the endodermal endocrine cells seem to be directly responding to Hedgehog signaling as these cells express both *Ptc1* and *Smo* (Thomas, 2000). Conversely, as discussed above, the earlier negative response to Hedgehog signaling appears to be indirect as the pancreatic endoderm in amniotes does not have Hedgehog signal transduction machinery at that stage. It remains to be determined whether the action of Hedgehog on the zebrafish pancreatic anlage is a direct response in the endoderm.

XVIII. Peripheral Nervous System

Schwann cells, a neural crest derivative, are responsible for synthesis of the myelin sheaths that insulate peripheral nerves (Mirsky *et al.,* 2002). Attention has focused on a second role that Schwann cells play in the Dhh-mediated organization of a perineurial nerve sheath that encases the peripheral nerve.

Dhh is expressed within the Schwann cell lineage from 11.5 dpc to at least 10 days postpartum (Bitgood and McMahon, 1995; Parmantier *et al.*, 1999). In contrast, *Ptc1* is expressed in the underlying perineurial mesenchyme, a fibroblast-like population of cells that converts to an epithelial sheath and forms a barrier preventing molecular and cellular infiltration into the nerve (Parmantier *et al.*, 1999). In the absence of *Dhh*, the perineurial sheath is thin and highly disorganized; perineurial cells are observed in the endoneurial space and cellular infiltration is observed within the nerve (Parmantier *et al.*, 1999). Thus, *Dhh* plays an essential role in regulating the connective tissue sheaths around peripheral nerves. Exactly how this is accomplished remains to be determined, but given the reduction in perineurial cells observed in *Dhh* mutants, Dhh may act, at least in part, to regulate proliferation and/or survival within this population. A report that Shh stimulates proliferation of sympathetic neural precursors and promotes tyrosine hydoxlase production in mature sympathetic neurons provides evidence of a second role for Hedgehog signaling in the peripheral nervous system (Williams *et al.*, 2000).

XIX. Pituitary Gland

The anterior pituitary arises from Rathke's pouch, an invagination of oral ectodermal epithelium. Distinct hormone-secreting cell types are generated at specific positions along its length (Marigo *et al.*, 1996a). As a result, the anterior pituitary is responsible for the hormonal control of a broad spectrum of physiological functions, including growth and reproduction. The expression of *Shh* in oral ectoderm adjacent to the ventral margin of Rathke's pouch and in overlying neural tissue suggests a possible role in the development of anterior pituitary cell types. Indeed, when Hedgehog signaling is blocked within the anterior pituitary primordium, a hypoplastic pituitary results. Conversely, *Shh* overexpression expands the anterior pituitary (Treier *et al.*, 2001). Interestingly, this expansion is primarily a result of increased numbers of gonadotropes and thyrotropes, indicating that there may be some regional specificity in the action of Shh signaling (Treier *et al.*, 2001). Although Shh appears to regulate precursor cell proliferation and dorsal–ventral polarity within the mammalian pituitary, it is unlikely to act alone. For example, FGF8 may act in conjunction with Shh to regulate proliferation, whereas *Bmp2*, a likely target of Shh in the ventral ectoderm, may play a more direct role in specifying ventral cell types (Treier *et al.*, 2001).

Pituitary formation has also been addressed in the zebrafish. A number of mutants with defects in midline development undergo a transdifferentiation of the pituitary anlagen (the adenohypophysis) to lens. This implicates midline signaling in specification of the pituitary. Interestingly, a similar phenotype is also observed in *Gli2* mutant zebrafish (Karlstrom *et al.*, 1999; Kondoh *et al.*, 2000). In the mouse, *Gli1$^{-/-}$/Gli2$^{-/-}$* compound mutants and *Shh* mutants entirely lack the pituitary (Park *et al.*, 2000; Treier *et al.*, 2001). Thus, Hedgehog signaling could play an early role in establishing the pituitary primordium.

XX. Prostate Gland

The prostate glands are paired, branched, secretory structures that bud from the urogenital sinus (at 17.5 dpc in the mouse) in a testosterone-dependent process. Indeed, the analogous region in the female urogenital sinus will also form prostatic structures if given testosterone at an appropriate period (Cunha *et al.*, 1980). Extensive branching of the prostate epithelium occurs within a mesenchymal stroma in the perinatal period, and the result is a multiductal organ with a number of openings to the urethra. *Shh* (and to a lesser extent *Ihh* [Bitgood and McMahon, 1995; Marigo *et al.*, 1995; Podlasek *et al.*, 1999]) is expressed within the epithelium of the urogenital sinus; expression is testosterone dependent. In the male *Shh* expression is reduced on withdrawal of testosterone, whereas in the female, *Shh* expression is elevated on addition of testosterone (Podlasek *et al.*, 1999). That Shh may be essential for normal ductal growth is indicated by antibody blocking experiments (Podlasek *et al.*, 1999). Presumably, Shh signaling plays an indirect role mediated through the stroma mesenchyme, as this appears to be the Hedgehog-responsive tissue. Several transcriptional regulators are known to regulate prostate development, including *HoxD13*, a target of Shh signaling in the limb bud, and *Nkx3.1*. *HoxD13* mutants display a ductal hypoplasia that is consistent with a diminished response to Shh (Podlasek *et al.*, 1997). In contrast, *Nkx3.1* mutants display a ductal hyperplasia, suggesting that *Nkx3.1* is a negative regulator of ductal growth (Bhatia-Gaur *et al.*, 1999). *Nkx3.1* expression is absent in *Shh* mutants (Schneider *et al.*, 2000) and, given what appear to be opposite roles for *Shh* and *Nkx3.1*, it seems unlikely that *Shh* is acting by an *Nkx3.1* dependent mechanism.

XXI. Sclerotome, Cartilage, and Bone

A. Sclerotome

The vertebrate skeleton arises from at least three distinct cellular sources: neural crest, paraxial mesoderm, and lateral mesoderm. As discussed earlier (see Section V, Face and Head), the skeleton of the head is formed from both neural crest cells emigrating from the dorsal neural tube and paraxial mesoderm. The vertebral column and ribs are exclusively derived from the sclerotomal component of the somitic mesoderm (for review see Christ *et al.*, 2000), whereas the sternum and long bones of the limbs derive from lateral plate mesoderm (see Section XII, Limb and Fin of Vertebrates). Despite the varied cellular origins of skeletal precursors, a common theme to their subsequent growth and differentiation emerges. Much of the skeleton (including bones of the face, limb, vertebrae, ribs, and sternum) develops from a cartilage template that later undergoes so-called endochondral ossification. In contrast, the clavicle and much of the cranium (the skull case) form bone directly from a mesenchymal condensate in the absence of a cartilage intermediate, a process termed intramembranous ossification (reviewed in Karsenty, 1999). Hedgehog signaling has been implicated in distinct aspects of vertebrate skeletogenesis: First, the induction, proliferation, and survival of sclerotomal progenitors of the axial vertebral skeleton and, later, coordination of proliferation and differentiation within the entire endochondral skeleton.

Development of the axial skeleton is dependent on early patterning events in the paraxial mesoderm of the somite, specifically the induction of a ventral sclerotomal population of cells. Over a period of several days, sclerotomal derivatives migrate from their site of origin to give rise to the vertebrae, intervertebral discs, and ribs. The precise contribution of these cells to specific components of these structures corresponds to their initial position within the somite. Thus, cells that are positioned more dorsal–medially in the sclerotome give rise to dorsal neural arches that wrap around the spinal cord, whereas the more ventromedial sclerotomal derivatives migrate around the notochord and form the vertebral bodies and intervertebral discs (reviewed in Christ *et al.*, 2000).

Extirpation of the neural tube and notochord (Christ, 1970; Rong *et al.*, 1992; Teillet and Le Douarin, 1983), or the physical separation of theses axial tissues from the early somite (Marcelle *et al.*, 1999), results in a complete loss of vertebrae and axial muscles. Re-engraftment of either the ventral neural tube or notochord rescues these structures (Rong *et al.*, 1992; Teillet *et al.*, 1998b). These approaches are complemented by experiments examining the patterning activity of notochord and floor plate when grafted adjacent to dorsal regions of the somite that demonstrate a role for notochord and floor plate signaling in somite patterning (Brand-Saberi *et al.*, 1993; Dietrich *et al.*, 1997; Goulding *et al.*, 1993; Pourquie *et al.*, 1993). Considerable evidence indicates that Shh secretion by these midline tissues is central to the induction of both the sclerotomal and myotomal components of the somite (see Section XV, Myogenesis).

In much of this work, sclerotomal induction is assayed by transcriptional activation of a transcriptional regulatory factor, *Pax1*. Pax1 is the earliest molecular marker of sclerotome and is essential for vertebral body and intervertebral disc formation (Wallin *et al.*, 1994). Limited studies with other sclerotomal markers have come to similar conclusions (Furumoto *et al.*, 1999; Kos *et al.*, 1998; Rohr *et al.*, 1999). Shh represses expression of *Pax3*, a uniformly expressed regulatory factor in the unpatterned somite, and induces expression of *Pax1* (C.-M. Fan *et al.*, 1995; C.-M. Fan and Tessier-Lavigne, 1994; Munsterberg *et al.*, 1995; Borycki *et al.*, 1998; Marcelle *et al.*, 1999; Murtaugh *et al.*, 1999; Teillet *et al.*, 1999). Induction correlates with the movement of Shh protein from the notochord, and later the floor plate, into the ventromedial somite (Gritli-Linde *et al.*, 2001) and the subsequent upregulation of general Shh response genes such as *Ptc1* and *Gli1* (Borycki *et al.*, 1998; Marigo *et al.*, 1996c). Further, antisense-mediated downregulation of *Shh* inhibits *Pax1* induction (Borycki *et al.*, 1998).

With the weight of evidence in support of Shh-mediated induction of sclerotome, the finding that *Pax1* is expressed in *Shh* mutants was a surprise (Chiang *et al.*, 1996). However, subsequent analysis of *Ptc1* expression revealed that a loss of

Shh alone does not abolish Hedgehog signaling in the ventral somite of the mouse embryo. This requires the removal of both Shh and gut-derived Ihh, indicating that these family members play semi-redundant roles in sclerotome induction (X. Zhang *et al.,* 2001). In the complete absence of any Hedgehog signaling in *Smo* mutants, *Pax1* expression is completely abolished (X. Zhang *et al.,* 2001). Thus, Hedgehog signaling is both necessary and sufficient for sclerotomal induction in vertebrates.

The molecular pathway by which Hedgehog signals induce a sclerotomal cell fate choice remains to be determined. At present, no direct, sclerotome-specific targets have been identified, although there are a number of candidates that include *Pax1*. There is the additional question of whether Shh induction is sufficient for sclerotome to progress to cartilage. This does not appear to be the case. Although Shh is a key step in the process, subsequent signaling by BMP members is required for sclerotomal cells to adopt chondrocyte fates (Murtaugh *et al.,* 1999).

After induction of sclerotome it is likely that Shh plays a continuing role in proliferation and survival of sclerotomal progenitors. As mentioned above, *Pax1*-expressing cells are lost at later stages, in *Shh* mutants most likely through cell death (Chiang *et al.,* 1996). Although the late loss of sclerotome may reflect a secondary consequence of the many abnormalities in *Shh* mutants, Teillet *et al.* (1998a) have demonstrated that Shh can rescue sclerotomal cell death induced on removal of the normal axial Shh signaling sources, the notochord and floor plate. Thus, Shh acts as a sclerotomal survival factor. Moreover, Shh induces proliferation of sclerotome both *in vivo* (Johnson *et al.,* 1994) and *in vitro* (C.-M. Fan *et al.,* 1995; C.-M. Fan and Tessier-Lavigne, 1994; Munsterberg *et al.,* 1995; Murtaugh *et al.,* 1999), indicating that Shh is also a mitogen for sclerotome cells.

B. Cartilage and Bone

Development of the endochondral skeleton is characterized by an orderly progression of events exemplified by long bone formation in the vertebrate (reviewed in Karsenty, 1999). After the initial condensation of chondrocyte precursors into a skeletal anlage, an outer layer of flattened cells, the perichondrium, forms around an inner core of undifferentiated, replicating chondrocytes. Growth of the cartilage element at this time is primarily by increasing cell number, through the general proliferative activity of the chondrocytes. In the next developmental stage, cells stop dividing at the center of the skeletal element and enter a postmitotic, prehypertrophic state. There follows a period of hypertrophic differentiation of postmitotic chondrocytes that results in a progressive enlargement of chondrocytes, and the accompanying synthesis of a mineralized extracellular matrix. Growth at this stage reflects the proliferative activity of chondrocytes within the two zones that lie closer to each of the prospective joint-forming regions at the ends of the long bones (the epiphyseal region), and the hypertrophy of more central, postmitotic chondrocytes.

The appearance of osteoblasts is closely coupled to the formation of hypertrophic chondrocytes. Osteoblasts are first detected by molecular and cellular analysis in the perichondrial/periosteal region, adjacent to hypertrophic chondrocyte where the osteoblasts lay down a periosteal, bone collar. Death of hypertrophic chondrocytes at the center of the element is accompanied by blood vessel and osteoblast invasion leading to a replacement of the cartilage matrix with bone and establishment of the bone marrow.

Considerable progress has been made in understanding how this sequence of events is controlled in the embryonic and early postnatal period. Ihh has emerged from these studies as a pivotal factor in coordinating, proliferation, and differentiation in the long bones, and other elements of the endochondral skeleton.

Ihh expression is largely restricted to early, postmitotic prehypertrophic chondrocytes (Bitgood and McMahon, 1995; St-Jacques *et al.*, 1999; Vortkamp *et al.*, 1996). Ectopic activation of Ihh signaling results in a pronounced increase in chondrocyte proliferation and blocks chondrocyte differentiation (Vortkamp *et al.*, 1996). This latter activity is shared with parathyroid hormone-related peptide (PTHrP), a regulatory factor synthesized predominantly, although not exclusively (Medill *et al.*, 2001), within the periarticular perichondrium (Vortkamp *et al.*, 1996). In the absence of either PTHrP or its receptor, which is expressed in proliferating chondrocytes, chondrocytes of mutant embryos undergo apositional hypertrophic differentiation, at a point closer to the articular surfaces than the chondrocytes of wild-type limbs (Amizuka *et al.*, 1994; Karaplis *et al.*, 1994; Lanske *et al.*, 1996; Lee *et al.*, 1996; Vortkamp *et al.*, 1996). Conversely, constitutive activation of the PTHrP pathway in chondroctyes blocks their normal differentiation (Schipani *et al.*, 1997; Weir *et al.*, 1996). Epistasis analysis demonstrates that Hedgehog signaling is unable to block apositional maturation of chondrocytes in the absence of the PTHrP pathway (Vortkamp *et al.*, 1996). Further, Ihh signaling is associated with an upregulation of PTHrP at the presumptive articular surfaces. These results support a model in which an Ihh–PTHrP signaling axis controls chondrocyte differentiation (Vortkamp *et al.*, 1996).

A number of studies have provided additional mechanistic insights. PTHrP signaling within proliferating chondrocytes (possibly through CREB- and AP1-dependent pathways [Ionescu *et al.*, 2001; Long *et al.*, 2001a]) inhibits further differentiation of chondrocytes. Some evidence indicates that PTHrP signaling directly inhibits *Ihh* expression in these cells (Yoshida *et al.*, 2001). The net result of the feedback mechanism is to prevent additional cells from entering a postmitotic state and thereby to maintain a proliferating zone of chondrocytes.

Analysis of *Ihh* mutants and chimeric embryos in which specific components of the Ihh and PTHrP pathways are mutated provides compelling support for the general model. *Ihh* mutants exhibit a dramatic apositional differentiation of chondrocytes that correlates with a loss of *PTHrP* expression (St-Jacques *et al.*, 1999). As predicted by the model, constitutive PTHrP signaling within chondrocytes blocks apositional differentiation in *Ihh* mutants (Karp *et al.*, 2000).

In chimeric studies, cartilage elements formed between wild-type and *PTHrP* receptor mutant cells display a cell-autonomous apositional differentiation of mutant cells, reflecting a loss of PTHrP responsiveness (Chung *et al.,* 1998). Consequently, *Ihh* is ectopically expressed as a result of the apositional differentiation of chondrocytes close to the articular surfaces (Chung *et al.,* 1998). Ectopic Ihh signaling leads to a resultant upregulation of *PTHrP,* and the increased PTHrP levels block differentiation of wild-type chondrocytes at positions where the cells would normally become postmitotic, leading to a lengthening of the columns of wild-type undifferentiated chondrocytes (Chung *et al.,* 1998). Removal of both *PTHrP* receptor and *Ihh* activities does not influence the apositional differentiation of hypertrophic chondrocytes. Rather, it restores the columns of wild-type chondrocytes to normal length, demonstrating that ectopic Ihh signaling is responsible for elevated levels of PTHrP production and the subsequent block of wild-type chondrocyte differentiation (Chung *et al.,* 2001).

Does Ihh directly regulate *PTHrP* expression? Analysis of *Ptc1* expression indicates an early, transient expression in the periarticular region. However, at later stages, upregulation of *Ptc1* is observed predominantly in a broad domain of proliferating chondrocytes and in the perichondrial/periosteal region (St-Jacques *et al.,* 1999). The activity pattern as visualized by *Ptc1* expression correlates well with the actual distribution of secreted Ihh (Gritli-Linde *et al.,* 2001). These data suggest an indirect signaling pathway and draws attention to the possible role of the perichondrium in the regulation of cartilage growth (Long *et al.,* 2001b).

Interestingly, several genes encoding BMPs and BMP receptors are expressed in the perichondrium (Solloway *et al.,* 1998; Zou *et al.,* 1997). Further, ectopic activation of BMP signaling can itself upregulate *PTHrP* expression and block chondrocyte differentiation (Zou *et al.,* 1997). Thus, Ihh could act indirectly, most likely through perichondrial expressed BMPs. Ihh upregulates expression of *Bmp2, Bmp4,* and *Bmp7* (Kameda *et al.,* 1999; Kawai and Sugiura, 2001; Macias *et al.,* 1997; Pathi *et al.,* 1999). However, the roles that these may play are complicated by the coinduction of their antagonists *Chordin* and *Noggin* (Pathi *et al.,* 1999). Interestingly, Noggin is unable to block the effects of *Ihh* overexpression in inhibiting chondrocyte differentiation. Further, addition of BMP2 cannot rescue apositional chondrogenesis that is the result of cyclopamine-mediated inhibition of Ihh signaling (Minina *et al.,* 2001). *Bmp2* expression is also reported to actually decrease *PTHrP* expression (Pateder *et al.,* 2000) whereas BMP7 appears to inhibit chondrocyte differentiation by a *PTHrP*-independent mechanism (Haaijman *et al.,* 1999). Clearly more work will be required to determine the precise roles of Ihh and BMP signaling in this important regulatory mechanism.

Chondrocyte differentiation is also blocked in proximal long bones of mice that lack the activity of Cbfa1, a transcriptional regulator of the runt family (Inada *et al.,* 1999). As the absence of differentiation here is accompanied by a loss of *PTHrP* receptor expression, Cbfa1 acts upstream of PTHrP/Ihh signaling (Inada *et al.,* 1999).

In addition to controlling differentiation through a signaling relay, Ihh regulates chondrocyte proliferation. The endochondral skeleton of *Ihh* mutants is greatly reduced; long bones are only 20% the length of those of wild-type littermates at birth, a phenotype that correlates with a pronounced reduction in proliferation (St-Jacques *et al.*, 1999). Reduced skeletal development is also observed in mice with mutations in *Gli* family members (Bai and Joyner, 2001; Mo *et al.*, 1997; Park *et al.*, 2000). Removal of Smo activity, and ectopic activation of Hedgehog signaling within undifferentiated chondrocytes, indicates that Ihh acts directly on chondrocytes to regulate chondrocyte proliferation, at least in part through the regulation of *cyclinD1* transcription (Long *et al.*, 2001b). Thus, Ihh is a mitogen for chondrocytes.

BMPs and FGFs have also been invoked in the control of chondrocyte proliferation. Growth plate expression of two distinct activating mutants of *Fgfr3* negatively regulate *Ihh* expression, leading to decreased chondrocyte proliferation in the postnatal animal (Naski *et al.*, 1998; Chen *et al.*, 2001). In contrast, a third activating allele actually stimulates early chondrocyte proliferation; whereas *Ihh* expression is not altered, *Ptc1* expression is elevated, suggesting that FGFR3 signaling acts downstream of *Ihh* (Iawata *et al.*, 2000). However, *Ptc1* expression is also reported to be elevated in mutants that lack functional *Fgfr3* (Naski *et al.*, 1998), a result that argues for an inhibitory role for FGFR3 on Hedgehog signaling. Consistent with this view, *Fgfr3* mutants display skeletal overgrowth —a phenotype opposite to that arising from the loss of Ihh signaling (Colvin *et al.*, 1996; Deng *et al.*, 1996). However, this phenotype is apparent only postnatally in the mouse. Therefore, FGFR3 is most likely to regulate Ihh action at postnatal stages, when Ihh signaling still appears to be operative (Farquharson *et al.*, 2001; Ferguson *et al.*, 1999).

Work indicates that BMP2 also stimulates chondrocyte proliferation, but only in the presence of Ihh signaling. Conversely, the proliferative effects of Ihh are blocked in the presence of Noggin (Minina *et al.*, 2001). These results suggest a synergy in the proliferative activities of Ihh and BMP signals. Interestingly, expression of *Bmp4*, a close relative of *Bmp2*, is increased in *Fgfr3* mutants (Naski *et al.*, 1998) suggesting that enhanced *Bmp* expression could stimulate chondrocyte proliferation.

Ihh has also been linked to bone formation. *Ihh* does not appear to be expressed by osteoblasts *in vivo*, so that it is likely that Ihh acts in a paracrine pathway. However, there is evidence of *Ihh* expression in cell lines with osteoblast potential (Murakami *et al.*, 1997). *Ihh* mutants lack a periosteal bone collar (St-Jacques *et al.*, 1999). Further, ectopic bone collars form in the perichondrium adjacent to apositionally differentiating chondrocytes that activate Ihh and their formation depends on Ihh signaling (Chung *et al.*, 1998, 2001). Thus, a strong correlation exists between Ihh signaling activity and periosteal bone formation in the endochondral skeleton. However, *Ihh* is insufficient to induce ectopic bone collars when expressed broadly in undifferentiated chondrocytes of transgenic mice (Long *et al.*, 2001b). One interpretation of these results is that Ihh, in combination with some other factor or factors normally produced by differentiating chondrocytes, induces periosteal osteoblasts: BMPs are clearly attractive candidates for this activity. In summary, Ihh secreted by differentiating chondrocytes, acting on the adjacent perichondrial region, couples cartilage differentiation to bone induction.

Ihh mutants also lack trabecular bone (St-Jacques *et al.,* 1999). A close association exists between vascular invasion of the mineralized cartilage and the first appearance of osteoblasts in the trabecular bone-forming region of the shaft of the long bones. *Ihh* mutant mice have limited vascular invasion of the long bones by the time of their death (shortly after birth). So, whether the absence of trabecular bone is a secondary consequence of the absence of vascular development, or whether it is linked to the absence of osteoblasts in the bone collar, or whether it reflects a distinct Ihh activity, remains to be clarified.

That *Ihh* may play a broader role in bone induction is suggested by the bone-inducing properties of Ihh on a number of mesenchymal cell lines (Kinto *et al.,* 1997; Nakamura *et al.,* 1997; Spinella-Jaegle *et al.,* 2001), on metatarsal explants in culture (Krishnan *et al.,* 2001), and after implantation of Hedgehog-secreting fibroblast cells into mice (Enomoto-Iwamoto *et al.,* 2000; Kinto *et al.,* 1997). In many of these assays, Ihh mirrors the activities of BMP members. Indeed, Hedgehog signaling synergizes with BMP signaling to induce osteogenic properties in many mesenchymal cell lines (Nakamura *et al.,* 1997; Spinella-Jaegle *et al.,* 2001).

Despite the strong evidence for Ihh in endochondral bone formation, intramembraneous bone formation is observed in the cranium and clavicle of *Ihh* mutants. Thus, there appears to be a position-specific, rather than an absolute, requirement for Ihh in osteoblast differentiation (St-Jacques *et al.,* 1999). However, bone formation is less advanced in the cranium of *Ihh* mutants, suggesting that Ihh is likely to play some role in intramembraneous bone development (St-Jacques *et al.,* 1999). *Shh* is reported to be expressed in the cranial sutures, the site where new cranial osteoblasts form (Kim *et al.,* 1998). Given the potential for cross-hybridization of RNA probes and the *Ihh* mutant phenotype, it is likely that Ihh, rather than Shh, is active in cranial sutures (Kim *et al.,* 1999). The localized expression of *Fgf* and *Bmp* family members within the sutures suggests a complex regulation that might parallel endochondral bone formation (Kim *et al.,* 1999).

Finally, given the central role that Ihh plays in coordinating skeletogenesis in the embryo, it is perhaps not surprising that *Ihh* and other regulatory signaling genes expressed in the early skeleton are re-expressed on fracture repair (Ferguson *et al.,* 1999; Ito *et al.,* 1999; Le *et al.,* 2001; Murakami and Noda, 2000; Vortkamp *et al.,* 1998). In comparing fracture repair with associated shear stress (through a cartilage-derived callus) to repair of stabilized fracture (predominantly by intramembraneous bone), only the former shows high levels of *Ihh* expression (Le *et al.,* 2001). The correlation with stress has interesting parallels with stress-induced activation of cartilage proliferation. In this model, Ihh production appears to be a key output of the stretch-induced stimulation of a mechanotransduction complex (Wu *et al.,* 2001).

XXII. Segments, Appendages, and Compartments of *Drosophila*

Nowhere has the role of Hedgehog signaling been more extensively analyzed than in the development of the *Drosophila* ectoderm, much of which gives rise to the exoskeleton of larval and adult forms of the animal. The exoskeleton is composed of a hard chitinous cuticle, rich in processes and sensory structures that provide convenient markers of the positional identity of the cells from with they derive. These markers have facilitated the detailed analysis of the roles played by Hh in controlling the identity, proliferation, polarity, and affinities of ectodermal cells. Each of these aspects of Hh function are considered in turn in the following sections.

A. Larval and Adult Abdominal Segments

Like all arthropods, the *Drosophila* body is overtly segmented. Segmental primordia are established early during embryogenesis at the cellular blastoderm stage, when an alternating pattern of segment-wide stripes of so-called pair-rule gene expression is activated in response to maternally deposited signals. Each segment is further subdivided into anterior (A) and posterior (P) compartments by the selective expression of the homeodomain protein Engrailed (En) in the posterior half of each segment. Pair-rule genes, such as *fushi tarazu* (*ftz*) and *even-skipped* (*eve*), define the future segmental and compartmental borders by activating transcription of the *en* gene in a narrow stripe of cells at the anterior boundary of their expression domains. Simultaneously, the same genes activate transcription of *hh* in the *en*-expressing cells and of *wg* (*Drosophila Wnt1*) in a similar stripe of cells immediately anterior and adjacent to those expressing *en/hh*. The interface between these two stripes of cells is known as the parasegment (PS) boundary: signaling by Hh and Wg emanating from each PS boundary is instrumental in maintaining and refining each *en* domain and in elaborating positional identity throughout each segment (reviewed by Ingham and Martinez-Arias, 1992).

Early studies of segment polarity gene function established an interdependence between the two stripes of ectodermal cells that express one or other signaling molecule, the activity of Hh being found to maintain transcription of *wg* in neighboring cells and vice versa (DiNardo *et al.*, 1988; Martinez Arias *et al.*, 1988). Strikingly, the range of both signals in this context seems to be highly restricted: only the cells immediately adjacent to those secreting either signal maintaining transcription of *wg* or *hh*. This property reinforced the notion that the positional identify of cells within each segment arises via a sequence of induction interactions between neighboring cells (Martinez Arias *et al.*, 1988). An alternative model, which envisages Hh acting at long range across the entire segment to specify cell identity in a dose-dependent manner, was advanced by Heemskirk and DiNardo

(1994). However, although consistent with the effects of modulating levels of Hh activity on the patterns of hairs that decorate the dorsal surface of the larval cuticle, these experiments could not distinguish between direct and indirect effects of Hh on individual cells.

One striking property of Hh-secreting cells not addressed by simple monotonic "gradient" models is that they signal bidirectionally along the A–P axis. Thus, not only does Hh pass anteriorly across the PS border to maintain *wg* transcription, it also passes posteriorly across the incipient segment border, as indicated by the upregulation of *ptc* transcription in these cells (Ingham *et al.*, 1991). Notably, however, this latter population does not activate *wg* transcription, indicating an underlying asymmetry in developmental potential within each segment. This property led to the suggestion that the segmental primordia are subdivided into populations that differ in their competence to respond to Hh activity. In this view, only cells in the posterior half of each anterior compartment are competent to activate *wg* transcription in response to Hh (Ingham *et al.*, 1991), a capacity that is conveyed by the expression of the *sloppy paired* (*slp*) gene, which encodes a FOX family transcription factor (Cadigan *et al.*, 1994). The importance of this asymmetric response to Hh is underscored by more recent studies of other Hh target genes, most notably *Serrate* (*Ser*), which encodes a ligand for the Notch (N) receptor and *rhomboid* (*rho*), which encodes a Golgi-resident protein required for the processing of epidermal growth factor receptor (EGFR) ligands. The localized expression of both genes is essential for the correct specification of cellular identify within each segment. Hh controls this localized expression by repressing the transcription of *Ser* in cells posterior to each segment boundary and subsequently activating expression of *rho* in the same cells (or their daughters) (Alexandre *et al.*, 1999). This establishes a series of discrete signal sources that seem to be sufficient to specify the identity of each row of cells along the A–P axis of the segment (Alexandre *et al.*, 1999). One important role of *rho* expression in cells immediately adjacent to the Hh domain is to promote the rapid endocytosis and lysosomal targeting of Wg, thereby preventing this signal from encroaching on cells of the anterior compartment (Dubois *et al.*, 2001). Thus, by activating *rho* transcription, Hh restricts the range of the Wg signal while at the same time maintaining its source by promoting *wg* transcription.

Another example of an asymmetric response to Hh is provided by the *stripe* (*sr*) gene, the product of which is necessary to promote muscle attachment at the segment boundary. Expression of *sr* is activated in cells just posterior to each Hh domain in response to Hh signaling but not in the Hh-responsive *wg*-expressing cells across the parasegment boundary. Analysis of the *sr* enhancer has identified an element necessary and sufficient for this spatially regulated expression; this contains binding sites for both Ci, the Hh-regulated transcription factor, and Pan (dTCF), the Wg-regulated transcription factor, which mediate positive and negative effects of the respective signals on its activity. Thus stimulation of *sr* expression by Hh is repressed in cells that also receive the Wg signal, leading to the observed asymmetric pattern of expression (Piepenburg *et al.*, 2000).

Histoblasts, the cells that will give rise to the adult abdominal segments, although contiguous with their larval counterparts, are segregated into so-called histoblast nests. These cells remain quiescent during the larval phases, resuming proliferation only after the larva has pupated. Like the larval segments, histoblast nests are subdivided into A and P compartments as exemplified by the restricted expression of *en* (Struhl *et al.*, 1997a). And as in the larva, Hh protein secreted by P cells plays a central role in organizing the pattern of each segment.

The extensive proliferation of histoblasts confers a distinct technical advantage when analyzing their development, because it affords the possibility of inducing mitotic recombination and thereby activating or inactivating gene activity in marked clones of cells at different developmental stages. Using this approach, it has been possible to analyze the effects of the localized ectopic expression of Hh and to investigate the consequences of eliminating or constitutively activating the Hh pathway in individual cells (Lawrence *et al.*, 1999a; Struhl *et al.*, 1997a). The results suggest that, in contrast to the situation in the larva, Hh acts directly to specify the identity of individual cells throughout the A compartment of each abdominal segment. As in the larva, cell identity can readily be assayed in the adult abdomen on the basis of cuticular morphology, such as pigmentation or the size and shape of hairs. In addition, each segment is decorated with an array of bristles—mechanosensory structures—of distinct sizes, shapes, and polarity that serve as convenient positional landmarks. Ectopic expression of Hh in clones of anterior compartment cells has a dramatic effect on the pattern of these structures, transforming the identity of both the Hh-expressing cells and that of their genetically wild-type neighbors. Cells within the clone activate En expression and adopt the morphological characteristics of posterior compartment cells whereas those outside the clone retain their anterior character but differentiate as though they were closely apposed to the compartment border (Struhl *et al.*, 1997a). Although this non-cell-autonomous effect of ectopic Hh is predicted if Hh itself acts as a morphogen, it is also consistent with the possibility that Hh normally acts to induce the expression of some other morphogen-like signal. Crucially, however, when the Hh pathway is ectopically activated in clones of anterior compartment cells by eliminating *ptc* or *pka* expression, transformations of cell fate occur within the clone while the identity of neighboring wild type cells is unaffected (Lawrence *et al.*, 1999a; Struhl *et al.*, 1997a). This strictly cell-autonomous effect of Hh pathway activation strongly implies that cell identity is specified in direct response to Hh activity. Consistent with this, when the ability of cells to respond to Hh is abrogated by removal of the *smo* gene, they adopt the character of anterior cells furthest away from the P compartment, irrespective of their actual location within the segment (Struhl *et al.*, 1997a).

Interestingly, as in the larva, the response of A compartment cells differs as a function of their location within the compartment. Thus, in response to high levels of Hh, cells in the anterior half of the compartment differentiate as though they were just posterior to the segment boundary, while those in the posterior

half differentiate as though they were located just anterior to the compartment
boundary. This difference in response is evident at the level of gene expression,
because posteriorly—but not anteriorly—located cells activate *en* in response to
Hh, reflecting the normal activation of *en* in A cells closest to the compartment
boundary (Lawrence *et al.*, 1999a). This induction of different cell types in response
to maximal Hh activity indicates that, as in the larva, the Hh signal is bidirectional,
with cells anterior and posterior to each domain of Hh expression receiving the
signal. Consistent with this, transcription of *ptc* is upregulated in cells on either side
of the posterior compartment (Struhl *et al.*, 1997a). Thus each segment contains a
U-shaped distribution of Hh activity, the lowest levels being reached in the center
of each anterior compartment.

Although the effects of loss or gain of Hh pathway activity on cell identity are
strictly cell autonomous, clones of cells lacking *ptc*, *pka*, or *smo* activity can influence the polarity of neighboring wild-type cells. For instance, bristles produced
by genetically wild-type cells in the vicinity of *ptc* or *pka* mutant clones are in-
variably oriented toward the center of the clone rather than pointing backward, as
is normally the case (Lawrence *et al.*, 1999a; Struhl *et al.*, 1997b). This suggests
that the planar polarity of abdominal cells is determined by a graded distribution
of some activity, the expression of which is controlled by Hh. In this view, the polarity signal would normally align cells such that the bristles point up the activity
gradient. At present, the identity of this signal remains elusive.

B. Adult Appendages

It is notable that in the abdominal segments, neither the inability to activate the
Hh pathway nor its constitutive activation has any effect on the size of mutant
clones, indicating that Hh activity is not required to promote cell proliferation
or survival of the histoblasts. By contrast, the activity of Hh is essential for the
growth of the adult appendages, such as the wings, antennae, and legs, which
derive from the head and thoracic segments of the larva. The appendage primordia
differ in a number of ways from the abdominal histoblasts. Unlike the latter, they
do not remain contiguous with their larval counterparts but instead form discrete
epithelial sheets of cells—the imaginal discs—that proliferate throughout the larval
stages of development. This proliferation depends critically on the activity of two
secreted proteins, Wingless (Wg) and Decapentaplegic (Dpp); and it is through
the regulation of transcription of the *wg* and *dpp* genes that Hh exerts its influence
on appendage growth (Basler and Struhl, 1994; Capdevila and Guerrero, 1994;
Ingham and Fietz, 1995; Tabata and Kornberg, 1994; Zecca *et al.*, 1995). The
role of Hh in appendage development has been characterized most extensively in
the wing, but similar principles apply to other imaginal discs: in essence, Hh acts
indirectly to stimulate proliferation and specify cell identity distant from its source
while acting directly to specify cell identity over a short range.

Like other segmentally derived structures, the wing is subdivided into A and P compartments: as in the abdomen, cells of the latter characteristically express *en* and *hh* (Lee *et al.*, 1992; Tabata *et al.*, 1992) while those of the former express *ci*. Along the A–P compartment boundary, A cells accumulate the activator form of Ci, Ci-A in response to Hh signaling, while throughout the rest of the compartment Ci is processed to yield the repressor form, Ci-R (Aza-Blanc *et al.*, 1997). Ci exerts dual control of *dpp* transcription via these two forms of the protein: thus in cells close to the compartment boundary, *dpp* is activated by Ci in response to Hh activity while in the rest of the compartment *dpp* transciption is repressed by Ci (Methot and Basler, 1999). The resultant stripe of Dpp-expressing cells acts as a localized source for a bell-shaped graded distribution of Dpp activity across the entire wing primordium: this not only promotes the growth of the disc but also specifies cell identity in a dose-dependent manner through the differential activation of Dpp target genes in both the A and P compartments (Lecuit *et al.*, 1996; Nellen *et al.*, 1996). This mechanism allows Hh to control indirectly the identity of cells in which it is itself expressed, contrasting with its direct mode of action in the abdominal segments. Co-opting Dpp in this way confers two distinct advantages: first, it couples the specification of cell identity and the stimulus for cell proliferation that underpins appendage outgrowth; second, it obviates the constraints on the range of Hh signaling imposed by the various factors that restrict the movement of Hh protein (see Ingham and McMahon, 2001).

Nevertheless, the appendages have retained a role for Hh as a direct inducer of cell identity closer to the compartment boundary, mirroring the role of Hh in the abdominal segments. This is best exemplified in the wing by the Hh-dependent regulation of two genes, *collier* (*col*) and *en*. Expression of *col*, which encodes a COE transcription factor, is essential for the differentiation of the central region of the wing blade, the so-called intervein region between veins L3 and L4 that run along the proximodistal axis of the wing, flanking the A–P compartment boundary. In the absence of Col function, the intervein region fails to form, cells instead differentiating enlarged L3 and L4 veins that fuse with one another. Consistent with this phenotype, expression of Col is restricted to a narrow stripe of A compartment cells that abut the A–P boundary, a pattern of expression that depends exclusively on Hh activity (Mohler *et al.*, 2000; Vervoort *et al.*, 1999). Significantly, the domains of Col and Dpp expression are adjacent, not coincident, implying that cells respond differentially to Hh activity as a function of their position. Various lines of evidence indicate that the levels of Hh activity experienced by a cell determine its response. This is clearly illustrated by the expression of *en*, which, as in the abdominal segments, is activated in A compartment cells closest to the A–P boundary in response to high levels of Hh activity (Guillen *et al.*, 1995; Mullor *et al.*, 1997; Strigini and Cohen, 1997). This expression of *en* is essential for the correct specification of cells in the central region of the wing, the activity suppressing the differentiation of innervated bristles along the wing margin between the L3 and L4 veins (Hidalgo, 1994).

In addition to these direct and dosage-dependent effects of Hh on cell fate, Hh also acts at short range in a nonautonomous manner to specify the differentiation of the L3 and L4 veins themselves. Interestingly, this depends on the modulation of EGFR activation, recalling the effects of local Hh signaling in the larval segments. In this case, however, Hh activates transcription of the EGFR ligand-encoding gene, *vein* (*vn*), in a subset of the cells that express *col* (Wessells *et al.*, 1999). Vn protein secreted by the intervein cells in turn promotes the specification of the L3 and L4 primordia, activating *rho* expression in these cells. The intervein cells appear unresponsive to Vn activity because of the Hh-dependent downregulation of EGFR expression mediated by Col (Mohler *et al.*, 2000).

Specification of the L3 primordium also requires the spatially restricted expression of the *araucan* (*ara*) and *caupolican* (*cap*) genes; these encode homeodomain transcription factors that promote expression of the proneural genes *acheaete* and *scute* genes, required for the development of the sensory organs specifically associated with L3. Both *ara* and *cap* are repressed by En induced in response to the high levels of Hh just anterior to the compartment border but are activated beyond this En domain in response to the combined inputs of Hh and Dpp signaling activity (Gomezskarmeta and Modolell, 1996). A similar combined input of Hh and the secondary signals that it regulates—in this case both Dpp and Wg—is also required for the positioning of sensory cell primordia in the leg imaginal discs. Here expression of the proneural repressor gene *hairy* (*h*) is activated in longitudinal stripes that overlap the domains of Ci-A accumulation in an Smo-dependent manner (Hays *et al.*, 1999). Dorsal and ventral enhancers that drive this expression have been identified and shown to be responsive both to Hh and to Dpp or Wg activity respectively (Hays *et al.*, 1999), reflecting the dorsoventral specific activation of *dpp* and *wg* in response to Hh signaling in the legs (Basler and Struhl, 1994).

C. Cell Affinities and Compartmental Boundaries

The lineage restriction between A and P compartments is defined by a remarkably straight line that marks the interface between Hh-expressing and nonexpressing cells. This contrasts with the random, wiggly boundaries that normally surround clones of marked wild-type cells within one or another compartment. The nature of the compartment boundary is taken to reflect differences in affinities between the two cell populations, presumed to be mediated by the differential expression of one or more cell recognition/adhesion molecules. Originally, it was assumed that this differential expression is controlled by the restricted activity of En in the P compartment; however, the effects of removing Smo activity from A compartment cells have revealed a requirement for Hh signaling in maintaining affinity differences at the A–P boundary. When Smo activity is removed from an A compartment cell close to the boundary, its progeny are displaced into the P compartment, where they form a smooth border with their sister cells in the A compartment (Blair and Ralston, 1997; Rodriguez and Basler, 1997). That this behavior reflects a transcriptional response of cells to Hh signaling is demonstrated by the finding that clones of cells lacking the activator form of Ci (Ci-A) behave in exactly the same way (Dahmann and Basler, 2000). In both cases, the cells retain their A identity—for instance, they differentiate characteristic sense organs and do not express *en*—but exhibit affinities similar to those of P cells. This dependence of affinity differences on the activator form of Ci could provide an explanation for the well-established finding that *en* clones cross into the A compartment, because in the absence of En activity, transcription of *ci* is derepressed in P compartment cells. Aberrant activation of *ci* in *en* mutant cells is not, however, sufficient to explain their behavior because P *en* cells that also lack all *ci* expression still sort out from their P neighbors. In this case, however, the cells form straight borders with both A and P cells, implying that En plays an additional, Ci-independent role in regulating affinity. Significantly, A cells that also lack all *ci* expression, and therefore lack both activator and repressor forms of the protein, behave in an identical manner. The simplest explanation of these findings is that cell affinities are controlled by the differential expression of a single adhesion molecule, the expression of which is repressed by En or Ci-R in P and A compartment cells, respectively, but upregulated by Ci-A in cells close to the compartment boundary (Dahmann and Basler, 2000). In this view, the A and P cell affinities would be fundamentally similar except at the compartment boundary, where an increased level of expression of the affinity molecule creates a barrier that prevents cell mixing between compartments. Because Hh activity forms a steep gradient close to the compartment border, it follows that there should be a gradient of cell affinities extending in an anterior direction away from the border. Such a gradient is difficult to detect by functional assays in the imaginal disc, but is apparent in the behavior of cells mutant for *ptc* in the abdominal segments, where the provenance and identity of individual cells can readily be scored. Here, clones of *ptc* mutant cells in which the Hh pathway is maximally activated progressively minimize contact with their neighbors the further they are from the Hh-expressing P compartment cells, eventually rounding up and detaching from the epithelium (Lawrence *et al.*, 1999b).

XXIII. Tongue and Taste

Taste in mammals is mediated in part by sensory taste papillae on the tongue. There are four main types in rodents: fungiform, circumvallate, foliate, and filiform. Only the first three give rise to gustatory sensory cells. Taste papillae are first observed on the tongue of the fetal mouse; taste receptors themselves are thought to form after sensory innervation of taste papillae, around the time of birth (Paulson *et al.*, 1985). Relatively little is known about the mechanisms that underlie papillary formation, innervation, taste cell differentiation, and the continued renewal of taste receptors throughout life. However, the expression of *Shh* and a number of other signaling factors suggests that they might play a role.

Shh is initially expressed broadly within the tongue epithelium at 12.5 dpc, but is limited to circumvallate and fungiform precursors by 14.5 dpc (Bitgood and McMahon, 1995; Hall *et al.*, 1999). Examination of *Ptc1* and *Gli1* expression in the embryonic tongue suggests that Shh signaling is occurring within the papilla epithelium, the site of sensory cell production, as well as in underlying mesenchyme (Hall *et al.*, 1999). Expression is observed before taste cell formation and continues in the taste papillae throughout life (Miura *et al.*, 2001). Thus, Shh signaling is well placed to play a role both in the initial induction and later renewal of taste receptors. Taste cells in the adult are maintained by sensory nerve input. Interestingly, denervation experiments in the adult that remove nerve IX-mediated support for taste cells result in a dramatic downregulation of Hedgehog signaling within taste papillae (Miura *et al.*, 2001). Whether there is a causal relationship between the loss of Shh expression and loss of taste cells remains to be determined.

In addition to *Shh*, *Bmp2* and *Bmp4* are also expressed in the taste papilla epithelium at 13.5 dpc (Bitgood and McMahon, 1995; Jung *et al.*, 1999), suggesting a possible link with *Shh*. However, the exact overlap in their specific expression profiles within distinct papillae has not been determined. Further, unlike *Shh*, expression of *Bmp4* is transient; no expression is observed after 15.5 dpc (Jung *et al.*, 1999).

XXIV. Tooth

As with hair and feathers, tooth formation requires a series of epithelial–mesenchymal interactions to establish the final differentiated organ. Tooth formation results from epithelial–mesenchymal interactions between ectodermal placodes on the oral surfaces of the mandibular and maxillary arches and underlying neural crest-derived dental mesenchyme. At the earliest placodal stage, before invagination of the ectodermal component of the tooth-into the dental mesenchyme, tooth-inducing properties appear to reside within the dental epithelium. After invagination, this property shifts to the dental papilla of the emerging tooth bud (for review see Maas and Bei, 1997). In the initial stages of tooth development the epithelium of the tooth anlagen forms a bud and then a cap. At this later stage, a population of nonreplicating ectodermal cells, the so-called enamel knot, appears toward the center of the molar epithelium. This structure has attracted considerable interest because of the expression of a large number of signaling molecules within the enamel knot (both primary and later arising secondary enamel knots) and the correlation of the appearance of these factors with specific cuspal patterning of the tooth germ (Jernvall *et al.*, 2000; Jernvall and Thesleff, 2000).

Further growth of the tooth leads to the bell stage, by which time the epithelial components form two distinct populations of cells: an inner and outer enamel epithelium, separated by the stellate reticulum. Localized cytodifferentiation of mesenchyme cells into polarized, pre-dentin-secreting odontoblasts is first observed at the bell stage in cells immediately facing the ectodermal inner dental epithelium. Shortly thereafter, preameloblasts, the precursors of enamel-secreting ameloblasts, within the inner dental epithelium. Thus, differentiation of these two critical cell types is closely coupled temporally and spatially (Thesleff and Jernvall, 1997; Thesleff and Sharpe, 1997). Depending on the particular tooth and species characteristics, the pattern of enamel production is tightly regulated.

Shh is expressed in the ectodermal component of the mammalian tooth from its initial appearance as a placodal thickening until the differentiation of ameloblasts (Bitgood and McMahon, 1995; Dassule and McMahon, 1998; Gritli-Linde *et al.,* 2001; Hardcastle *et al.,* 1998; Koyama *et al.,* 1996). The initial restriction of *Shh* to the tooth anlagen is probably achieved in part by Wnt7b-mediated repression initiated by adjacent nondental oral epithelium (Sarkar *et al.,* 2000). *Wnt10b,* a second *Wnt* member, is actually expressed in the dental placode, but may play a later role in modulating *Shh* expression within the epithelium of the tooth (Dassule and McMahon, 1998). In addition, there is indirect evidence for *Prx1/Prx2*-dependent mesenchymal signaling in the positive regulation of *Shh* expression (ten Berge *et al.,* 2001).

By the cap stage, *Shh* expression becomes restricted to the enamel knot. However, Shh protein extends a considerable distance in the basement membrane that underlies the prospective inner, and much of the outer enamel epithelium and into the dental papilla (Gritli-Linde *et al.,* 2001). Analysis of *Ptc1* and *Gli1* expression suggests that Shh acts as a planar signal within the ectoderm and as a paracrine signal to the dental mesenchyme at these early stages (Dassule and McMahon, 1998; Hardcastle *et al.,* 1998).

By the bell stage, *Shh* is expressed broadly throughout the undifferentiated inner enamel epithelium, but is downregulated on ameloblast differentiation. However, expression is observed in a cuboidal, epithelial population that overlies the differentiated ameloblasts, the stratum intermedium (Gritli-Linde *et al.,* 2001; Koyama *et al.,* 2001). Although *Shh* is no longer expressed in ameloblasts themselves, Shh is supplied by cells of the stratum intermedium. The upregulation of *Ptc1* in postmitotic ameloblasts in the postnatal tooth supports a continued signaling role for Shh in these cells (Gritli-Linde *et al.,* 2001). In addition, the strong upregulation of *Ptc1* observed in the stellate reticulum implies that Shh signaling is also active in this population, consistent with Shh protein localization studies (Gritli-Linde *et al.,* 2001). As at all earlier stages, Shh signaling is maintained in the dental papilla (Gritli-Linde *et al.,* 2001).

The changing pattern of Shh signaling in the tooth, from induction to differentiation, suggests that *Shh* may play direct roles at different stages of tooth development in both ectodermal and mesenchymal components of the tooth. Cobourne *et al.* (2001) have attempted to address the earliest activity of Shh, by antibody- and drug (forskolin)-mediated inhibition of signaling in cultures of mandibular explants. Their study suggests that Shh is not required for tooth placode formation per se, but Shh does regulate placodal growth and epithelial survival over the limited period of the assay (to bud stages). Further, *Gli2/Gli3* compound mutants form incisors that arrest development at early bud stages (Hardcastle *et al*, 1998). Growth of the tooth germ is also disrupted on ectopic application of Shh, a result consistent with a role for *Shh* in the control of epithelial invagination at these stages (Hardcastle *et al.,* 1998). Interestingly, although birds lost dentition some 60 million years ago, they develop in the oral regions of the mouth simple epithelium placodes that resemble the earliest stages of tooth placode formation. Their placodes do not express *Shh,* but can be induced to do so by a combination of FGF and BMP signals. Activation of *Shh* is associated with ingrowth of the epithelial placode, although whether Shh is actually necessary for ingrowth has not been addressed (Chen *et al.,* 2000).

The complete removal of Shh signaling after the bud stage leads to a dramatic phenotype in the mouse (Dassule *et al.,* 2000). Both incisors and molars fail to invaginate fully, remaining as small, superficial structures within the jaws, reflecting an abnormal morphogenesis that results from decreased growth of both ectodermal and mesenchymal populations (incisors being more severely affected than molars). Interestingly, although polarization of both ameloblast and odontoblast precursors is abnormal in these mutants, dentin- and enamel-specific markers are expressed. Hence, Shh signaling is not essential for these aspects of ameloblast and odontoblast differentiation (Dassule *et al.,* 2000).

How exactly Shh coordinates growth and morphogenesis of distinct cell populations remains to be determined. One hypothesis is that Shh may act indirectly through the regulation of other signals as observed elsewhere. However, the initial examination of a number of signaling factors coexpressed with *Shh* in the tooth (including members of the *Bmp*, *Fgf*, and *Wnt* families) suggests that *Shh* acts independently of these factors (Dassule *et al.,* 2000). Analysis of Shh signaling in *Msx1* mutants suggests that *Msx1* expression in the mesenchyme is necessary for a normal Shh-mediated response at early stages (Y. Zhang *et al.,* 1999). Further analysis suggests that *Msx1*-mediated regulation of mesenchymal *Bmp* expression maintains *Shh* expression in the overlying epithelium (Y. Zhang *et al.,* 2000; Zhao *et al.,* 2000). Yet, this finding is difficult to reconcile with the apparent inhibitory action attributed to ectopic BMP signaling on *Shh* expression in wild-type embryos (Y. Zhang *et al.,* 2000; Zhao *et al.,* 2000). Clearly, new approaches will be required to elucidate distinct temporal and spatial aspects of *Shh* action in tooth development.

XXV. Visual System

The eyes of both insects and vertebrates derive from a single medially located anlagen that is separated into bilateral primordia in response to secreted signals. Remarkably, many of the genes that define the anlagen in vertebrates, such as *Otx*,

Six3, and *Pax6*, have orthologs that are similarly deployed in the embryonic eye field of *Drosophila*.

In vertebrates, Shh secreted by the prechordal plate plays a critical role in partitioning the eye field into two distinct optic primordia. Cells that are closest to the midline activate *Pax2* expression and form the optic stalk while more laterally located cells express *Pax6* and give rise to the retina. Overexpression of *Shh* in zebrafish embryos was found to activate *Pax2* throughout the eye field and suppress retinal development, suggesting that high levels of Shh might be responsible for the specification of optic stalk fate (Ekker *et al.*, 1995b; Macdonald *et al.*, 1995). In line with this interpretation, mutation of *Shh* in both mouse and human causes the reciprocal effect, the entire eye field forming a single retina, the so-called cyclopic phenotype (Belloni *et al.*, 1996; Chiang *et al.*, 1996; Roessler *et al.*, 1996).

Strikingly similar cyclopic phenotypes have been found to occur when dorsal midline signaling, in this case mediated by Dpp, is attenuated in *Drosophila* embryos (Chang *et al.*, 2001). Such embryos develop a single optic lobe and larval eye along the midline in place of the head ectoderm that normally separates the bilateral eye primordia. Given that *dpp* is a target gene of *hh* activity in various other contexts in *Drosophila*, this phenotype could imply an analogous role for Hh signaling in partitioning the eye field in insects as well as vertebrates. However, midline expression of *dpp* precedes expression of Hh in the head of the *Drosophila* embryo and there is no evidence of an interdependence of the expression of either gene in this context. Instead, inactivation of *hh* results in a loss of both the larval eye and adult eye primordia (Chang *et al.*, 2001; Suzuki and Saigo, 2000a), indicating a role for Hh signaling in specifying both structures after the initial partitioning of the eye field by Dpp. Notably, failure to specify the adult eye primordium is reflected by the absence of *eye/Pax6* expression. Thus in *Drosophila hh* functions as an activator of Pax6, in contrast to the repressive effect of Shh on Pax6 in the vertebrate eye field.

The cells of the *Drosophila* adult eye primordium become incorporated into the eye–antennal imaginal disc—a sacklike structure composed of a monolayer epithelium of undifferentiated cells—where they continue to proliferate throughout the larval stages of development. This proliferation depends on the localized activation of the transmembrane receptor protein Notch along the dorsoventral boundary of the developing retinal field (Dominguez and de Celis, 1998). Establishment of this boundary—which subsequently also plays a key role in patterning the eye as it differentiates—depends on the restricted expression of genes of the Iroquois Complex (IRO-C) in the dorsal half of the primordium. Removal of *smo* from clones of cells within the eye primordium results in a loss of IRO-C expression from most of the clone, whereas the same genes are ectopically expressed in and around clones of cells that lack *ptc* activity (Cavodeassi *et al.*, 1999). These findings imply that the spatially restricted expression of IRO-C is regulated by a signal whose expression is in turn regulated by Hh activity. Interestingly, expression of Hh during the first two larval instars occurs predominantly in the cells of the peripodial membrane that is apposed to the main disc epithelium (Cho *et al.*, 2000).

Hh protein produced by these cells appears to signal to the disc epithelial cells via long processes that extend across the lumen of the disc. The expression of Hh is highly dynamic during these stages, shifting from ventral to dorsal domains and back again. As well as regulating IRO-C in dorsal cells, its activity is required for the correct spatial regulation of the Notch ligands Delta and Serrate within the disc primordium (Cho *et al.*, 2000).

Despite their obvious morphological differences, there are some commonalities in the differentiation of the *Drosophila* compound eye and its vertebrate counterpart that suggest the possibility that at least some of the underlying mechanisms may be conserved. In both cases, distinct neural cell types are generated from common progenitors, although the highly ordered sequence of cell recruitment that characterizes the insect eye (Tomlinson and Ready, 1987) contrasts with the overlapping generation of different cell types in the vertebrate retina (Cepko *et al.*, 1996). Nevertheless, in both cases, extrinsic signals clearly play a pivotal role in controlling differentiation and, in both cases, Hh proteins have been implicated as one such signal.

The compound eyes of the adult *Drosophila* are composed of about 750 clusters of individual units called ommatidia, each of which consists of a precise number of photoreceptor, pigment, and lens cells arranged in a stereotypic pattern. Ommatidia arise progressively within the eye primordium as a wave of differentiation passes in a posterior-to-anterior direction across the imaginal disc epithelium. At the wave front, cells arrest in G_1 and undergo a characteristic apical constriction that causes a physical indentation in the disc known as the morphogenetic furrow (reviewed by Tomlinson and Ready, 1987). Expression of the proneural gene *atonal* is activated in cells just anterior to the furrow (Dominguez and Hafen, 1997) and as cells enter the furrow they commence differentiation, the incipient photoreceptors initiating the expression of the neuron-specific Elav protein. In addition, the photoreceptors activate expression of Hh, the activity of which influences the differentiation both of future photoreceptors and of target neurons within the brain.

The functional significance of Hh expression in differentiating photoreceptors was first addressed by Ma *et al.* (1993) and Heberlein *et al.* (1993). Both groups showed that inactivation of Hh during the third larval instar—the developmental stage during which differentiation commences—blocks the progression of the morphogenetic furrow, resulting in a dramatic reduction in the size of the eye. Conversely, driving ectopic expression of Hh in the proliferating cells anterior to the furrow, was found to be sufficient to induce their precocious differentiation: moreover, these cells initiate the formation of a new furrow, which in turn promotes further differentiation (Heberlein *et al.*, 1995). Taken together, these observations imply that in normal development, Hh protein secreted by photoreceptor cells posterior to the furrow induces the differentiation of anterior cells; these in turn secrete Hh protein, a reiterative process that drives the wave of differentiation across the disc epithelium. Such a progressive expansion of the Hh expression domain is a unique feature of eye development, contrasting with the situation in the developing appendages, where the source of Hh signaling remains restricted to a specific lineage compartment (see Section XXII). In common with Hh activity in the appendages, however, the photoreceptor-derived Hh activates the expression of the *dpp* gene within the furrow, raising the possibility that the secreted Dpp protein relays the Hh-dependent differentiation signal. Consistent with this, loss of Dpp activity also blocks furrow progression; however, although ectopic Dpp activity can induce ectopic furrows, this effect is restricted to certain regions of the disc (Pignoni and Zipursky, 1997), implying that additional Hh-dependent input is normally required for neuronal differentiation.

To test whether Hh itself might act directly on cells to promote their differentiation, Strutt and Mlodzik (1997) used mutations of *smo* to block the ability of anterior cells to respond to Hh. Although Smo activity is required for the initial expression of *atonal* in cells anterior to the furrow, implying that Hh may act directly across multiple cell diameters to regulate early *atonal* activation (Dominguez, 1999), cells lacking Smo activity are nevertheless capable of differentiating, provided they are in close proximity to wild-type cells (Strutt and Mlodzik, 1997). This suggests that some other signal regulated by Hh acts over a short range to induce photoreceptor differentiation in concert with Dpp. One possible candidate for such a signal is the Notch ligand Delta, which is required for photoreceptor differentiation and which, when coexpressed with Dpp, can mimic the effects of ectopic Hh activity, triggering neural differentiation in cells at any location anterior to the furrow (Baonza and Freeman, 2001). Other data, however, implicate an as yet unidentified Raf-dependent signal in this process (Greenwood and Struhl, 1999). Notably, cells lacking the Dpp receptor are still able to differentiate into photoreceptors provided they can receive Hh, indicating that Hh itself can substitute for Dpp. Such differentiation, is, however, significantly delayed relative to wild type, suggesting that Dpp acts by priming cells to respond to the second Hh-dependent signal, thus accelerating the rate of differentiation (Greenwood and Struhl, 1999).

The vertebrate retina consists of a multilayered sheet of neural cells composed of photoreceptors, interneurons, output neurons, and glia, all of which arise progressively during embryonic and perinatal development from a common pool of precursor cells (Cepko *et al.*, 1996). The different cell types are born in a defined and overlapping temporal sequence that has been partially conserved during evolution. In most species, the first cells to be born (beginning at stage E10 in the mouse) are the retinal ganglion cells (RGCs); photoreceptors and interneurons begin to appear almost immediately afterward and continue to be born for some time beyond the point at which RGC differentiation ceases (Cepko *et al.*, 1996).

hh genes have been found to be expressed both in the retinal pigmented epithelium that surrounds the retina as well as in the retina itself of chick, mouse, rat, and zebrafish (Jensen and Wallace, 1997; Levine *et al.*, 1997; Neumann and Nüsslein-Volhard, 2000; Stenkamp *et al.*, 2000; Takabatake *et al.*, 1997; Wallace and Raff, 1999; Zhang and Yang, 2001). Wallace and Raff (1999) reported Shh to be first detectable in the developing ganglion cell layer of the mouse retina at E14, some time after the first RGCs are born but corresponding to the peak period of RGC production. Subsequent studies of the zebrafish and chick have found the onset of *Shh* expression to be somewhat earlier, coinciding in time and space with the first RGCs to differentiate (Neumann and Nüsslein-Volhard, 2000; X. Zhang and Yang, 2001). Expression then expands outward, mirroring the wave of RGC differentiation that spreads across the retina. In chick, the incipient RGCs ahead of the Shh wave have been shown to upregulate *Ptc-1* transcription, consistent with the notion that these cells are responding to Shh secreted by the differentiated RGCs (X. Zhang and Yang, 2001). In line with this interpretation, although expression of *Shh* is initiated in the retinas of *sonic you* (*syu*) mutant zebrafish embryos, which lack wild-type Shh activity, it fails to spread across the retina and RGCs fail to differentiate. Compelling evidence that Shh propagates the wave comes from genetic mosaic experiments in which donor wild-type cells can activate *Shh* expression in *syu* mutant host retina, provided they are located at the point at which the wave normally initiates (Neumann and Nüsslein-Volhard, 2000). As the wave of differentiation propagates outward, *Ptc-1* upregulation becomes restricted to the medial region of the retina, where retinal progenitor cells continue to proliferate, suggesting that these proliferating cells are also responding to Shh. This could imply that Shh secreted by the RGCs has a dual role, acting both to promote and inhibit RCG differentiation, depending on the location of the target cells and the levels of Shh to which they are exposed. Consistent with this interpretation, overexpression of *Shh* in the chick retina results in a reduction in RGC production whereas attenuation of Shh activity, mediated by the function-blocking monoclonal antibody 5E1, causes an overproduction of RGCs (X. Zhang and Yang, 2001). Exposure of *in vitro* cultures of rat retinal cells to Shh, by contrast, was found to have no effect on RGC differentiation, yet stimulated retinal progenitor cell proliferation and caused a 2- to 10-fold increase in the numbers of photoreceptors that differentiated compared with control cultures (Levine *et al.*, 1997). Although this lack of effect on RCGs may reflect the stage at which the cells were explanted, the increase in photoreceptor differentiation induced by Shh treatment is consistent with the finding that abrogation of *twhh* and *shh* activity, using antisense oligonucleotides, disrupts photoreceptor differentiation in the zebrafish retina (Stenkamp *et al.*, 2000). These effects correlate with the expression of Shh and Twhh in the retinal pigment epithelium (RPE) of zebrafish (Stenkamp *et al.*, 2000). In rodents the RPE expresses *Ihh*; although a mutation in the mouse *Ihh* gene has been described (St-Jacques *et al.*, 1999), no effects of the loss of Ihh function in the rodent eye have been reported.

A novel role for Shh in regulating pathfinding by RGC axons has been suggested in the study by Trousse *et al.*, (2001). They noted that expression of *Shh* in the ventral brain is specifically downregulated in the region of the future optic chiasm, where retinal axons enter the brain and cross the midline. Overexpression of *Shh* in this region suppresses midline crossing, suggesting that Shh induces the expression of some repulsive factor or acts itself as such a factor. In line with the latter possibility, purified Shh can induce growth cone arrest of RGC axons in explant cultures.

The differentiation of target neurons in the visual center of the *Drosophila* brain is matched to the number of photoreceptors by the interaction between incoming retinal axons and lamina precursor cells in the brain. The arrival of axons triggers the G_1-arrested lamina precursor cells (LPCs) to undergo a final division and differentiate into neurons. In addition, the incoming axons elicit the migration and maturation of glial cells that associate with the lamina neurons to form postsynaptic cartridge units. Glia that are in close proximity with retinal axons show an upregulation of *ptc* expression (Huang and Kunes, 1998), suggesting that they respond to Hh signaling. Hh protein is not expressed within the brain at this stage but is found along the retinal axons within the optic stalk and brain, making it a good candidate for a signal mediating the interaction between the incoming axons and the LPCs and glia. Consistent with this, inactivation of Hh after photoreceptor differentiation but before axon outgrowth blocks the division and differentiation of the LPCs, whereas misexpression of Hh in the brain is sufficient to initiate LPC differentiation in the absence of incoming axons (Huang and Kunes, 1996). Surprisingly, however, inactivation of Hh has no effect on glial migration and maturation; thus, although glia can respond to axonally derived Hh, as revealed by their upregulation of *ptc* transcription, their axon-dependent behavior is not mediated by Hh (Huang and Kunes, 1998). In the mouse there is evidence that Shh is similarly translocated along RGC axons into the optic nerve: in this case the axonally derived Shh induces upregulation of *Ptc-1* transcription in astrocytes and stimulates their proliferation (Wallace and Raff, 1999).

XXVI. Human Congenital Malformations

Humans are more closely screened genetically than any other organism. Indeed, virtually every individual with even very minor developmental defects is brought to medical attention in the developed nations and familial inheritance patterns are vigorously explored and followed up. As increasing numbers of human genetic studies pinpoint the exact genetic lesions responsible for malformations, the acquired information will provide structure–function data based on the genotype–phenotype correlations, which will in turn help us to better understand how the genes function in their developmental roles and how they interact with genetic modifiers of their activity. At the same time, study of the homologous genes in model organisms continues to provide fundamental insights into the etiology of human congenital malformations.

Given their importance in regulating many different aspects of embryonic patterning, growth, and cell type specification, it is not surprising that mutations in human *hedgehog* genes and genes involved in mediating or regulating their activity have been linked to a myriad of human congenital malformations.

As discussed above, *Ihh* plays critical roles in the growth and differentiation of cartilage and bone, and loss-of-function mutations of *Ihh* in mice result in, among other things, shortening of the skeletal elements. Mutations in human *IHH* have been shown to be responsible for a dominantly inherited brachydactyly type A-1, characterized by shortening of the long bones and a shortening or absence of the middle phalanges (Gao *et al.*, 2001). The *IHH* locus (Marigo *et al.*, 1995) has also been tightly linked to inheritance of syndactyly type 1 (Bosse *et al.*, 2000). This malformation is distinct from brachydactyly, involving fusion of phalanges, but could be due to a different *IHH* allele or to the effect of distinct modifying loci segregating in these families.

Mutations in *SHH* have also been shown to cause limb defects, consistent with the important role of *Shh* in patterning the developing limb bud. As discussed above, Shh is normally produced in the posterior of the developing limb bud, where it influences both the number and morphology of the digits. Ectopic *Shh* induced experimentally in the anterior of the limb bud results in varying degrees of digit duplication, characterized as preaxial polydactyly. Moreover, in most genetic mouse models of preaxial polydactyly, ectopic *Shh* expression is observed in the anterior limb bud (e. g., in the *Xt*, *Rim4*, *lst*, *lx*, *Xpl*, and *Ssq* mutants) (Masuya *et al.*, 1995, 1997; Sharpe *et al.*, 1999). This strongly suggests that mutations in genes that lead to misregulation of *SHH* are likely to be responsible for many cases of human preaxial polydactyly as well. Preaxial polydactyly is the most frequent limb malformation and one of the most common of all developmental malformations, affecting approximately 15 in 10,000 births (Ivy, 1957; Sesgin and Stark, 1961). Obviously, although misregulation of *SHH* in humans cannot be tested directly, once genes responsible for repression of *Shh* in the anterior limb bud in mice are uncovered, these become candidates for human preaxial polydactyly. In at least one case, a mutation leading to ectopic *Shh* expression in the mouse limb bud appears to be in a regulatory element of *Shh* itself (Sharpe *et al.*, 1999). Similar genetic lesions may be involved in humans as well. Indeed, a number of human polydactylies, including cases of preaxial polydactyly type II and type III, complex polysyndactyly, and all mapped cases of polydactyly involving triphallangial thumb (Heutink *et al.*, 1994; Hing *et al.*, 1995; Radhakrishna *et al.*, 1996; Tsukurov *et al.*, 1994; Zguricas *et al.*, 1999), map to a locus that cannot be separated from the genetic location of *SHH* (Marigo *et al.*, 1995).

Polydactyly can also be generated by mutations of other genes in the Hedgehog pathway. In the limb, the major mediator of Shh activity appears to be Gli3 (see Ingham and McMahon, 2001). In the posterior of the limb, *Gli3* is transcriptionally downregulated by Shh (Marigo *et al.*, 1996b) and, in addition, the repressor form of Gli3 is posttranslationally blocked by Shh signaling (Wang *et al.*, 2000). The result is a gradient of the processed repressor form of Gli3 running from anterior (high) to posterior (low) (Wang *et al.*, 2000), thus leading to the derepression of Shh targets in the posterior. Partial loss of Gli3 activity in the anterior is therefore equivalent, in principle, to ectopic Shh expression in the anterior limb bud.

Loss-of-funtion mutations in *GLI3* cause Greig cephalopolysyndactyly syndrome (Jones, 1997; Kalff-Suske *et al.*, 1999; Vortkamp *et al.*, 1991; Wild *et al.*, 1997). This dominantly inherited syndrome includes predominantly preaxial polydactyly, syndactyly, and abnormally broad first digits (thumb and big toe). In addition, Greig patients exhibit facial anomalies including hypertelorism and frontal bossing, likely reflecting other regions of Hedgehog signaling where levels of GLI3 protein are critical. *GLI3* mutations are also responsible for Pallister–Hall syndrome, which is also dominantly inherited (Jones, 1997; Kang *et al.*, 1997). This distinct syndrome includes postaxial or central polydactyly and syndactyly, as well as hypothalmic hamartoma, imperforate anus, anteverted nares, occasional malformations of the axial skeleton, and in some instances holoprosencephaly (a condition considered separately below). Mutations in *GLI3* are also responsible for dominant inheritance of postaxial polydactyly type A, preaxial polydactyly type IV, and postaxial polydactyly type A/B (Radhakrishna *et al.*, 1997, 1999). Because the molecular nature of the lesions in the *GLI3* gene does not appear to correlate with the different phenotypic presentations listed above (Kalff-Suske *et al.*, 1999; Radhakrishna *et al.*, 1999), the differences in outcome likely are due to the differential activity of alleles of modifier genes also impinging on the Hedgehog pathway.

It is also worth noting in this context that mutations in the transcriptional coactivator *CBP* produce a dominant phenotype called Rubinstein–Taybi syndrome (Blough *et al.*, 2000; Petrij *et al.*, 1995) that is similar to the spectrum of malformations caused by *GLI3* mutations. The *Drosophila* CBP homolog is known to act as a cofactor during transcriptional activation by Ci (Akimaru *et al.*, 1997), and the vertebrate CBP protein also binds to Gli3, although not to Gli1 (Dai *et al.*, 1999). Thus, CBP may play a key role in the activator function of Gli3, which opposes the activity of the Gli3 repressor form. Thus, if the dominantly inherited *CBP* mutations responsible for Rubinstein–Taybi syndrome were dominant gain-of-function rather than haploinsufficiency lesions, then they would be expected to have similar effects as *GLI3* loss of function in the limb bud, phenocopying ectopic Hedgehog activity.

Although *GLI3* is the only one of the *GLI* genes directly implicated thus far in human congenital malformations, the roles of all three have been explored by targeted mutational analysis of mice, including analysis of the individual mutants, compound heterozygotes, and double homozygous mutants. It has been noted that the spectrum of defects observed in the various *Gli* mutant mice bears a striking resemblance to the human VACTERL association of malformations, which include limb abnormalities, vertebral defects, anal atresia, tracheoesophageal fistula, esophageal atresia, renal dysplasia, and cardiac anomalies (reviewed in Kim *et al.*, 2001). These parallels suggest that at least some instances of VACTERL could be caused by defects in genes involved in the Hedgehog signaling pathway.

Another alteration in the Hedgehog pathway that can result in activation of Hedgehog targets is the loss of the Hedgehog receptor Patched (Ptc1 in higher vertebrates). Patched acts on downstream effectors to prevent activation of target gene transcription. In the presence of Hedgehog protein, this negative influence is abrogated and target genes are transcribed. The loss of Patched produces the equivalent result, releasing the signaling system from repression (see Ingham and McMahon, 2001, for a review of the biochemistry of Hedgehog signaling). Thus, partial loss of Ptc1 activity, like partial loss of Gli3 activity, should be functionally equivalent to ectopic Shh signaling in the anterior of the limb bud and elsewhere in the embryo where levels of Hedgehog activity are critical.

Accordingly, individuals who inherit a single copy of a defective *PTCH1* gene display a range of congenital defects as part of what is termed Gorlin syndrome (Hahn *et al.,* 1996; Johnson *et al.,* 1996). Defects seen in Gorlin syndrome include preaxial polydactyly, immobile thumbs, short metacarpals, broad facies, rib defects, and dental abnormalities (Bale *et al.,* 1991; Gorlin, 1995; Jones, 1997). For reasons that are discussed in a separate section (see Section XXVII, Cancer), patients with Gorlin syndrome, also called "nevoid basal cell carcinoma syndrome" or "basal cell nevus syndrome," have a high predisposition to certain forms of cancer in addition to the congenital defects described above.

As discussed elsewhere in this review, another important location of *Shh* expression is along the midline (i.e., in the notochord, the anterior midline mesoderm known as the prechordal plate, and the floor plate of the neural tube). Midline Shh plays an essential role in patterning the ventral neural tube and surrounding axial mesoderm (discussed elsewhere in this review). The importance of Shh in patterning the midline is underscored by the finding that haploinsufficiency for *SHH* in humans is a cause of holoprosencephaly (Belloni *et al.,* 1996; Nanni *et al.,* 1999; Roessler *et al.,* 1996, 1997), a midline patterning defect characterized by incomplete separation of the ventral forebrain into distinct cerebral hemispheres and associated craniofacial anomalies such as single maxillary incisor, cleft lip and palate, a proboscis-like nasal structure, and cyclopia. Each of these phenotypes can be understood as a consequence of defective midline signaling. For example, the vertebrate eye first develops as a single developmental field until the inductive influence of Shh from the underlying anterior midline prechordal plate induces a split in the eye field to form two distinct organ primordia (see Section XXV, Visual System). Holoprosencephaly has an incidence as high as 1 in 250 conceptuses; however, because of the severe lethality associated with this defect, holoprosencephaly is seen in only 1 in 16,000 live births, which is still a considerable number (Matsunaga and Shiota, 1977; Roach *et al.,* 1975). Although *SHH* mutations have been shown to cause both sporadic and familial cases of holoprosencephaly, other cases of this malformation have been mapped to genes not directly involved in the Hedgehog pathway, highlighting the complex polygenic nature of this malformation.

One alternative genetic context that can lead to holoprosencephaly is Smith–Lemli–Opitz syndrome (SLOS). SLOS includes a number of features also seen in mice and humans with defects in the Hedgehog pathway, including polydactyly, syndactyly, anteverted nares, CNS hypoplasia, holoprosencephaly, and cryptorchidism (Donnai *et al.,* 1987; Kelley *et al.,* 1996; Tint *et al.,* 1994). SLOS is caused by mutations in the 7-dehydrocholesterol reductase gene, which acts in the final step of cholesterol biosynthesis (Fitzky *et al.*, 1998; Wassif *et al.*, 1998; Waterham *et al.*, 1998). Moreover, treating experimental animals with cholesterol biosynthesis inhibitors (Lanoue *et al.*, 1997; Reppeto *et al.*, 1990; Roux *et al.*, 1979) produces an array of phenotypes that not only parallels SLOS, but is also extremely similar to the phenotype of mice carrying homozygous mutations in *Shh* (Chiang *et al.*, 1996). Although it remains to be determined whether the phenotypes resulting from defects in cholesterol metabolism, such as SLOS, are entirely due to their impact on Hedgehog signaling, there is good reason to believe that it is a large part of the story. Cholesterol plays several important roles in the biogenesis, transport, and reception of Hedgehog signaling (reviewed in Ingham and McMahon, 2001). For example, Hedgehog proteins are first synthesized as precursors that are autoproteolytically cleaved to generate an N-terminal fragment that contains the signaling activity (Marti *et al.*, 1995; Porter *et al.*, 1995). In this catalytic reaction, the active Hedgehog fragment becomes covalently linked to cholesterol, which is important for its subsequent activity and localization (Porter *et al.*, 1996). In addition, treatment of cells with steroidal alkaloids that act as distal inhibitors of cholesterol biosynthesis inhibit the ability of cells to respond to Hedgehog signals even when added exogenously in an active form (Cooper *et al.*, 1998; Incardona *et al.*, 1998).

Because Hedgehog proteins are covalently coupled to cholesterol, the signaling proteins are anchored to the membrane of secretory cells, which would be expected to restrict their range of action. Yet, as discussed elsewhere in this review, Hedgehog proteins have developmental roles that in some cases depend on their ability to act over multiple cell diameters. This implies that there must exist specific mechanisms for transporting Hedgehog proteins intercellularly (reviewed in Ingham and McMahon, 2001). Indeed, in *Drosophila,* movement of cholesterol-modified Hh is dependent on the activity of a gene called *tout-velu* (*ttv*) that encodes a glycosyltransferase involved in the synthesis of heparan sulfate proteoglycans (Bellaiche *et al.*, 1998), which are presumably directly involved in Hh trafficking. The human homologs of *ttv* are the *EXT* genes. Mutations in *EXT1* or *EXT2* cause hereditary multiple exostoses (HME) (Ahn *et al.*, 1995; Stickens *et al.*, 1996; Wuyts *et al.*, 1997), a dominant disorder characterized by benign growths or exostoses forming along the epiphyses of long bones. HME patients also have severe skeletal malformations, short stature, and limb length inequalities. The expression patterns of *EXT1* and *EXT2* in mice are consistent with a role in mediating *IHH* transport similar to that of *ttv* in *Drosophila,* which would be consistent with the phenotypes observed (Stickens *et al.*, 2000). In addition, other *ttv/EXT* homologs exist in the human genome that are not associated with HME (Van Hui *et al.*, 1998; Wise *et al.*, 1997; Wuyts *et al.*, 1997). These could be differentially expressed and play equivalent roles in the transport of other Hedgehog gene products in humans.

XXVII. Cancer

As discussed throughout this review, the vertebrate *hedgehog* genes exhibit many activities during embryogenesis, including the ability to promote proliferation in a variety of settings. As such, it is not surprising that inappropriate activation of this pathway postnatally can contribute to the development of neoplasia. Moreover, because of the way in which Hedgehog signal transduction pathway is regulated, this can be accomplished both by genetic alterations that promote the activity of the *hedgehog* genes themselves or their downstream, positively-acting effectors, and by mutations that result in the loss of negatively acting regulators in the pathway. It is in this latter context that the link between Hedgehog signaling and cancer was first appreciated.

In the absence of Hedgehog protein, the Hedgehog receptor Patched acts to repress the signaling activity of the downstream effector molecule Smoothened. Conversely, binding of Hedgehog to Patched relieves this suppression, resulting in activation of target genes by the Gli family of transcription factors. Significantly, loss of Patched has the same effect as Hedgehog binding, because in its absence there is no suppression and Smoothened is free to initiate downstream signaling. As discussed above (see Section XXVI, Human Congenital Malformations), mutations in the human Patched gene, *PTCH1*, result in a condition known as Gorlin syndrome, basal cell nevus syndrome, or nevoid basal call carcinoma syndrome (Hahn *et al.*, 1996; Johnson *et al.*, 1996), in which *PTCH1* heterozygotes inherit a spectrum of developmental defects due to a ligand-independent activation of Hedgehog signaling. In addition, these patients are predisposed to developing various forms of cancer including multiple basal cell carcinomas, medulloblastomas, ovarian fibromas, and less frequently rhabdomyosarcomas, meningiomas, fibrosarcomas, and cardiac fibromas (Bale *et al.*, 1991; Gorlin, 1995; Kimonis *et al.*, 1997). The tumors arise when a mutagenic event inactivates the single wild-type *PTCH1* allele in these cells, leaving them with no *PTCH1* and hence fully activated Hedgehog signaling. Thus, *PTCH1* functions as a classic tumor suppressor, and individuals with germ line mutations in this gene develop multiple basal cell carcinomas.

Whereas all Gorlin syndrome patients develop basal cell carcinomas, the other types of tumors associated with this disease occur much less frequently. For example, only 5% of such individuals develop medulloblastoma, although a high frequency of medulloblastoma is seen in *Ptc1* heterozygous mice (Goodrich *et al.*, 1997a; Hahn *et al.*, 1998). Interestingly, although the *Ptc1* heterozygous mice do develop a wide range of tumors, they have not been reported to develop basal cell carcinoma. Nonetheless, activating the Hedgehog pathway in transgenic mice by expressing either *Shh* or an activated form of *Smo* in the skin does result in formation of skin lesions histologically indistinguishable from human basal cell carcinoma (H. Fan *et al.*, 1997; Oro *et al.*, 1997; Xie *et al.*, 1998). The reasons for these intriguing differences in susceptibility to different tumor types in the different species are currently unclear.

Whereas Gorlin syndrome is comparatively rare, sporadic basal cell carcinoma is the most common form of human cancer, representing one-third of all tumors diagnosed and affecting one in six individuals in their lifetime (Landis *et al.*, 1998). Significantly, *PTCH1* mutations have been identified in up to 40% of sporadic basal cell carcinomas (Aszterbaum *et al.*, 1998; Gailani *et al.*, 1996; Unden *et al.*, 1997; Wolter *et al.*, 1997). Moreover, loss of heterozygosity in markers tightly linked to *PTCH1* suggests that the frequency of *PTCH1* involvement in sporadic basal cell carcinoma may actually be greater than 50% (Gailani *et al.*, 1992). Exposure to sunlight is a key predisposing factor for basal cell carcinoma. Indeed, mutations in *p53*, which occur in the majority of basal cell carcinoma, almost always contain molecular lesions consistent with UVB damage (Ziegle *et al.*, 1993). However, less than 50% of the *PTCH1* mutations in basal cell carcinoma carry the signature C-to-A transitions indicative of UVB damage (Gailani *et al.*, 1996). In contrast, basal cell carcinomas arising in patients with xeroderma pigmentosum have *PTCH1* mutations that almost always carry the UVB signature (Bodak *et al.*, 1999; D'Errico *et al.*, 2000; X. Zhang *et al.*, 2001).

PTCH1 mutations have also been found to be associated with a wide variety of other sporadic tumor types. These include medulloblastoma (Pietsch *et al.*, 1997; Raffel *et al.*, 1997; Vorechovsky *et al.*, 1997; Wolter *et al.*, 1997), squamous cell esophageal carcinoma (Maesawa *et al.*, 1998), meningioma (Xie *et al.*, 1997), transitional cell bladder carcinoma (McGarvey *et al.*, 1998), benign skin trichoepitheliomas (Vorechovsky *et al.*, 1997), and various noninflammatory cysts (Levanat *et al.*, 2000). It is worth noting that all of these tumors are derived from tissues in which Hedgehog signaling plays an endogenous role. For example, basal cell carcinoma is believed to be derived from stem cells in the bulge of the hair follicle (see Section VIII, Hair and Feather Morphogenesis, above). A full, up-to-date catalog of all *PTCH1* mutations identified in human tumors can be found at the *PTCH1* Mutation Database on the Web at http://www.cybergene.se/PTCH.

Activating mutations in *SMOH* would also be expected to have the same downstream consequences as inactivation of *PTCH1*. Indeed, constitutively activating mutations in *SMOH* have been identified in up to 20% of the sporadic basal cell carcinomas surveyed (Lam *et al.*, 1999; Reifenberger *et al.*, 1998; Xie *et al.*, 1998). *SMOH* mutations have also been found to be associated with medulloblastoma (Reifenberger *et al.*, 1998).

The *Gli* genes encode transcription factors that mediate Hedgehog signaling, but in addition, all three vertebrate *Gli* genes are themselves transcriptionally regulated by Hedgehog activity. In particular, *Gli1* expression is induced in response to Hedgehog signaling (see Ingham and McMahon, 2001). Specifically, whereas *PTCH1* and *SMOH* mutations are detected only in a total of approximately 60% of sporadic basal cell carcinomas, a high level of *GLI1* transcriptional activation is a consistent feature of all such tumors (Dahmane *et al.*, 1997; Ghali *et al.*, 1999), suggesting that activation of the Hedgehog signaling pathway, and hence of Hedgehog target genes by *GLI1*, may be a requisite step in the formation of these tumors. Consistent with this, transcription of *PTCH1*, which is itself a universal target of Hedgehog signaling, is also upregulated in all cases of familial and sporadic basal cell carcinoma examined to date (Unden *et al.*, 1997).

In addition to stimulation of the Hedgehog pathway, an alternative way to increase *Gli1* expression in a cell, and hence to activate Hedgehog target genes, is through gene amplification. Indeed, the *GLI1* gene was originally identified on the basis of its amplification in gliomas (Kinzler *et al.*, 1987). *GLI1* gene amplification has also been reported in some lymphomas, osteosarcomas, and rhabdomyosarcomas (Roberts *et al.*, 1989; Werner *et al.*, 1997). Although the clinical relevance of this *GLI1* DNA amplification is obscured by the fact that other significant genes might be coamplified with *GLI1* in these cells, it has been shown that the degree of *GLI1* gene expression in sarcomas carrying *GLI1* amplification does indeed correlate with the pathological tumor grade (Stein *et al.*, 1999).

An important issue that is not completely resolved is the functional significance of Hedgehog signaling in these examples of tumorgenesis. One possibility is that Hedgehog signaling promotes the invasiveness of cells, allowing them to establish macroscopic growth. Indeed, in at least one *in vitro* invasion assay, the level of Hedgehog signaling in cells, as assayed by levels of *PTCH1* expression, was correlated with greater invasive properties (Saldanha *et al.*, 1998). However, it is likely that the most significant impact of ectopic Hedgehog signaling in cancer is at the level of proliferation. In particular, in both basal cell carcinoma and medulloblastoma, the two tumor types most consistently linked with activation of the Hedgehog pathway, *SHH* is believed to stimulate proliferation of the putative tumor progenitor cell (Chiang *et al.*, 1999; Dahmane and Ruiz-i-Altaba, 1999; St-Jacques *et al.*, 1998; Wallace, 1999; Wechsler-Reya and Scott, 1999). However, the molecular mechanisms underlying this are still unclear. It has been shown that in an epithelial cell model of basal cell carcinoma, Shh expression prevents cells from exiting S and G_2/M phases in response to differentiation signals and also blocks p21-induced growth arrest, thus allowing long-term proliferation (H. Fan and Khavari, 1999). Therefore, the inappropriate activation of Hedgehog targets likely promotes cell proliferation instead of allowing normal differentiation and exit from the cell cycle. Alternatively, it has been suggested that PTCH1 might, in the absence of Hedgehog signaling, directly affect the subcellular localization of cyclin B1, thereby altering the G_2/M checkpoint by changing the localization of MPK (Barnes *et al.*, 2001). This seems unlikely, however, because a mechanism based on direct physical interaction between PTCH1 and cyclin B1 would not explain how an identical phenotype could be mediated by an activated form of SMOH, nor would it make sense of the upregulation and amplification of *GLI1* seen in other tumors. Because *GLI1* upregulation is a consistent finding in all basal cell carcinomas, transcriptional changes would seem to be the most likely mechanism by which the pathway affects tumorigenesis. As such, it opens up the possibility that pharmacological agents that inhibit the Hedgehog pathway downstream of PTCH1 and SMOH might provide important and novel therapeutics for treating these diseases.

A final link between the Hedgehog pathway and cancer is seen in malignant tumors arising in patients with hereditary multiple exostoses. As discussed above (see Section XXVI, Human Congenital Malformations), this common hereditary skeletal dysplasia is caused by haploinsufficiency for *EXT1* or *EXT2*, the human homologs of the *Drosophila ttv* gene, which is required for intracellular transport of Hh. The bony benign tumors called exostoses, which are characteristic of this disorder, occasionally develop into malignant sarcomas. This transition has been linked to loss of heterozygosity at the *EXT* loci, such that the unaffected copy of the gene is lost (Hecht *et al.,* 1995; Raskind *et al.,* 1995). Thus, like *PTCH1*, *EXT1* and *EXT2* can be considered tumor suppressor genes. However, in contrast to *PTCH1*, the *EXT* genes promote Hedgehog signaling. Thus, if the etiology of these tumors is related to the role of *EXT1* and *EXT2* in Hedgehog transport, then the loss of IHH-mediated regulation of the bone differentiation pathway may be directly responsible for this neoplasia.

Acknowledgments

We are particularly indebted to the tireless efforts of Paul Cassidy in all aspects of the preparation of this manuscript. We also acknowledge the helpful comments of Amel Gritli-Linde, Penny Rashbass and Anne-Gaelle Borycki, and the support of Terri Broderick. Finally, we thank our families for their tolerance of our efforts. Work in Andrew P. McMahon's laboratory is supported by a grant from the NIH (NS 33642), work in Philip W. Ingham's laboratory is supported by Wellcome Trust Programme Grant 051824/Z/97, and work in Cliff Tabin's laboratory is supported by NIH Grant HD32433.

References

Afzelius, B. A. (1976). A human syndrome caused by immotile cilia. *Science* **193,** 317–319.
Agarwala, S., Sanders, T. A., and Ragsdale, C. W. (2001). Sonic hedgehog control of size and shape in midbrain pattern formation. *Science* **291,** 2147–2150.
Ahlgren, S. C., and Bronner-Fraser, M. (1999). Inhibition of sonic hedgehog signaling in vivo results in craniofacial neural crest cell death. *Curr Biol.* **9,** 1304–1314.
Ahlgren, U., Pfaff, S. L., Jessell, T. M., Edlund, T., and Edlund, H. (1977). Independent requirement for ISLI information of pancreatic mesenchyme and islet cells. *Nature* **385,** 257–260.
Ahn, J., Ludecke, H. J., Lindow, S., Horton, W. A., Lee, B., *et al.* (1995). Cloning of the putative tumour suppressor gene for hereditary multiple exotoses (EXT1). *Nat. Genet.* **11,** 137–143.
Akimaru, H., Chen, Y., Dai, P., Hou, D. X., Nonaka, M., *et al.* (1997). *Drosophila* CBP is a co-activator of cubitus interruptus in hedgehog signaling. *Nature (London)* **397,** 735–738.
Akimenko, M. A., and Ekker, M. (1995). Anterior duplication of the Sonic hedgehog expression pattern in the pectoral fin buds of zebrafish treated with retinoic acid. *Dev. Biol.* **170,** 243–247.

Alberta, J. A., Park, S. K., Mora, J., Yuk, D., Pawlitzky, I., *et al.* (2001). Sonic hedgehog is required during an early phase of oligodendrocyte development in mammalian brain. *Mol. Cell. Neurosci.* **18,** 434–441.

Alescio, T., and Cassini, A. (1962). Induction in vitro of tracheal buds by pulmonary mesenchyme grafted onto tracheal epithelium. *J. Exp. Zool.* **150,** 83–94.

Alexandre, C., Lecourtois, M., and Vincent, J.-P. (1999). Wingless and Hedgehog pattern *Drosophila* denticle belts by regulating the production of short-range signals. *Development* **126,** 5689–5698.

Amizuka, N., Warshawsky, H., Henderson, J. E., Goltzman, D., and Karaplis, A. C. (1994). Parathyroid hormone-related peptide-depleted mice show abnormal epiphyseal cartilage development and altered endochondral bone formation. *J. Cell Biol.* **126,** 1611–1623.

Apelqvist, A., Ahlgren, U., and Edlund, H. (1997). Sonic hedgehog directs specialised mesoderm differentiation in the intestine and pancreas. *Curr. Biol.* **7,** 801–804.

Arkell, R., and Beddington, R. S. (1997). BMP-7 influences pattern and growth of the developing hindbrain of mouse embryos. *Development* **124,** 1–12.

Aszterbaum, M., Rothman, A., Johnson, R. L., Fisher, M., Xie, J., *et al.* (1998). Identification of mutations in the human PATCHED gene in sporadic basal cell carcinomas and in patients with the basal cell nevus syndrome. *J. Invest. Dermatol.* **110,** 885–888.

Aza-Blanc, P., Ramirezweber, F., Laget, M., Schwartz, C., and Kornberg, T. (1997). Proteolysis that is inhibited by hedgehog targets cubitus-interruptus protein to the nucleus and converts it to a repressor. *Cell* **89,** 1043–1053.

Azpiazu, N., Lawrence, P. A., Vincent, J. P., and Frasch, M. (1996). Segmentation and specification of the *Drosophila* mesoderm. *Genes Dev.* **10,** 3183–3194.

Bai, C. B., and Joyner, A. L. (2001). Gli1 can rescue the in vivo function of Gli2. *Development* **128,** 5161–5172.

Bale, S. J., Amos, C. I., Parry, D. M., and Bale, A. E. (1991). Relationship between head circumference and height in normal adults and in the nevoid basal cell carcinoma syndrome and neurofibromatosis type I. *Am. J. Med. Genet.* **40,** 206–210.

Baonza, A., and Freeman, M. (2001). Notch signalling and the initiation of neural development in the *Drosophila* eye. *Development* **128,** 3889–3898.

Barnes, E. A., Kong, M., Ollendorff, V., and Donoghue, D. J. (2001). Patched1 interacts with cycin B1 to regulate cell cycle progression. *EMBO J.* **20,** 2214–2223.

Barresi, M., Stickney, H., and Devoto, S. (2000). The zebrafish slow-muscle-omitted gene product is required for Hedgehog signal transduction and the development of slow muscle identity. *Development* **127,** 2189–2199.

Barth, K. A., and Wilson, S. W. (1995). Expression of zebrafish nk2.2 is influenced by sonic hedgehog/vertebrate hedgehog-1 and demarcates a zone of neuronal differentiation in the embryonic forebrain. *Development* **121,** 1755–1768.

Basler, K., and Struhl, G. (1994). Compartment boundaries and the control of *Drosophila* limb pattern by hedgehog protein. *Nature (London)* **368,** 208–214.

Beattie, C. E., Hatta, K., Halpern, M. E., Liu, H., Eisen, J. S., and Kimmel, C. B. (1997). Temporal separation in the specification of primary and secondary motoneurons in zebrafish. *Dev. Biol.* **187,** 171–182.

Belaoussoff, M., Farrington, S. M., and Baron, M. H. (1998). Hematopoietic induction and respecification of A-P identity by visceral endoderm signaling in the mouse embryo. *Development* **125,** 5009–5018.

Bellaiche, Y., The, I., and Perrimon, N. (1998). Tout-velu is a *Drosophila* homologue of the putative tumour suppressor EXT-1 and is needed for Hh diffusion. *Nature (London)* **394,** 85–88.

Belloni, E., Muenke, M., Roessler, E., Traverso, G., Siegel-Bartelt, J., *et al.* (1996). Identification of *Sonic hedgehog* as a candidate gene responsible for holoprosencephaly. *Nat. Genet.* **14,** 353–356.

Bellusci, S., Henderson, R., Winnier, G., Oikawa, T., and Hogan, B. L. (1996). Evidence from normal expression and targeted misexpression that bone morphogenetic protein (Bmp-4) plays a role in mouse embryonic lung morphogenesis. *Development* **122,** 1693–1702.

Bellusci, S., Furuta, Y., Rush, M. G., Henderson, R., Winnier, G., and Hogan, B. L. (1997a). Involvement of Sonic hedgehog (Shh) in mouse embryonic lung growth and morphogenesis. *Development* **124,** 53–63.

Bellusci, S., Grindley, J., Emoto, H., Itoh, N., and Hogan, B. L. (1997b). Fibroblast growth factor 10 (FGF10) and branching morphogenesis in the embryonic mouse lung. *Development* **124,** 4867–4878.

Bhardwaj, G., Murdoch, B., Wu, D., Baker, D. P., Williams, K. P., *et al.* (2001). Sonic hedgehog induces the proliferation of primitive human hematopoietic cells via BMP regulation. *Nat. Immunol.* **2,** 172–180.

Bhat, K. M. (1996). The patched signaling pathway mediates repression of gooseberry allowing neuroblast specification by wingless during *Drosophila* neurogenesis. *Development* **122,** 2921–2932.

Bhatia-Gaur, R., Donjacour, A. A., Sciavolino, P. J., Kim, M., Desai, N., *et al.* (1999). Roles for Nkx3.1 in prostate development and cancer. *Genes Dev.* **13,** 966–977.

Bilder, D., and Scott, M. P. (1998). Hedgehog and wingless induce metameric pattern in the *Drosophila* visceral mesoderm. *Dev. Biol.* **201,** 43–56.

Bingham, S., Nasevicius, A., Ekker, S. C., and Chandrasekhar, A. (2001). Sonic hedgehog and tiggy-winkle hedgehog cooperatively induce zebrafish branchiomotor neurons. *Genesis* **30,** 170–174.

Bitgood, M. J., and McMahon, A. P. (1995). Hedgehog and Bmp genes are coexpressed at many diverse sites of cell–cell interaction in the mouse embryo. *Dev. Biol.* **172,** 126–138.

Bitgood, M. J., Shen, L., and McMahon, A. P. (1996). Sertoli cell signaling by Desert hedgehog regulates the male germline. *Curr. Biol.* **6,** 298–304.

Blader, P., Fischer, N., Gradwohl, G., Guillemont, F., and Strahle, U. (1997). The activity of neurogenin1 is controlled by local cues in the zebrafish embryo. *Development* **124,** 4557–4569.

Blagden, C., Currie, P., Ingham, P., and Hughes, S. (1997). Notochord induction of Zebrafish Slow Muscle is mediated by Sonic Hedgehog. *Genes Dev.* **11,** 2163–2175.

Blair, S., and Ralston, A. (1997). *Smoothened*-mediated hedgehog signaling is required for the maintenance of the anterior-posterior lineage restriction in the developing wing of *Drosophila*. *Development* **124,** 4053–4063.

Blough, R. I., Petrij, F., Dauwerse, J. G., Milatovich-Cherry, A., Weiss, L., *et al.* (2000). Variation in microdeletions of the cyclic AMP-responsive element-binding protein gene at chromosome band 16p13.3 in the Rubinstein-Taybi syndrome. *Am. J. Med. Genet.* **90,** 29–34.

Bodak, N., Queille, S., Avril, M. F., Bouadjar, B., Drougard, C., *et al.* (1999). High levels of patched gene mutations in basal-cell carcinomas from patients with xeroderma pigmentosum. *Proc. Natl. Acad. Sci. USA* **96,** 5117–5122.

Boettger, T., Wittler, L., and Kessel, M. (1999). FGF8 functions in the specification of the right body side of the chick. *Curr. Biol.* **9,** 277–280.

Borjigin, J., Deng, J., Wang, M. M., Li, X., Blackshaw, S., and Snyder, S. H. (1999). Circadian rhythm of patched1 transcription in the pineal regulated by adrenergic stimulation and cAMP. *J. Biol. Chem.* **274,** 35012–35015.

Borycki, A., Brunk, B., Tajbakhsh, S., Buckingham, M., Chiang, C., and Emerson, C. (1999). Sonic hedgehog controls epaxial muscle determination through *myf5* activation. *Development* **126,** 4053–4063.

Borycki, A. G., Mendham, L., and Emerson, C. P., Jr. (1998). Control of somite patterning by Sonic hedgehog and its downstream signal response genes. *Development* **125,** 777–790.

Bosse, K., Betz, R. C., Lee, Y. A., Wienker, T. F., Reis, A., *et al.* (2000). Localization of a gene for syndactyly type 1 to chromosome 2q34-q36. *Am. J. Hum. Genet.* **67,** 492–497.

Brand, M., Heisenberg, C. P., Warga, R. M., Pelegri, F., Karlstrom, R. O., *et al.* (1996). Mutations affecting development of the midline and general body shape during zebrafish embryogenesis. *Development* **123,** 129–142.

Brand-Saberi, B., Ebensperger, C., Wilting, J., Balling, R., and Christ, B. (1993). The ventralizing effect of the notochord on somite differentiation in chick embryos. *Anat. Embryol. (Berl.)* **188**, 239–245.

Brewster, R., Lee, J., and Ruiz i Altaba, A. (1998). Gli/Zic factors pattern the neural plate by defining domains of cell differentiation. *Nature (London)* **393**, 579–583.

Briscoe, J., and Ericson, J. (2001). Specification of neuronal fates in the ventral neural tube. *Curr. Opin. Neurobiol.* **11**, 43–49.

Briscoe, J., Sussel, L., Serup, P., Hartigan-O'Connor, D., Jessell, T. M., *et al.* (1999). Homeobox gene Nkx2.2 and specification of neuronal identity by graded Sonic hedgehog signalling. *Nature (London)* **398**, 622–627.

Briscoe, J., Pierani, A., Jessell, T. M., and Ericson, J. (2000). A homeodomain protein code specifies progenitor cell identity and neuronal fate in the ventral neural tube. *Cell* **101**, 435–445.

Briscoe, J., Chen, Y., Jessell, T. M., and Struhl, G. (2001). A hedgehog-insensitive form of patched provides evidence for direct long-range morphogen activity of sonic hedgehog in the neural tube. *Mol. Cell* **7**, 1279–1291.

Buffinger, N., and Stockdale, F. (1995). Myogenic specification of somites is mediated by diffusible factors. *Dev. Biol.* **169**, 96–108.

Burke, R., Nellen, D., Bellotto, M., Hafen, E., Senti, K. A., *et al.* (1999). Dispatched, a novel sterol-sensing domain protein dedicated to the release of cholesterol-modified hedgehog from signaling cells. *Cell* **99**, 803–815.

Buscher, D., Bosse, B., Heymer, J., and Ruther, U. (1997). Evidence for genetic control of Sonic hedgehog by Gli3 in mouse limb development. *Mech. Dev.* **62**, 175–182.

Byrd, N., Becker, S., Maye, P., Narasimhaiah, R., St-Jacques, B., *et al.* (2002). Hedgehog is required for murine yolk sac angiogenesis. *Development* **129**, 361–372.

Cadigan, K. M., Grossniklaus, U., and Gehring, W. J. (1994). Localized expression of *sloppy paired* protein maintains the polarity of *Drosophila* parasegments. *Genes Dev.* **8**, 899–913.

Cann, G., Lee, J., and Stockdale, F. (1999). Sonic hedgehog enhances somite cell viability and formation of primary slow muscle fibers in avian segmented mesoderm. *Anat. Embryol. (Berl.)* **200**, 239–252.

Capdevila, J., and Guerrero, I. (1994). Targeted expression of the signalling molecule *decapentaplegic* induces pattern duplications and growth alterations in *Drosophila* wings. *EMBO J.* **13**, 4459–4468.

Capdevila, J., and Johnson, R. L. (1998). Endogenous and ectopic expression of noggin suggests a conserved mechanism for regulation of BMP function during limb and somite patterning. *Dev. Biol.* **197**, 205–217.

Capdevila, J., Tsukui, T., Rodriquez Esteban, C., Zappavigna, V., and Izpisua Belmonte, J. C. (1999). Control of vertebrate limb outgrowth by the proximal factor Meis2 and distal antagonism of BMPs by Gremlin. *Mol. Cell* **4**, 839–849.

Capdevila, J., Vogan, K. J., Tabin, C. J., and Izpisua Belmonte, J. C. (2000). Mechanisms of left–right determination in vertebrates. *Cell* **101**, 9–21.

Cardoso, W. V. (2001). Molecular regulation of lung development. *Annu. Rev. Physiol.* **63**, 471–494.

Cardoso, W. V., Itoh, A., Nogawa, H., Mason, I., and Brody, J. S. (1997). FGF-1 and FGF-7 induce distinct patterns of growth and differentiation in embryonic lung epithelium. *Dev. Dyn.* **208**, 398–405.

Carl, M., and Wittbrodt, J. (1999). Graded interference with FGF signalling reveals its dorsoventral asymmetry at the mid–hindbrain boundary. *Development* **126**, 5659–5667.

Cavodeassi, F., Diez del Corral, R., Campuzano, S., and Domínguez, M. (1999). Compartments and organising boundaries in the *Drosophila* eye: The role of the homeodomain Iroquois proteins. *Development* **126**, 4933–4942.

Cepko, C. L., Austin, C. P., Yang, X., Alexiades, M., and Ezzeddine, D. (1996). Cell fate determination in the vertebrate retina. *Proc. Natl. Acad. Sci. USA* **93**, 589–595.

Chan, Y. M., and Jan, Y. N. (1999). Conservation of neurogenic genes and mechanisms. *Curr. Opin. Neurobiol.* **9**, 582–588.

Chandrasekhar, A., Warren, J. T., Jr., Takahashi, K., Schauerte, H. E., van Eeden, F. J., et al. (1998). Role of sonic hedgehog in branchiomotor neuron induction in zebrafish. Mech. Dev. 76, 101–115.

Chang, D. T., Lopez, A., von Kessler, D. P., Chiang, C., Simandl, B. K., et al. (1994). Products, genetic linkage and limb patterning activity of a murine hedgehog gene. Development 120, 3339–3353.

Chang, T., Mazotta, J., Dumstrei, K., Dumitrescu, A., and Hartenstein, V. (2001). Dpp and Hh signaling in the Drosophila embryonic eye field. Development 128, 4691–4704.

Charite, J., de Graaff, W., Shen, S., and Deschamps, J. (1994). Ectopic expression of Hoxb-8 causes duplication of the ZPA in the forelimb and homeotic transformation of axial structures. Cell 78, 589–601.

Charite, J., McFadden, D. G., and Olson, E. N. (2000). The bHLH transcription factor dHAND controls Sonic hedgehog expression and establishment of the zone of polarizing activity during limb development. Development 127, 2461–2470.

Charrier, J. B., Lapointe, F., Douarin, N. M., and Teillet, M. A. (2001). Anti-apoptotic role of Sonic hedgehog protein at the early stages of nervous system organogenesis. Development 128, 4011–4020.

Chen, J. N., van Eeden, F. J., Warren, K. S., Chin, A., Nusslein-Volhard, C., et al. (1997). Left–right pattern of cardiac BMP4 may drive asymmetry of the heart in zebrafish. Development 124, 4373–4382.

Chen, W., Burgess, S., and Hopkins, N. (2001). Analysis of the zebrafish smoothened mutant reveals conserved and divergent functions of hedgehog activity. Development 128, 2385–2396.

Chen, Y., and Struhl, K. (1996). Dual roles for patched in sequestering and transducing Hedgehog. Cell 87, 553–563.

Chen, Y., Zhang, Y., Jiang, T. X., Barlow, A. J., St Amand, T. R., et al. (2000). Conservation of early odontogenic signaling pathways in Aves. Proc. Natl. Acad. Sci. USA 97, 10044–10049.

Chiang, C., Litingtung, Y., Lee, E., Young, K. E., Corden, J. L., et al. (1996). Cyclopia and defective axial patterning in mice lacking Sonic hedgehog gene function. Nature (London) 383, 407–413.

Chiang, C., Swan, R. Z., Grachtchouk, M., Bolinger, M., Litingtung, Y., et al. (1999). Essential role for Sonic hedgehog during hair follicle morphogenesis. Dev. Biol. 205, 1–9.

Chiang, C., Litingtung, Y., Harris, M. P., Simandl, B. K., Li, Y., et al. (2001). Manifestation of the limb prepattern: Limb development in the absence of sonic hedgehog function. Dev. Biol. 236, 421–435.

Cho, K.-O., Chern, J., Izaddoost, S., and Choi, K.-W. (2000). Novel signaling from the peripodial membrane is essential for eye disc patterning in Drosophila. Cell 103, 331–342.

Choi, K., Kennedy, M., Kazarov, A., Papadimitriou, J. C., and Keller, G. (1998). A common precursor for hematopoietic and endothelial cells. Development 125, 725–732.

Christ, B. (1970). Experimente zur Lageentwicklung der Somiten. Anat. Anz. Erg-H Bd. 126, 555–564.

Christ, B., Huang, R., and Wilting, J. (2000). The development of the avian vertebral column. Anat. Embryol. (Berl.) 202, 179–194.

Chuang, P. T., and McMahon, A. P. (1999). Vertebrate Hedgehog signalling modulated by induction of a Hedgehog-binding protein. Nature (London) 397, 617–621.

Chung, S., McLean, M. R., and Rymond, B. C. (1999). Yeast ortholog of the Drosophila crooked neck protein promotes spliceosome assembly through stable U4/U6.U5 snRNP addition. RNA 5, 1042–1054.

Chung, U. I., Lanske, B., Lee, K., Li, E., and Kronenberg, H. (1998). The parathyroid hormone/parathyroid hormone-related peptide receptor coordinates endochondral bone development by directly controlling chondrocyte differentiation. Proc. Natl. Acad. Sci. USA 95, 13030–13035.

Chung, U. I., Schipani, E., McMahon, A. P., and Kronenberg, H. M. (2001). Indian hedgehog couples chondrogenesis to osteogenesis in endochondral bone development. J. Clin. Invest. 107, 295–304.

Chuong, C. M., Widelitz, R. B., Ting-Berreth, S., and Jiang, T. X. (1996). Early events during avian skin appendage regeneration: Dependence on epithelial–mesenchymal interaction and order of molecular reappearance. *J. Invest. Dermatol.* **107,** 639–646.

Chuong, C. M., Patel, N., Lin, J., Jung, H. S., and Widelitz, R. B. (2000). Sonic hedgehog signaling pathway in vertebrate epithelial appendage morphogenesis: Perspectives in development and evolution. *Cell. Mol. Life Sci.* **57,** 1672–1681.

Clark, A. M., Garland, K. K., and Russell, L. D. (2000). Desert hedgehog (Dhh) gene is required in the mouse testis for formation of adult-type Leydig cells and normal development of peritubular cells and seminiferous tubules. *Biol. Reprod.* **63,** 1825–1838.

Cobourne, M. T., Hardcastle, Z., and Sharpe, P. T. (2001). Sonic hedgehog regulates epithelial proliferation and cell survival in the developing tooth germ. *J. Dent. Res.* **80,** 1974–1979.

Collignon, J., Varlet, I., and Robertson, E. J. (1996). Relationship between asymmetric nodal expression and the direction of embryonic turning. *Nature* (*London*) **381,** 155–158.

Colvin, J. S., Bohne, B. A., Harding, G. W., McEwen, D. G., and Ornitz, D. M. (1996). Skeletal overgrowth and deafness in mice lacking fibroblast growth factor receptor 3. *Nat. Genet.* **12,** 390–397.

Concordet, J.-P., Lewis, K., Moore, J., Goodrich, L. V., Johnson, R. L., *et al.* (1996). Spatial regulation of a Zebrafish *patched* Homologue reflects the roles of sonic hedgehog and Protein Kinase A in neural tube and somite patterning. *Development* **122,** 2835–2846.

Cooper, M. K., Porter, J. A., Young, K. E., and Beachy, P. A. (1998). Teratogen-mediated inhibition of target tissue response to Shh signaling. *Science* **280,** 1603–1607.

Cotsarelis, G., Sun, T. T., and Lavker, R. M. (1990). Label-retaining cells reside in the bulge area of pilosebaceous unit: Implications for follicular stem cells, hair cycle, and skin carcinogenesis. *Cell* **61,** 1329–1337.

Coutelle, O., Blagden, C., Hampson, R., Halai, C., Rigby, P., and Hughes, S. (2001). Hedgehog signalling is required for maintenance of myf5 and myoD expression and timely terminal differentiation in zebrafish adaxial myogenesis. *Dev. Biol.* **236,** 136–150.

Crossley, P. H., Minowada, G., MacArthur, C. A., and Martin, G. R. (1996). Roles for FGF8 in the induction, initiation, and maintenance of chick limb development. *Cell* **84,** 127–136.

Cunha, G. R., Chung, L. W., Shannon, J. M., and Reese, B. A. (1980). Stromal–epithelial interactions in sex differentiation. *Biol. Reprod.* **22,** 19–42.

Currie, P. D., and Ingham, P. W. (1996). Induction of a specific muscle cell type by a hedgehog-like protein in zebrafish. *Nature* (*London*) **382,** 452–455.

Dahmane, N., and Ruiz-i-Altaba, A. (1999). Sonic hedgehog regulates the growth and patterning of the cerebellum. *Development* **126,** 3089–3100.

Dahmane, N., Lee, J., Robins, P., Heller, P., and Ruiz-i-Altaba, A. (1997). Activation of the transcription factor GliI and the Sonic hedgehog signalling pathway in skin tumours. *Nature* (*London*) **389,** 876–881.

Dahmann, C., and Basler, K. (2000). Opposing transcriptional outputs of Hedgehog signaling and engrailed control compartmental cell sorting at the *Drosophila* A/P boundary. *Cell* **100,** 411–422.

Dahn, R. D., and Fallon, J. F. (1999). Limbiting outgrowth: BMP's as negative regulators in limb development. *Bioessays* **21,** 721–725.

Dai, P., Akimaru, H., Tanaka, Y., Maekawa, T., Nakafuku, M., and Ishii, S. (1999). Sonic Hedgehog-induced activation of the GliI promoter is mediated by GLI3. *J. Biol. Chem.* **274,** 8143–8152.

Dale, J. K., Vesque, C., Lints, T. J., Sampath, T. K., Furley, A., *et al.* (1997). Cooperation of BMP7 and SHH in the induction of forebrain ventral midline cells by prechordal mesoderm. *Cell* **90,** 257–269.

Dale, K., Sattar, N., Heemskerk, J., Clarke, J. D., Placzek, M., and Dodd, J. (1999). Differential patterning of ventral midline cells by axial mesoderm is regulated by BMP7 and chordin. *Development* **126,** 397–408.

Danos, M. C., and Yost, H. J. (1995). Linkage of cardiac left–right asymmetry and dorsal–anterior development in *Xenopus*. *Development* **121,** 1467–1474.

Danos, M. C., and Yost, H. J. (1996). Role of notochord in specification of cardiac left–right orientation in zebrafish and *Xenopus*. *Dev. Biol.* **177,** 96–103.

Dassule, H. R., and McMahon, A. P. (1998). Analysis of epithelial–mesenchymal interactions in the initial morphogenesis of the mammalian tooth. *Dev. Biol.* **202,** 215–227.

Dassule, H. R., Lewis, P., Bei, M., Maas, R., and McMahon, A. P. (2000). Sonic hedgehog regulates growth and morphogenesis of the tooth. *Development* **127,** 4775–4785.

Davies, J. E., and Miller, R. H. (2001). Local sonic hedgehog signaling regulates oligodendrocyte precursor appearance in multiple ventricular zone domains in the chick metencephalon. *Dev. Biol.* **233,** 513–525.

Davis, A. P., and Capecchi, M. R. (1994). Axial homeosis and appendicular skeleton defects in mice with a targeted disruption of HoxD-11. *Development* **120,** 2187–2198.

Davis, A. P., Witte, D. P., Hsieh-Li, H. M., Potter, S. S., and Capecchi, M. R. (1995). Absence of radius and ulna in mice lacking HoxA-11 and HoxD-11. *Nature (London)* **375,** 791–795.

Deng, C., Wynshaw-Boris, A., Zhou, F., Kuo, A., and Leder, P. (1996). Fibroblast growth factor receptor 3 is a negative regulator of bone growth. *Cell* **84,** 911–921.

D'Errico, M., Calcagnile, A., Canzona, F., Didona, B., Posteraro, P., *et al.* (2000). UV mutation signature in tumor suppressor genes involved in skin carcinogenesis in xeroderma pigmentosum patients. *Oncogene* **19,** 463–467.

Deshpande, G., Swanhart, L., Chiang, P., and Schedl, P. (2001). Hedgehog signaling in germ cell migration. *Cell* **106,** 759–769.

Detmer, K., Walker, A. N., Jenkins, T. M., Steele, T. A., and Dannawi, H. (2000). Erythroid differentiation in vitro is blocked by cyclopamine, an inhibitor of hedgehog signaling. *Blood Cells Mol. Dis.* **26,** 360–372.

Dhouailly, D. (1973). Dermo–epidermal interactions between birds and mammals: Differentiation of cutaneous appendages. *J. Embryol. Exp. Morphol.* **30,** 587–603.

Dhouailly, D. (1975). *Wilhelm Roux Arch.* **117,** 323–340.

Dickinson, M. E., Krumlauf, R., and McMahon, A. P. (1994). Evidence for a mitogenic effect of Wnt-1 in the developing mammalian central nervous system. *Development* **120,** 1453–1471.

Dietrich, S., Schubert, F. R., and Lumsden, A. (1997). Control of dorsoventral pattern in the chick paraxial mesoderm. *Development* **124,** 3895–3908.

DiNardo, S., Sher, E., Heemskerk-Jongens, J., Kassis, J., and O'Farrell, P. (1988). Two tiered regulation of spatially patterned *engrailed* gene expression during *Drosophila* embryogenesis. *Nature (London)* **332,** 604–609.

Ding, Q., Motoyama, J., Gasca, S., Mo, R., Sasaki, H., *et al.* (1998). Diminished Sonic hedgehog signaling and lack of floor plate differentiation in Gli2 mutant mice. *Development* **125,** 2533–2543.

Dominguez, M. (1999). Dual role for Hedgehog in the regulation of the proneural gene atonal during ommatidia development. *Development* **126,** 2345–2353.

Dominguez, M., and de Celis, J. (1998). A dorsal/ventral boundary established by Notch controls growth and polarity in the *Drosophila* eye. *Nature (London)* **396,** 276–278.

Dominguez, M., and Hafen, E. (1997). Hedgehog directly controls initiation and propagation of retinal differentiation in the *Drosophila* eye. *Genes Dev.* **11,** 3254–3264.

Donnai, D., Hughes, H. E., and Winter, R. M. (1987). Postaxial acrofacial dysostosis (Miller) syndrome. *J. Med. Genet.* **24,** 422–425.

Drossopoulou, G., Lewis, K. E., Sanz-Ezquerro, J. J., Nikbakht, N., McMahon, A. P., *et al.* (2000). A model for anteroposterior patterning of the vertebrate limb based on sequential long- and short-range Shh signaling and Bmp signaling. *Development* **127,** 1337–1348.

Du, S., Devoto, S., Westerfield, M., and Moon, R. (1997). Positive and negative regulation of muscle cell identity by members of the hedgehog and TGF-β gene families. *J. Cell Biol.* **139,** 145–156.

Dubois, L., Lecourtois, M., Alexandre, C., Hirst, E., and Vincent, J. (2001). Regulated endocytic routing modulates wingless signaling in *Drosophila* embryos. *Cell* **105,** 613–624.

Duman-Scheel, M., Li, X., Orlov, I., Noll, M., and Patel, N. H. (1997). Genetic separation of the neural and cuticular patterning functions of gooseberry. *Development* **124,** 2855–2865.

Duprez, D., Fournier-Thibault, C., and Le Douarin, N. (1998). Sonic Hedgehog induces proliferation of committed skeletal muscle cells in the chick limb. *Development* **125,** 495–505.

Duprez, D. M., Kostakopoulou, K., Francis-West, P. H., Tickle, C., and Brickell, P. M. (1996). Activation of Fgf-4 and HoxD gene expression by Bmp-2 expressing cells in the developing chick limb. *Development* **122,** 1821–1828.

Dutton, R., Yamada, T., Turnley, A., Bartlett, P. F., and Murphy, M. (1999). Sonic hedgehog promotes neuronal differentiation of murine spinal cord precursors and collaborates with neurotrophin 3 to induce Islet-1. *J. Neurosci.* **19,** 2601–2608.

Dyer, M. A., Farrington, S. M., Mohn, D., Munday, J. R., and Baron, M. H. (2001). Indian hedgehog activates hematopoiesis and vasculogenesis and can respecify prospective neurectodermal cell fate in the mouse embryo. *Development* **128,** 1717–1730.

Echelard, Y., Epstein, D. J., St-Jacques, B., Shen, L., Mohler, J., *et al.* (1993). Sonic hedgehog, a member of a family of putative signaling molecules, is implicated in the regulation of CNS polarity. *Cell* **75,** 1417–1430.

Eggenschwiler, J. T., and Anderson, K. V. (2000). Dorsal and lateral fates in the mouse neural tube require the cell-autonomous activity of the open brain gene. *Dev. Biol.* **227,** 648–660.

Eggenschwiler, J. T., Espinoza, E., and Anderson, K. V. (2001). Rab23 is an essential negative regulator of the mouse Sonic hedgehog signalling pathway. *Nature (London)* **412,** 194–198.

Ekker, S., McGrew, L., Lai, C., Lee, J., von Kessler, D., *et al.* (1995a). Distinct expression and shared activities of members of the hedgehog gene family of *Xenopus laevis. Development* **121,** 2337–2347.

Ekker, S., Ungar, A., Greenstein, P., von Kessler, D., Porter, J., *et al.* (1995b). Patterning activities of vertebrate hedgehog proteins in the developing eye and brain. *Curr. Biol.* **5,** 944–955.

Ellis, T., Gambardella, L., Horcher, M., Tschanz, S., Capol, J., *et al.* (2001). The transcriptional repressor CDP (Cutl1) is essential for epithelial cell differentiation of the lung and the hair follicle. *Genes Dev.* **15,** 2307–2319.

Endo, T., Yokoyama, H., Tamura, K., and Ide, H. (1997). Shh expression in developing and regenerating limb buds of *Xenopus laevis. Dev. Dyn.* **209,** 227–232.

Enomoto-Iwamoto, M., Nakamura, T., Aikawa, T., Higuchi, Y., Yuasa, T., *et al.* (2000). Hedgehog proteins stimulate chondrogenic cell differentiation and cartilage formation. *J. Bone. Miner. Res.* **15,** 1659–1668.

Epstein, D. J., McMahon, A. P., and Joyner, A. L. (1999). Regionalization of Sonic hedgehog transcription along the anteroposterior axis of the mouse central nervous system is regulated by Hnf3-dependent and -independent mechanisms. *Development* **126,** 281–292.

Ericson, J., Muhr, J., Placzek, M., Lints, T., Jessell, T. M., and Edlund, T. (1995). Sonic hedgehog induces the differentiation of ventral forebrain neurons: A common signal for ventral patterning within the neural tube. *Cell* **81,** 747–756.

Ericson, J., Morton, S., Kawakami, A., Roelink, H., and Jessell, T. M. (1996). Two critical periods of Sonic Hedgehog signaling required for the specification of motor neuron identity. *Cell* **87,** 661–673.

Ericson, J., Rashbass, P., Schedl, A., Brenner-Morton, S., Kawakami, A., *et al.* (1997). Pax6 controls progenitor cell identity and neuronal fate in response to graded Shh signaling. *Cell* **90,** 169–180.

Essner, J. J., Vogan, K. J., Wagner, M. K., Tabin, C. J., Yost, H. J., and Brueckner, M. (2002). Conserved function for embryonic nodal cilia. *Nature* **418,** 37–38.

Etheridge, L. A., Wu, T., Liang, J. O., Ekker, S. C., and Halpern, M. E. (2001). Floor plate develops upon depletion of tiggy-winkle and sonic hedgehog. *Genesis* **30,** 164–169.

Fallon, J. F., Lopez, A., Ros, M. A., Savage, M. P., Olwin, B. B., and Simandl, B. K. (1994). FGF-2: Apical ectodermal ridge growth signal for chick limb development. *Science* **264**, 104–107.

Fan, C.-M., and Tessier-Lavigne, M. (1994). Patterning of mammalian somites by surface ectoderm and notochord: Evidence for sclerotome induction by a hedgehog homolog. *Cell* **79**, 1175–1186.

Fan, C.-M., Porter, J. A., Chiang, C., Chang, D. T., Beachy, P. A., and Tessier-Lavigne, M. (1995). Long-range sclerotome induction by sonic hedgehog: Direct role of the amino-terminal cleavage product and modulation by the cyclic AMP signaling pathway. *Cell* **81**, 457–465.

Fan, C.-M., Lee, C., and Tessier-Lavigne, M. (1997). A role for WNT proteins in induction of dermomyotome. *Dev. Biol.* **191**, 160–165.

Fan, H., and Khavari, P. A. (1999). Sonic hedgehog opposes epithelial cell cycle arrest. *J. Chem. Biol.* **147**, 71–76.

Fan, H., Oro, A. E., Scott, M. P., and Khavari, P. A. (1997). Induction of basal cell carcinoma features in transgenic human skin expressing Sonic hedgehog. *Nat. Med.* **3**, 788–792.

Farquharson, C., Jefferies, D., Seawright, E., and Houston, B. (2001). Regulation of chondrocyte terminal differentiation in the postembryonic growth plate: The role of the PTHrP–Indian hedgehog axis. *Endocrinology* **142**, 4131–4140.

Farrington, S. M., Belaoussoff, M., and Baron, M. H. (1997). Winged-helix, Hedgehog and Bmp genes are differentially expressed in distinct cell layers of the murine yolk sac. *Mech. Dev.* **62**, 197–211.

Favier, B., Le Meur, M., Chambon, P., and Dolle, P. (1995). Axial skeleton homeosis and forelimb malformtions in HoxD-11 mutant mice. *Proc. Natl. Acad. Sci. USA* **92**, 310–314.

Ferguson, C., Alpern, E., Miclau, T., and Helms, J. A. (1999). Does adult fracture repair recapitulate embryonic skeletal formation? *Mech. Dev.* **87**, 57–66.

Fernandez-Teran, M., Piedra, M. E., Kathiriya, I. S., Srivastava, D., Rodriguez-Rey, J. C., and Ros, M. A. (2000). Role of dHAND in the anterior–posterior polarization of the limb bud: Implications for the Sonic hedgehog pathway. *Development* **127**, 2133–2142.

Fitzky, B. U., Witsch-Baumgartner, M., Erdel, M., Lee, J. N., Paik, Y. K., *et al.* (1998). Mutations in the Δ^7-sterol reductase gene in patients with the Smith-Lemli-Opitz syndrome. *Proc. Natl. Acad. Sci. USA* **95**, 8181–8186.

Forbes, A. J., Liu, H., Ingham, P. W., and Spradling, A. C. (1996a). *Hedgehog* is required for the proliferation and specification of somatic cells during egg chamber assembly in *Drosophila* oogenesis. *Development* **122**, 1125–1135.

Forbes, A. J., Spradling, A. C., Ingham, P. W., and Lin, H. (1996b). The role of segment polarity genes during early oogenesis in *Drosophila*. *Development* **122**, 3283–3294.

Franco, P. G., Paganelli, A. R., Lopez, S. L., and Carrasco, A. E. (1999). Functional association of retinoic acid and hedgehog signaling in *Xenopus* primary neurogenesis. *Development* **126**, 4257–4265.

Fuchs, E., and Segre, J. A. (2000). Stem cells: A new lease on life. *Cell* **100**, 143–155.

Furthauer, M., Thisse, C., and Thisse, B. (1997). A role for FGF-8 in the dorsoventral patterning of the zebrafish gastrula. *Development* **124**, 4253–4264.

Furumoto, T. A., Miura, N., Akasaka, T., Mizutani-Koseki, Y., Sudo, H., *et al.* (1999). Notochord-dependent expression of MFH1 and PAX1 cooperates to maintain the proliferation of sclerotome cells during the vertebral column development. *Dev. Biol.* **210**, 15–29.

Gaiano, N., Kohtz, J. D., Turnbull, D. H., and Fishell, G. (1999). A method for rapid gain-of-function studies in the mouse embryonic nervous system. *Nat. Neurosci.* **2**, 812–819.

Gailani, M. R., Bale, S. J., Leffell, D. J., DiGiovanna, J. J., Peck, G. L., *et al.* (1992). Developmental defects in Gorlin syndrome related to a putative tumor suppressor gene on chromosome 9. *Cell* **69**, 111–117.

Gailani, M. R., Stahle-Backdahl, M., Leffell, D. J., Glynn, M., Zaphiropoulos, P. G., *et al.* (1996). The role of the human homologue of *Drosophila* patched in sporadic basal cell carcinomas. *Nat. Genet.* **14**, 78–81.

Gao, B., Guo, J., She, C., Shu, A., Yang, M., *et al.* (2001). Mutations in IHH, encoding Indian hedgehog, cause brachydactyly type A-1. *Nat. Genet.* **28,** 386–388.

Garber, B., and Moscona, A. A. (1964). Reconstruction of skin in the chorioallantoic membrane from suspensions of chick and mouse skin cells. *J. Exp. Zool.* **155,** 179–202.

Garber, B., Kollar, E. J., and Moscona, A. A. (1968). Aggregation in vivo of dissociated cells. 3. Effect of state of differentiation of cells on feather development in hybrid aggregates of embryonic mouse and chick skin cells. *J. Exp. Zool.* **168,** 455–472.

Gat, U., DasGupta, R., Degenstein, L., and Fuchs, E. (1998). De novo hair follicle morphogenesis and hair tumors in mice expressing a truncated β-catenin in skin. *Cell* **95,** 605–614.

Ghali, L., Wong, S. T., Green, J., Tidman, N., and Quinn, A. G. (1999). Gli1 protein is expressed in basal cell carcinomas, outer root sheath keratinocytes and a subpopulation of mesenchymal cells in normal human skin. *J. Invest. Dermatol.* **113,** 595–599.

Glazer, L., and Shilo, B. Z. (2001). Hedgehog signaling patterns the tracheal branches. *Development* **128,** 1599–1606.

Goldstein, A. M., Ticho, B. S., and Fishman, M. C. (1998). Patterning the heart's left–right axis: From zebrafish to man. *Dev. Genet.* **22,** 278–287.

Gomezskarmeta, J., and Modolell, J. (1996). *Araucan* and *caupolican* provide a link between compartment subdivisions and patterning of sensory organs and veins in the *Drosophila* wing. *Genes Dev.* **10,** 2935–2945.

Goodrich, L. V., Johnson, R. L., Milenkovic, L., McMahon, J. A., and Scott, M. P. (1996). Conservation of the hedgehog/patched signaling pathway from flies to mice: Induction of a mouse patched gene by Hedgehog. *Genes Dev.* **10,** 301–312.

Goodrich, L. V., Milenkovic, L., Higgins, K. M., and Scott, M. P. (1997). Altered neural cell fates and medulloblastoma in mouse patched mutants. *Science* **277,** 1109–1113.

Gorlin, R. J. (1995). Nevoid basal cell carcinoma syndrome. *Dermatol. Clin.* **13,** 113–125.

Goulding, M. D., Lumsden, A., and Gruss, P. (1993). Signals from the notochord and floor plate regulate the region-specific expression of two Pax genes in the developing spinal cord. *Development* **117,** 1001–1016.

Grandel, H., and Schulte-Merker, S. (1998). The development of the paired fins in the zebrafish (*Danio rerio*). *Mech. Dev.* **79,** 99–120.

Grandel, H., Draper, B. W., and Schulte-Merker, S. (2000). dackel acts in the ectoderm of the zebrafish pectoral fin bud to maintain AER signaling. *Development* **127,** 4169–4178.

Greenwood, S., and Struhl, G. (1999). Progression of the morphogenetic furrow in the *Drosophila* eye: The roles of Hedgehog, Decapentaplegic and the Raf pathway. *Development* **126,** 5795–5808.

Grieshammer, U., Minowada, G., Pisenti, J. M., Abbott, U. K., and Martin, G. R. (1996). The chick limbless mutation causes abnormalities in limb bud dorsal–ventral patterning: Implications for the mechanism of apical ridge formation. *Development* **122,** 3851–3861.

Grindley, J. C., Bellusci, S., Perkins, D., and Hogan, B. L. (1997). Evidence for the involvement of the Gli gene family in embryonic mouse lung development. *Dev. Biol.* **188,** 337–348.

Gritli-Linde, A., Lewis, P., McMahon, A. P., and Linde, A. (2001). The whereabouts of a morphogen: Direct evidence for short- and graded long-range activity of hedgehog signaling peptides. *Dev. Biol.* **236,** 364–386.

Guillen, I., Mullor, J. L., Capdevila, J., Sanchez-Herrero, E., Morata, G., and Guerrero, I. (1995). The function of *engrailed* and the specification of *Drosophila* wing pattern. *Development* **121,** 3447–3456.

Gunther, T., Struwe, M., Aguzzi, A., and Schughart, K. (1994). Open brain, a new mouse mutant with severe neural tube defects, shows altered gene expression patterns in the developing spinal cord. *Development* **120,** 3119–3130.

Gustafsson, M. K., Pan, H., Pinney, D. F., Liu, Y., Lewandowski, A., *et al.* (2002). Myf5 is a direct target of long-range Shh signaling and Gli regulation for muscle specification. *Genes Dev.* **16,** 114–126.

Haaijman, A., Karperien, M., Lanske, B., Hendriks, J., Lowik, C. W., *et al.* (1999). Inhibition of terminal chondrocyte differentiation by bone morphogenetic protein 7 (OP-1) in vitro depends on the periarticular region but is independent of parathyroid hormone-related peptide. *Bone* **25,** 397–404.

Haffen, K., Lacroix, B., Kedinger, M., and Simon-Assmann, P. M. (1983). Inductive properties of fibroblastic cell cultures derived from rat intestinal mucosa on epithelial differentiation. *Differentiation* **23,** 226–233.

Hahn, H., Wicking, C., Zaphriopoulou, P. G., Gailani, M. R., Shanley, S., *et al.* (1996). Mutations of the human homolog of *Drosophila* patched in the nevoid basal cell carcinoma syndrome. *Cell* **85,** 841–851.

Hahn, H., Wojnowski, L., Zimmer, A. M., Hall, J., Miller, G., and Zimmer, A. (1998). Rabdomyosarcomas and radiation hypersensitivity in a mouse model of Gorlin syndrome. *Nat. Med.* **4,** 619–622.

Hall, J. M., Hooper, J. E., and Finger, T. E. (1999). Expression of sonic hedgehog, patched, and Gli1 in developing taste papillae of the mouse. *J. Comp. Neurol.* **406,** 143–155.

Halpern, M. E., Ho, R. K., Walker, C., and Kimmel, C. (1993). Induction of muscle pioneers and floor plate is distinguished by the zebrafish *no tail* mutation. *Cell* **75,** 99–111.

Hammerschmidt, M., Bitgood, M. J., and McMahon, A. P. (1996). Protein kinase A is a common negative regulator of Hedgehog signaling in the vertebrate embryo. *Genes Dev.* **10,** 647–658.

Haraguchi, R., Mo, R., Hui, C. C., Motoyama, J., Makino, S., *et al.* (2001). Unique functions of Sonic hedgehog signaling during external genitalia development. *Development* **128,** 4241–4250.

Hardcastle, Z., Mo, R., Hui, C. C., and Sharpe, P. T. (1998). The Shh signalling pathway in tooth development: Defects in Gli2 and Gli3 mutants. *Development* **125,** 2803–2811.

Hardy, M. H. (1992). The secret life of the hair follicle. *Trends Genet.* **8,** 55–61.

Harrison, K. A., Thaler, J., Pfaff, S. L., Gu, H., and Kehrl, J. H. (1999). Pancreas dorsal lobe agenesis and abnormal islets of Langerhans in Hlxb9-deficient mice. *Nat. Genet.* **23,** 71–75.

Hayashi, T., and Murakami, R. (2001). Left–right asymmetry in *Drosophila melanogaster* gut development. *Dev. Growth Differ.* **43,** 239–246.

Hays, R., Buchanan, K. T., Neff, C., and Orenic, T. V. (1999). Patterning of *Drosophila* leg sensory organs through combinatorial signaling by Hedgehog, Decapentaplegic and Wingless. *Development* **126,** 2891–2899.

Heberlein, U., Wolff, T., and Rubin, G. M. (1993). The TGFβ homolog dpp and the segment polarity gene hedgehog are required for propagation of a morphogenetic wave in the *Drosophila* Retina. *Cell* **75,** 913–926.

Heberlein, U., Singh, C. M., Luk, A. Y., and Donohoe, T. J. (1995). Growth and differentiation in the *Drosophila* eye coordinated by *hedgehog*. *Nature (London)* **373,** 709–711.

Hebrok, M., Kim, S. K., and Melton, D. A. (1998). Notochord repression of endodermal Sonic hedgehog permits pancreas development. *Genes Dev.* **12,** 1705–1713.

Hecht, J. T., Hogue, D., Strong, L. C., Hansen, M. F., Blanton, S. H., and Wagner, M. (1995). Hereditary multiple exostosis and chrondrosarcoma: Linkage to chromosome II and loss of heterozygosity for EXT-linked markers on chromosomes II and 8. *Am. J. Hum. Genet.* **56,** 1125–1131.

Heemskerk, J., and Dinardo, S. (1994). *Drosophila hedgehog* acts as a morphogen in cellular patterning. *Cell* **76,** 449–460.

Helms, J. A., Kim, C. H., Eichele, G., and Thaller, C. (1996). Retinoic acid signaling is required during early chick limb development. *Development* **122,** 1385–1394.

Helms, J. A., Kim, C. H., Hu, D., Minkoff, R., Thaller, C., and Eichele, G. (1997). Sonic hedgehog participates in craniofacial morphogenesis and is down-regulated by teratogenic doses of retinoic acid. *Dev. Biol.* **187,** 25–35.

Heutink, P., Zguricas, J., van Oosterhout, L., Breedveld, G. J., Testers, L., *et al.* (1994). The gene for triphalangeal thumb maps to the subtelomeric region of chromosome 7q. *Nat. Genet.* **6,** 287–292.

Hidalgo, A. (1994). Three distinct roles for the engrailed gene in *Drosophila* wing development. *Curr. Biol.* **4,** 1087–1098.

Hing, A. V., Helms, C., Slaugh, R., Burgess, A., Wang, J. C., *et al.* (1995). Linkage of preaxial polydactyly type 2 to 7q36. *Am. J. Med. Genet.* **58,** 128–135.

Hogan, B. L. (1999). Morphogenesis. *Cell* **96,** 225–233.

Honig, S. (1981). Positional signal transmssion in the developing chick limb. *Nature (London)* **291,** 72–73.

Hornbruch, A., and Wolpert, L. (1986). Positional signalling by Hensen's node when grafted to the chick limb bud. *J. Embryol. Exp. Morphol.* **94,** 257–265.

Hu, D., and Helms, J. A. (1999). The role of sonic hedgehog in normal and abnormal craniofacial morphogenesis. *Development* **126,** 4873–4884.

Huang, Z., and Kunes, S. (1996). Hedgehog, transmitted along retinal axons, triggers neurogenesis in the developing visual centers of the *Drosophila* brain. *Cell* **86,** 411–422.

Huang, Z., and Kunes, S. (1998). Signals transmitted along retinal axons in *Drosophila*: Hedgehog signal reception and the cell circuitry of lamina cartridge assembly. *Development* **125,** 3753–3764.

Huelsken, J., Vogel, R., Erdmann, B., Cotsarelis, G., and Birchmeier, W. (2001). β-Catenin controls hair follicle morphogenesis and stem cell differentiation in the skin. *Cell* **105,** 533–545.

Hui, C. C., and Joyner, A. L. (1993). A mouse model of Greig cephalopolysyndactyly syndrome: The extra-toesJ mutation contains an intragenic deletion of the Gli3 gene. *Nat. Genet.* **3,** 241–246.

Hui, C. C., Slusarski, D., Platt, K. A., Holmgren, R., and Joyner, A. L. (1994). Expression of three mouse homologs of the *Drosophila* segment polarity gene cubitus interruptus, Gli, Gli-2, and Gli-3, in ectoderm- and mesoderm-derived tissues suggests multiple roles during postimplantation development. *Dev. Biol.* **162,** 402–413.

Hynes, M., Porter, J. A., Chiang, C., Chang, D., Tessier-Lavigne, M., *et al.* (1995). Induction of midbrain dopaminergic neurons by Sonic hedgehog. *Neuron* **15,** 35–44.

Hynes, M., Stone, D. M., Dowd, M., Pitts-Meek, S., Goddard, A., *et al.* (1997). Control of cell pattern in the neural tube by the zinc finger transcription factor and oncogene Gli-1. *Neuron* **19,** 15–26.

Hynes, M., Ye, W., Wang, K., Stone, D., Murone, M., *et al.* (2000). The seven-transmembrane receptor smoothened cell-autonomously induces multiple ventral cell types. *Nat. Neurosci.* **3,** 41–46.

Imokawa, Y., and Yoshizato, K. (1997). Expression of Sonic hedgehog gene in regenerating newt limb blastemas recapitulates that in developing limb buds. *Proc. Natl. Acad. Sci. USA* **94,** 9159–9164.

Inada, M., Yasui, T., Nomura, S., Miyake, S., Deguchi, K., *et al.* (1999). Maturational disturbance of chondrocytes in Cbfa1-deficient mice. *Dev. Dyn.* **214,** 279–290.

Incardona, J. P., Gaffield, W., Kapur, R. P., and Roelink, H. (1998). The teratogenic *Veratrum* alkaloid cyclopamine inhibits sonic hedgehog signal transduction. *Development* **125,** 3553–3562.

Incardona, J. P., Gaffield, W., Lange, Y., Cooney, A., Pentchev, P. G., *et al.* (2000). Cyclopamine inhibition of Sonic hedgehog signal transduction is not mediated through effects on cholesterol transport. *Dev. Biol.* **224,** 440–452.

Ingham, P., and Fietz, M. (1995). Quantitative effects of *hedgehog* and *decapentaplegic* activity on the patterning of the *Drosophila* wing. *Curr. Biol.* **5,** 432–441.

Ingham, P. W., and Martinez-Arias, A. (1992). Boundaries and fields in early embryos. *Cell* **68,** 221–235.

Ingham, P. W., and McMahon, A. P. (2001). Hedgehog signaling in animal development: Paradigms and principles. *Genes. Dev.* **15,** 3059–3087.

Ingham, P. W., Taylor, A. M., and Nakano, Y. (1991). Role of the *Drosophila* patched gene in positional signaling. *Nature (London)* **353,** 184–187.

Inoue, Y., Niwa, N., Mito, T., Ohuchi, H., Yoshioka, H., and Noji, S. (2002). Expression patterns of hedgehog, wingless, and decapentaplegic during gut formation of *Gryllus bimaculatus* (cricket). *Mech. Dev.* **110,** 245–248.

Ionescu, A. M., Schwarz, E. M., Vinson, C., Puzas, J. E., Rosier, R., *et al.* (2001). PTHrP modulates chondrocyte differentiation through AP-1 and CREB signaling. *J. Biol. Chem.* **276,** 11639–11647.

Iseki, S., Araga, A., Ohuchi, H., Nohno, T., Yoshioka, H., *et al.* (1996). Sonic hedgehog is expressed in epithelial cells during development of whisker, hair, and tooth. *Biochem. Biophys. Res. Commun.* **218,** 688–693.

Ito, H., Akiyama, H., Shigeno, C., Iyama, K., Matsuoka, H., and Nakamura, T. (1999). Hedgehog signaling molecules in bone marrow cells at the initial stage of fracture repair. *Biochem. Biophys. Res. Commun.* **262,** 443–451.

Ivy, R. H. (1957). Congenital anomalies as recorded on birth certificates in the Division of Vital Statistics of the Pennsylvania Department of Health for the period 1951–1955 inclusive. *Plast. Reconst. Surg.* **20,** 400–411.

Izpisua-Belmonte, J. C., Tickle, C., Dolle, P., Wolpert, L., and Duboule, D. (1991). Expression of the homeobox Hox-4 genes and the specification of position in chick wing development. *Nature (London)* **350,** 585–589.

Izraeli, S., Lowe, L. A., Bertness, V. L., Good, D. J., Dorward, D. W., *et al.* (1999). The SIL gene is required for mouse embryonic axial development and left–right specification. *Nature (London)* **399,** 691–694.

Jagla, K., Frasch, M., Jagla, T., Dretzen, G., Bellard, F., and Bellard, M. (1997). Ladybird, a new component of the cardiogenic pathway in *Drosophila* required for diversification of heart precursors. *Development* **124,** 3471–3479.

Jensen, A., and Wallace, V. (1997). Expression of sonic-hedgehog and its putative role as a precursor cell mitogen in the developing mouse retina. *Development* **124,** 363–371.

Jernvall, J., and Thesleff, I. (2000). Reiterative signaling and patterning during mammalian tooth morphogenesis. *Mech. Dev.* **92,** 19–29.

Jernvall, J., Keranen, S. V., and Thesleff, I. (2000). Evolutionary modification of development in mammalian teeth: Quantifying gene expression patterns and topography. *Proc. Natl. Acad. Sci. USA* **97,** 14444–14448.

Jessell, T. M. (2000). Neuronal specification in the spinal cord: Inductive signals and transcriptional codes. *Nat. Rev. Genet.* **1,** 20–29.

Johnson, R. L., Laufer, E., Riddle, R. D., and Tabin, C. (1994). Ectopic expression of Sonic hedgehog alters dorsal-ventral patterning of somites. *Cell* **79,** 1165–1173.

Johnson, R. L., Rothman, A. L., Xie, J., Goodrich, L. V., Bare, J. W., *et al.* (1996). Human homolog of patched, a candidate gene for the basal cell nevus syndrome. *Science* **272,** 1668–1671.

Jones, K. L. (1997). "Smith's Recognizable Patterns of Human Malformations." W. B. Saunders, Philadelphia.

Jung, H. S., Oropeza, V., and Thesleff, I. (1999). Shh, Bmp-2, Bmp-4 and Fgf-8 are associated with initiation and patterning of mouse tongue papillae. *Mech. Dev.* **81,** 179–182.

Kalff-Suske, M., Wild, A., Topp, J., Wesslig, M., Jacobsen, E. M., *et al.* (1999). Point mutations throughout the GLI3 gene cause Greig cephalopolysyndactyly syndrome. *Hum. Mol. Genet.* **8,** 1769–1777.

Kalyani, A. J., Piper, D., Mujtaba, T., Lucero, M. T., and Rao, M. S. (1998). Spinal cord neuronal precursors generate multiple neuronal phenotypes in culture. *J. Neurosci.* **18,** 7856–7868.

Kameda, T., Koike, C., Saitoh, K., Kuroiwa, A., and Iba, H. (1999). Developmental patterning in chondrocytic cultures by morphogenic gradients: BMP induces expression of Indian hedgehog and noggin. *Genes Cells* **4,** 175–184.

Kang, S., Graham, J. M., Jr., Olney, A. H., and Biesecker, L. G. (1997). GLI3 frameshift mutations cause autosomal dominant Pallister–Hall syndrome. *Nat. Genet.* **15,** 266–268.

Karaplis, A. C., Luz, A., Glowacki, J., Bronson, R. T., Tybulewicz, V. L., *et al.* (1994). Lethal skeletal dysplasia from targeted disruption of the parathyroid hormone-related peptide gene. *Genes Dev.* **8,** 277–289.

Karlsson, L., Bondjers, C., and Betsholtz, C. (1999). Roles for PDGF-A and sonic hedgehog in development of mesenchymal components of the hair follicle. *Development* **126,** 2611–2621.

Karlstrom, R. O., Talbot, W. S., and Schier, A. F. (1999). Comparative synteny cloning of zebrafish you-too: Mutations in the Hedgehog target gli2 affect ventral forebrain patterning. *Genes. Dev.* **13,** 388–393.

Karp, S. J., Schipani, E., St-Jacques, B., Hunzelman, J., Kronenberg, H., and McMahon, A. P. (2000). Indian hedgehog coordinates endochondral bone growth and morphogenesis via parathyroid hormone related-protein-dependent and -independent pathways. *Development* **127,** 543–548.

Karsenty, G. (1999). The genetic transformation of bone biology. *Genes Dev.* **13,** 3037–3051.

Kawai, S., and Sugiura, T. (2001). Characterization of human bone morphogenetic protein (BMP)-4 and -7 gene promoters: Activation of BMP promoters by Gli, a sonic hedgehog mediator. *Bone* **29,** 54–61.

Kedinger, M., Simon-Assmann, P. M., Lacroix, B., Marxer, A., Hauri, H. P., and Haffen, K. (1986). Fetal gut mesenchyme induces differentiation of cultured intestinal endodermal and crypt cells. *Dev. Biol.* **113,** 474–483.

Kedinger, M., Duluc, I., Fritsch, C., Lorentz, O., Plateroti, M., and Freund, J. N. (1998). Intestinal epithelial–mesenchymal cell interactions. *Ann. N.Y. Acad. Sci.* **859,** 1–17.

Kelley, R. L., Roessler, E., Hennekam, R. C., Feldman, G. L., Kosaki, K., *et al.* (1996). Holoprosencephaly in RSH/Smith-Lemli-Opitz syndrome: Does abnormal cholesterol metabolism affect the function of Sonic hedgehog? *Am. J. Med. Genet.* **66,** 478–484.

Kenney, A. M., and Rowitch, D. H. (2000). Sonic hedgehog promotes G_1 cyclin expression and sustained cell cycle progression in mammalian neuronal precursors. *Mol. Cell. Biol.* **20,** 9055–9067.

Kim, H. J., Rice, D. P., Kettunen, P. J., and Thesleff, I. (1998). FGF-, BMP- and Shh-mediated signalling pathways in the regulation of cranial suture morphogenesis and calvarial bone development. *Development* **125,** 1241–1251.

Kim, I. S., Otto, F., Zabel, B., and Mundlos, S. (1999). Regulation of chondrocyte differentiation by Cbfa1. *Mech. Dev.* **80,** 159–170.

Kim, J., Kim, P., and Hui, C. C. (2001). The VACTERL association: Lessons from the Sonic hedgehog pathway. *Clin. Genet.* **59,** 306–315.

Kim, S. K., and Melton, D. A. (1998). Pancreas development is promoted by cyclopamine, a hedgehog signaling inhibitor. *Proc. Natl. Acad. Sci. USA* **95,** 13036–13041.

Kimonis, V. E., Goldstein, A. M., Pastakia, B., Yang, M. L., Kase, R., *et al.* (1997). Clinical manifestations in 105 persons with nevoid basal cell carcinoma syndrome. *Am. J. Med. Genet.* **69,** 299–308.

King, F. J., Skakmary, A., Cox, D. N., and Lin, H. (2001). Yb modulates the divisions of both germline and somatic stem cells through piwi- and hh-mediated mechanisms in the *Drosophila* ovary. *Mol. Cell.* **7,** 497–508.

King, T., Beddington, R. S., and Brown, N. A. (1998). The role of the brachyury gene in heart development and left-right specification in the mouse. *Mech. Dev.* **79,** 29–37.

Kinto, N., Iwamoto, M., Enomoto-Iwamoto, M., Noji, S., Ohuchi, H., *et al.* (1997). Fibroblasts expressing Sonic hedgehog induce osteoblast differentiation and ectopic bone formation. *FEBS Lett.* **404,** 319–323.

Kinzler, K. W., Bigner, S. H., Bigner, D. D., Trent, J. M., Law, M. L., *et al.* (1987). Identification of an amplified, highly expressed gene in a human glioma. *Science* **236,** 70–73.

Kohtz, J. D., Baker, D. P., Corte, G., and Fishell, G. (1998). Regionalization within the mammalian telencephalon is mediated by changes in responsiveness to Sonic Hedgehog. *Development* **125,** 5079–5089.

Kohtz, J. D., Lee, H. Y., Gaiano, N., Segal, J., Ng, E., *et al.* (2001). N-terminal fatty-acylation of sonic hedgehog enhances the induction of rodent ventral forebrain neurons. *Development* **128,** 2351–2363.

Kondoh, H., Uchikawa, M., Yoda, H., Takeda, H., Furutani-Seiki, M., and Karlstrom, R. O. (2000). Zebrafish mutations in Gli-mediated hedgehog signaling lead to lens transdifferentiation from the adenohypophysis anlage. *Mech. Dev.* **96,** 165–174.

Kos, L., Chiang, C., and Mahon, K. A. (1998). Mediolateral patterning of somites: Multiple axial signals, including Sonic hedgehog, regulate Nkx-3.1 expression. *Mech. Dev.* **70,** 25–34.

Koyama, E., Yamaai, T., Iseki, S., Ohuchi, H., Nohno, T., *et al.* (1996). Polarizing activity, Sonic hedgehog, and tooth development in embryonic and postnatal mouse. *Dev. Dyn.* **206,** 59–72.

Koyama, E., Wu, C., Shimo, T., Iwamoto, M., Ohmori, T., *et al.* (2001). Development of stratum intermedium and its role as a Sonic hedgehog-signaling structure during odontogenesis. *Dev. Dyn.* **222,** 178–191.

Kraus, P., Fraidenraich, D., and Loomis, C. A. (2001). Some distal limb structures develop in mice lacking Sonic hedgehog signaling. *Mech. Dev.* **100,** 45–58.

Krauss, S., Concordet, J. P., and Ingham, P. W. (1993). A functionally conserved homolog of the *Drosophila* segment polarity gene hh is expressed in tissues with polarizing activity in zebrafish embryos. *Cell* **75,** 1431–1444.

Krishnan, V., Ma, Y., Moseley, J., Geiser, A., Friant, S., and Frolik, C. (2001). Bone anabolic effects of Sonic/Indian hedgehog are mediated by bmp-2/4-dependent pathways in the neonatal rat metatarsal model. *Endocrinology* **142,** 940–947.

Kruger, M., Mennerich, D., Fees, S., Schafer, R., Mundlos, S., and Braun, T. (2001). Sonic hedgehog is a survival factor for hypaxial muscles during mouse development. *Development* **128,** 743–752.

Lacaud, G., Robertson, S., Palis, J., Kennedy, M., and Keller, G. (2001). Regulation of hemangioblast development. *Ann. N.Y. Acad. Sci.* **938,** 96–107; discussion 8.

Laforest, L., Brown, C. W., Poleo, G., Geraudie, J., Tada, M., *et al.* (1998). Involvement of the sonic hedgehog, patched 1 and bmp2 genes in patterning of the zebrafish dermal fin rays. *Development* **125,** 4175–4184.

Lam, C. W., Xie, J., To, K. F., Ng, H. K., Lee, K. C., *et al.* (1999). A frequent activated smoothened mutation in sporadic basal cell carcinomas. *Oncogene* **18,** 833–836.

Landis, S. H., Murray, T., Bolden, S., and Wingo, P. A. (1998). Cancer studies 1998. *CA Cancer J. Clin.* **48,** 6–29.

Lanoue, L., Dehart, D. B., Hinsdale, M. E., Maeda, N., Tint, G. S., and Sulik, K. K. (1997). Limb, genital, CNS, and facial malformations result from gene/environment-induced cholesterol deficiency: Further evidence for a link to sonic hedghog. *Am. J. Med. Genet.* **73,** 24–31.

Lanske, B., Karaplis, A. C., Lee, K., Luz, A., Vortkamp, A., *et al.* (1996). PTH/PTHrP receptor in early development and Indian hedgehog-regulated bone growth. *Science* **273,** 663–666.

Laufer, E., Nelson, C. E., Johnson, R. L., Morgan, B. A., and Tabin, C. (1994). Sonic hedgehog and Fgf-4 act through a significant cascade and feedback loop to integrate growth and patterning of the developing limb bud. *Cell* **79,** 993–1003.

Lawrence, P., Casal, J., and Struhl, G. (1999a). hedgehog and engrailed: Pattern formation and polarity in the *Drosophila* abdomen. *Development* **126,** 2431–2439.

Lawrence, P., Casal, J., and Struhl, G. (1999b). The hedgehog morphogen and gradients of cell affinity in the abdomen of *Drosophila*. *Development* **126,** 2441–2449.

Le, A. X., Miclau, T., Hu, D., and Helms, J. A. (2001). Molecular aspects of healing in stabilized and non-stabilized fractures. *J. Orthop. Res.* **19,** 78–84.

Lebeche, D., Malpel, S., and Cardoso, W. V. (1999). Fibroblast growth factor interactions in the developing lung. *Mech. Dev.* **86,** 125–136.

Lecuit, T., Brook, W., Ng, M., Calleja, M., Sun, H., and Cohen, S. (1996). Two distinct mechanisms for long-range patterning by *decapentaplegic* in the *Drosophila* wing. *Nature (London)* **381,** 387–393.

Lee, C., Buttitta, L. A., May, N. R., Kispert, A., and Fan, C.-M. (2000). SHH-N upregulates Sfrp2 to mediate its competitive interaction with WNT1 and WNT4 in the somitic mesoderm. *Development* **127,** 109–118.

Lee, C., Buttitta, L. A., and Fan, C.-M. (2001). Evidence that the WNT-inducible growth arrest-specific gene 1 encodes an antagonist of sonic hedgehog signaling in the somite. *Dev. Biol.* **98,** 11347–11352.

Lee, J., Platt, K. A., Censullo, P., and Ruiz i Altaba, A. (1997). Gli1 is a target of Sonic hedgehog that induces ventral neural tube development. *Development* **124,** 2537–2552.

Lee, J. J., von Kessler, D. P., Parks, S., and Beachy, P. A. (1992). Secretion and localised transcription suggests a role in positional signalling for products of the segmentation gene hedgehog. *Cell* **70,** 777–789.

Lee, K., Lanske, B., Karaplis, A. C., Deeds, J. D., Kohno, H., *et al.* (1996). Parathyroid hormone-related peptide delays terminal differentiation of chondrocytes during endochondral bone development. *Endocrinology* **137,** 5109–5118.

Lee, K. J., and Jessell, T. M. (1999). The specification of dorsal cell fates in the vertebrate central nervous system. *Annu. Rev. Neurosci.* **22,** 261–294.

Levanat, S., Pavelic, B., Crnic, I., Oreskovic, S., and Manojlovic, S. (2000). Involvement of PTCH gene in various noninflammatory cysts. *J. Mol. Med.* **78,** 140–146.

Levin, M., and Mercola, M. (1998). Gap junctions are involved in the early generation of left–right asymmetry. *Dev. Biol.* **203,** 90–105.

Levin, M., and Mercola, M. (1999). Gap junction-mediated transfer of left–right patterning signals in the early chick blastoderm is upstream of Shh asymmetry in the node. *Development* **126,** 4703–4714.

Levin, M., Johnson, R. L., Stern, C. D., Kuehn, M., and Tabin, C. (1995). A molecular pathway determining left–right asymmetry in chick embryogenesis. *Cell* **82,** 803–814.

Levin, M., Roberts, D. J., Holmes, L. B., and Tabin, C. (1996). Laterality defects in conjoined twins. *Nature (London)* **384,** 321.

Levin, M., Pagan, S., Roberts, D. J., Cooke, J., Kuehn, M. R., and Tabin, C. J. (1997). Left/right patterning signals and the independent regulation of different aspects of situs in the chick embryo. *Dev. Biol.* **189,** 57–67.

Levine, E., Roelink, H., Turner, J., and Reh, T. (1997). Sonic hedgehog promotes rod photoreceptor differentiation in mammalian retinal cells in-vitro. *J. Neurosci.* **17,** 6277–6288.

Lewis, K. E., and Eisen, J. S. (2001). Hedgehog signaling is required for primary motoneuron induction in zebrafish. *Development* **128,** 3485–3495.

Lewis, K. E., Currie, P. D., Roy, S., Schauerte, H., Haffter, P., and Ingham, P. W. (1999). Control of muscle cell-type specification in the zebrafish embryo by Hedgehog signalling. *Dev. Biol.* **216,** 469–480.

Lewis, M. T. (2001). Hedgehog signaling in mouse mammary gland development and neoplasia. *J. Mammary Gland Biol. Neoplasia* **6,** 53–66.

Lewis, M. T., Ross, S., Strickland, P. A., Sugnet, C. W., Jimenez, E., *et al.* (1999). Defects in mouse mammary gland development caused by conditional haploinsufficiency of Patched-1. *Development* **126,** 5181–5193.

Lewis, P. M., Dunn, M. P., McMahon, J. A., Logan, M., Martin, J. F., *et al.* (2001). Cholesterol modification of sonic hedgehog is required for long-range signaling activity and effective modulation of signaling by Ptc1. *Cell* **105,** 599–612.

Li, H., Arber, S., Jessell, T. M., and Edlund, H. (1999). Selective agenesis of the dorsal pancreas in mice lacking homeobox gene Hlxb9. *Nat Genet* **23,** 67–70.

Liem, K. F., Jr., Tremml, G., Roelink, H., and Jessell, T. M. (1995). Dorsal differentiation of neural plate cells induced by BMP-mediated signals from epidermal ectoderm. *Cell* **82,** 969–979.

Liem, K. F., Jr., Jessell, T. M., and Briscoe, J. (2000). Regulation of the neural patterning activity of sonic hedgehog by secreted BMP inhibitors expressed by notochord and somites. *Development* **127,** 4855–4866.

Litingtung, Y., and Chiang, C. (2000). Specification of ventral neuron types is mediated by an antagonistic interaction between Shh and Gli3. *Nat. Neurosci.* **3,** 979–985.

Litingtung, Y., Lei, L., Westphal, H., and Chiang, C. (1998). Sonic hedgehog is essential to foregut development. *Nat. Genet.* **20,** 58–61.

Lohr, J. L., Danos, M. C., and Yost, H. J. (1997). Left–right asymmetry of a nodal-related gene is regulated by dorsoanterior midline structures during *Xenopus* development. *Development* **124,** 1465–1472.

Long, F., Schipani, E., Asahara, H., Kronenberg, H., and Montminy, M. (2001a). The CREB family of activators is required for endochondral bone development. *Development* **128,** 541–550.

Long, F., Zhang, X. M., Karp, S., Yang, Y., and McMahon, A. P. (2001b). Genetic manipulation of hedgehog signaling in the endochondral skeleton reveals a direct role in the regulation of chondrocyte proliferation. *Development* **128,** 5099–5108.

Lopez-Martin, A., Chang, D. T., Chiang, C., Porter, J. A., Ros, M. A., *et al.* (1995). Limb-patterning activity and restricted posterior localization of the amino-terminal product of Sonic hedgehog cleavage. *Curr. Biol.* **5,** 791–796.

Lu, H.-C, Revelli, J.-P., Goering, L., Thaler, C., and Eichele, G. (1997). Retinoid signaling is required for the establishment of a ZPA and for the expression of HoxB-8, a mediator of ZPA formation. *Development* **124,** 1643–1651.

Lu, M. F., Cheng, H. T., Kern, M. J., Potter, S. S., Tran, B., *et al.* (1999). prx-1 functions cooperatively with another paired-related homeobox gene, prx-2, to maintain cell fates within the craniofacial mesenchyme. *Development* **126,** 495–504.

Lu, Q. R., Yuk, D., Alberta, J. A., Zhu, Z., Pawlitzky, I., *et al.* (2000). Sonic hedgehog regulated oligodendrocyte lineage genes encoding bHLH proteins in the mammalian central nervous system. *Neuron* **25,** 317–329.

Ma, C., Zhou, Y., Beachy, P. A., and Moses, K. (1993). The segment polarity gene hedgehog is required for progression of the morphogenetic furrow in the developing *Drosophila* eye. *Cell* **75,** 927–938.

Maas, R., and Bei, M. (1997). The genetic control of early tooth development. *Crit. Rev. Oral. Biol. Med.* **8,** 4–39.

MacCabe, J. A., Errick, J., and Saunders, J. W., Jr. (1974). Ectodermal control of the dorsoventral axis in the leg bud of the chick embryo. *Dev. Biol.* **39,** 69–82.

Macdonald, R., Barth, K. A., Xu, Q., Holder, N., Mikkola, I., and Wilson, S. W. (1995). Midline signalling is required for Pax gene regulation and patterning of the eyes. *Development* **121,** 3267–3278.

Macias, D., Ganan, Y., Sampath, T. K., Piedra, M. E., Ros, M. A., and Hurle, J. M. (1997). Role of BMP-2 and OP-1 (BMP-7) in programmed cell death and skeletogenesis during chick limb development. *Development* **124,** 1109–1117.

Maesawa, C., Tamura, G., Iwaya, T., Ogaswara, S., Ishida, K., *et al.* (1998). Mutations in the human homologue of the *Drosophila* patched gene in esophageal squamous cell carcinoma. *Genes Chromosomes Cancer* **21,** 276–279.

Mahlapuu, M., Enerback, S., and Carlsson, P. (2001). Haploinsufficiency of the forkhead gene Foxf1, a target for sonic hedgehog signaling, causes lung and foregut malformations. *Development* **128,** 2397–2406.

Marcelle, C., Ahlgren, S., and Bronner-Fraser, M. (1999). In vivo regulation of somite differentiation and proliferation by sonic hedgehog. *Dev. Biol.* **214,** 277–287.

Marigo, V., and Tabin, C. J. (1996). Regulation of patched by sonic hedgehog in the developing neural tube. *Proc. Natl. Acad. Sci. USA* **93,** 9346–9351.

Marigo, V., Roberts, D. J., Lee, S. M., Tsukurov, O., Levi, T., *et al.* (1995). Cloning, expression, and chromosomal location of SHH and IHH: Two human homologues of the *Drosophila* segment polarity gene hedgehog. *Genomics* **28,** 44–51.

Marigo, V., Davey, R. A., Zuo, Y., Cunningham, J. M., and Tabin, C. J. (1996a). Biochemical evidence that patched is the Hedgehog receptor. *Nature (London)* **384,** 176–179.

Marigo, V., Johnson, R. L., Vortkamp, A., and Tabin, C. J. (1996b). Sonic hedgehog differentially regulates expression of GLI and GLI3 during limb development. *Dev. Biol.* **180,** 273–283.

Marigo, V., Scott, M. P., Johnson, R. L., Goodrich, L. V., and Tabin, C. J. (1996c). Conservation in hedgehog signaling: Induction of a chicken patched homolog by Sonic hedgehog in the developing limb. *Development* **122,** 1225–1233.

Maroto, M., Reshef, R., Munsterberg, A. E., Koester, S., Goulding, M., and Lassar, A. B. (1997). Ectopic Pax-3 activates MyoD and Myf-5 expression in embryonic mesoderm and neural tissue. *Cell* **89,** 139–148.

Marti, E., Bumcrot, D. A., Takada, R., and McMahon, A. P. (1995). Requirement of 19K form of Sonic hedgehog for induction of distinct ventral cell types in CNS explants. *Nature (London)* **375,** 322–325.

Martinez Arias, A., Baker, N. E., and Ingham, P. W. (1988). Role of segment polarity gene in the definition and maintenance of cell states in the *Drosophila* embryo. *Development* **103,** 157–170.

Martinez-Morales, J. R., Barbas, J. A., Marti, E., Bovolenta, P., Edgar, D., and Rodriguez-Tebar, A. (1997). Vitronectin is expressed in the ventral region of the neural tube and promotes the differentiation of motor neurons. *Development* **124,** 5139–5147.

Masuya, H., Sagai, T., Wakana, S., Moriwaki, K., and Shiroishi, T. (1995). A duplicated zone of polarizing activity in polydactylous mouse mutants. *Genes Dev.* **9,** 1645–1653.

Masuya, H., Sagai, T., Moriwaki, K., and Shiroishi, T. (1997). Multigenic control of the localization of the zone of polarizing activity in limb morphogenesis in the mouse. *Dev. Biol.* **182,** 42–51.

Matise, M. P., Epstein, D. J., Park, H. L., Platt, K. A., and Joyner, A. L. (1998). Gli2 is required for induction of floor plate and adjacent cells, but not most ventral neurons in the mouse central nervous system. *Development* **125,** 2759–2770.

Matsunaga, E., and Shiota, K. (1977). Holoprosencephaly in human embryos: Epidemiologic studies of 150 cases. *Teratology* **16,** 261–272.

Matsuzaki, M., and Saigo, K. (1996). Hedgehog signaling independent of engrailed and wingless required for post-S1 neuroblast formation in *Drosophila* CNS. *Development* **122,** 3567—3575.

McDonald, J. A., and Doe, C. Q. (1997). Establishing neuroblast-specific gene expression in the *Drosophila* CNS: Huckebein is activated by Wingless and Hedgehog and repressed by Engrailed and Gooseberry. *Development* **124,** 1079–1087.

McGarvey, T. W., Maruta, Y., Tomaszewski, J. E., Linnenbach, A. J., and Malkowicz, S. B. (1998). PTCH gene mutations in invasive transitional cell carcinoma of the bladder. *Oncogene* **17,** 1167–1172.

McMahon, J. A., Takada, S., Zimmerman, L. B., Fan, C. M., Harland, R. M., and McMahon, A. P. (1998). Noggin-mediated antagonism of BMP signaling is required for growth and patterning of the neural tube and somite. *Genes Dev.* **12,** 1438–1452.

Medill, N. J., Praul, C. A., Ford, B. C., and Leach, R. M. (2001). Parathyroid hormone-related peptide expression in the epiphyseal growth plate of the juvenile chicken: Evidence for the origin of the parathyroid hormone-related peptide found in the epiphyseal growth plate. *J. Cell Biochem.* **80,** 504–511.

Melloy, P. G., Ewart, J. L., Cohen, M. F., Desmond, M. E., Kuehn, M. R., and Lo, C. W. (1998). No turning, a mouse mutation causing left–right and axial patterning defects. *Dev. Biol.* **193,** 77–89.

Meno, C., Saijoh, Y., Fujii, H., Ikeda, M., Yokoyama, T., *et al.* (1996). Left–right asymmetric expression of the TGF β-family member lefty in mouse embryos. *Nature (London)* **381,** 151–155.

Meno, C., Shimono, A., Saijoh, Y., Yashiro, K., Mochida, K., *et al.* (1998). lefty-1 is required for left–right determination as a regulator of lefty-2 and nodal. *Cell* **94,** 287–297.

Merino, R., Rodriguez-Leon, J., Macias, D., Ganan, Y., Economides, A. N., and Hurle, J. M. (1999). The BMP antagonist Gremlin regulates outgrowth, chondrogenesis and programmed cell death in the developing limb. *Development* **126,** 5515–5522.

Methot, N., and Basler, K. (1999). Hedgehog controls limb development by regulating the activities of distinct transcriptional activator and repressor forms of Cubitus interruptus. *Cell* **96,** 819–831.

Methot, N., and Basler, K. (2001). An absolute requirement for Cubitus interruptus in Hedgehog signaling. *Development* **128,** 733–742.

Metzger, R. J., and Krasnow, M. A. (1999). Genetic control of branching morphogenesis. *Science* **284,** 1635–1639.

Meyers, E. N., and Martin, G. R. (1999). Differences in left–right axis pathways in mouse and chick: Functions of FGF8 and SHH. *Science* **285,** 403–406.

Miao, N., Wang, M., Ott, J. A., D'Alessandro, J. S., Woolf, T. M., *et al.* (1997). Sonic hedgehog promotes the survival of specific CNS neuron populations and protects these cells from toxic insult in vitro. *J. Neurosci.* **17,** 5891–5899.

Min, H., Danilenko, D. M., Scully, S. A., Bolon, B., Ring, B. D., *et al.* (1998). Fgf-10 is required for both limb and lung development and exhibits striking functional similarity to *Drosophila* branchless. *Genes. Dev.* **12,** 3156–3161.

Minina, E., Wenzel, H. M., Kreschel, C., Karp, S., Gaffield, W., *et al.* (2001). BMP and Ihh/PTHrP signaling interact to coordinate chondrocyte proliferation and differentiation. *Development* **128,** 4523–4534.

Minoo, P., Su, G., Drum, H., Bringas, P., and Kimura, S. (1999). Defects in tracheoesophageal and lung morphogenesis in Nkx2.1$^{-/-}$ mouse embryos. *Dev. Biol.* **209,** 60–71.

Mirsky, R., Jessen, K. R., Brennan, A., Parkinson, D., Dong, Z., *et al.* (2002). Schwann cells as regulators of nerve development. *J. Physiol. Paris.* **96,** 17–24.

Miura, H., Kusakabe, Y., Sugiyama, C., Kawamatsu, M., Ninomiya, Y., *et al.* (2001). Shh and Ptc are associated with taste bud maintenance in the adult mouse. *Mech. Dev.* **106,** 143–145.

Mo, R., Freer, A. M., Zinyk, D. L., Crackower, M. A., Michaud, J., *et al.* (1997). Specific and redundant functions of Gli2 and Gli3 zinc finger genes in skeletal patterning and development. *Development* **124,** 113–123.

Mohler, J. (1988). Requirements for hedgehog, a segmental polarity gene, in patterning larval and adult cuticle of *Drosophila*. *Genetics* **120,** 1061–1072.

Mohler, J., and Vani, K. (1992). Molecular organization and embryonic expression of the hedgehog gene involved in cell–cell communication in segmental patterning of *Drosophila*. *Development* **115,** 957–971.

Mohler, J., Seecoomar, M., Agarwal, S., Bier, E., and Hsai, J. (2000). Activation of knot (kn) specifies the 3–4 intervein region in the *Drosophila* wing. *Development* **127,** 55–63.

Monsoro-Burq, A., and Le Douarin, N. (2000). Left–right asymmetry in BMP4 signalling pathway during chick gastrulation. *Mech. Dev.* **97,** 105–108.

Monsoro-Burq, A., and Le Douarin, N. M. (2001). BMP4 plays a key role in left–right patterning in chick embryos by maintaining Sonic Hedgehog asymmetry. *Mol. Cell.* **7,** 789–799.

Moore, L. A., Broihier, H. T., Van Doren, M., Lunsford, L. B., and Lehmann, R. (1998). Identification of genes controlling germ cell migration and embryonic gonad formation in *Drosophila*. *Development* **125,** 667–678.

Morgan, B. A., Orkin, R. W., Noramly, S., and Perez, A. (1998). Stage-specific effects of sonic hedgehog expression in the epidermis. *Dev. Biol.* **201,** 1–12.

Motoyama, J., Heng, H., Crackower, M. A., Takabatake, T., Takeshima, K., *et al.* (1998a). Overlapping and non-overlapping Ptch2 expression with Shh during mouse embryogenesis. *Mech. Dev.* **78,** 81–84.

Motoyama, J., Liu, J., Mo, R., Ding, Q., Post, M., and Hui, C. C. (1998b). Essential function of Gli2 and Gli3 in the formation of lung, trachea and oesophagus. *Nat. Genet.* **20,** 54–57.

Motoyama, J., Takabatake, T., Takeshima, K., and Hui, C. (1998c). Ptch2, a second mouse Patched gene is co-expressed with Sonic hedgehog. *Nat. Genet.* **18,** 104–106.

Mullor, J., Calleja, M., Capdevila, J., and Guerrero, I. (1997). Hedgehog activity, independent of decapentaplegic, participates in wing disc patterning. *Development* **124,** 1227–1237.

Munoz-Sanjuan, I., Cooper, M. K., Beachy, P. A., Fallon, J. F., and Nathans, J. (2001). Expression and regulation of chicken fibroblast growth factor homologous factor (FHF)-4 during craniofacial morphogenesis. *Dev. Dyn.* **220,** 238–245.

Munsterberg, A., Kitajewski, J., Bumcrot, D. A., McMahon, A. P., and Lassar, A. B. (1995). Combinatorial signaling by Sonic hedgehog and WNt family members induces myogenic bHLH gene expression in the somite. *Genes. Dev.* **9,** 2911–2922.

Murakami, S., and Noda, M. (2000). Expression of Indian hedgehog during fracture healing in adult rat femora. *Calcif. Tissue Int.* **66,** 272–276.

Murakami, S., Nifuji, A., and Noda, M. (1997). Expression of Indian hedgehog in osteoblasts and its posttranscriptional regulation by transforming growth factor-β. *Endocrinology* **138,** 1972–1978.

Murtaugh, L. C., Chyung, J. H., and Lassar, A. B. (1999). Sonic hedgehog promotes somitic chondrogenesis by altering the cellular response to BMP signaling. *Genes Dev.* **13,** 225–237.

Nakagawa, Y., Kaneko, T., Ogura, T., Suzuki, T., Torii, M., *et al.* (1996). Roles of cell-autonomous mechanisms for differential expression of region-specific transcription factors in neuroepithelial cells. *Development* **122,** 2449–2464.

Nakamura, T., Aikawa, T., Iwamoto-Enomoto, M., Iwamoto, M., Higuchi, Y., *et al.* (1997). Induction of osteogenic differentiation by hedgehog proteins. *Biochem. Biophys. Res. Commun.* **237,** 465–469.

Nanni, L., Ming, J. E., Bocian, M., Steinhaus, K., Bianchi, D. W., *et al.* (1999). The mutational spectrum of the sonic hedgehog gene in holoprosencephaly: SHH mutations cause a significant proportion of autosomal dominant holoprosencephaly. *Hum. Mol. Genet.* **8,** 2479–2488.

Narita, T., Ishii, Y., Nohno, T., Noji, S., and Yasugi, S. (1998). Sonic hedgehog expression in developing chicken digestive organs is regulated by epithelial–mesenchymal interactions. *Dev. Growth Differ.* **40,** 67–74.

Naski, M. C., Colvin, J. S., Coffin, J. D., and Ornitz, D. M. (1998). Repression of hedgehog signaling and BMP4 expression in growth plate cartilage by fibroblast growth factor receptor 3. *Development* **125,** 4977–4988.

Nellen, D., Burke, R., Struhl, G., and Basler, K. (1996). Direct and long-range action of a *dpp* morphogen gradient. *Cell* **85,** 357–368.

Nery, S., Wichterle, H., and Fishell, G. (2001). Sonic hedgehog contributes to oligodendrocyte specification in the mammalian forebrain. *Development* **128,** 527–540.

Neumann, C., and Nüsslein-Volhard, C. (2000). Patterning of the zebrafish retina by a wave of sonic hedgehog activity. *Science* **289,** 2137–2139.

Neumann, C. J., Grandel, H., Gaffield, W., Schulte-Merker, S., and Nusslein-Volhard, C. (1999). Transient establishment of anteroposterior polarity in the zebrafish pectoral fin bud in the absence of sonic hedgehog activity. *Development* **126,** 4817–4826.

Niswander, L., and Martin, G. R. (1992). Fgf-4 expression during gastrulation, myogenesis, limb and tooth development in the mouse. *Development* **114,** 755–768.

Niswander, L., Tickle, C., Vogel, R., Booth, I., and Martin, G. R. (1993). FGF-4 replaces the apical ectodermal ridge and directs outgrowth and patterning of the limb. *Cell* **75,** 579–587.

Niswander, L., Jeffrey, S., Martin, G. R., and Tickle, C. (1994). A positive feedback loop coordinates growth and patterning in the vertebrate limb. *Cell* **75,** 579–587.

Nohno, T., Noji, S., Koyama, E., Ohyama, K., Myokai, F., *et al.* (1991). Involvement of the Chox-4 chicken homeobox genes in determination of anteroposterior axial polarity during limb development. *Cell* **64,** 1197–1205.

Nohno, T., Kawakami, Y., Ohuchi, H., Fujiwara, A., Yoshioka, H., and Noji, S. (1995). Involvement of the Sonic hedgehog gene in chick feather formation. *Biochem. Biophys. Res. Commun.* **206,** 33–39.

Noji, S., Nohno, T., Koyama, E., Muto, K., Ohyama, K., *et al.* (1991). Retinoic acid induces polarizing activity but is unlikely to be a morphogen in the chick limb bud. *Nature (London)* **350,** 83–86.

Nonaka, S., Tanaka, Y., Okada, Y., Takeda, S., Harada, A., *et al.* (1998). Randomization of left–right asymmetry due to loss of nodal cilia generating leftward flow of extraembryonic fluid in mice lacking KIF3B motor protein. *Cell* **95,** 829–837.

Normaly, S., Pisenti, J. M., Abbott, U. K., and Morgan, B. A. (1996). Gene expression in the limbless mutant: Polarized gene expression in the absence of Shh and an AER. *Dev. Biol.* **179,** 339–346.

Nusslein-Volhard, C., and Weischhaus, E. (1980). Mutations affecting segment numbers and polarity in *Drosophila. Nature* **287,** 795–801.

Odenthal, J., van Eeden, F. J., Haffter, P., Ingham, P. W., and Nusslein-Volhard, C. (2000). Two distinct cell populations in the floor plate of the zebrafish are induced by different pathways. *Dev. Biol.* **219,** 350–63.

Offield, M. F., Jetton, T. L., Labosky, P. A., Ray, M., Stein, R., Magnuson, M. A., Hogan, B. L. M., and Wright, C. V. E. (1996). PDX-1 is required for pancreatic outgrowth and differentiation of the rostral duodenum. *Development* **122,** 983–995.

Oh, S. P., and Li, E. (1997). The signaling pathway mediated by the type IIB activin receptor controls axial patterning and lateral asymmetry in the mouse. *Genes Dev.* **11,** 1812–1826.

Oppenheim, R. W., Homma, S., Marti, E., Prevette, D., Wang, S., *et al.* (1999). Modulation of early but not later stages of programmed cell death in embryonic avian spinal cord by sonic hedgehog. *Mol. Cell. Neurosci.* **13,** 348–361.

Orentas, D. M., Hayes, J. E., Dyer, K. L., and Miller, R. H. (1999). Sonic hedgehog signaling is required during the appearance of spinal cord oligodendrocyte precursors. *Development* **126,** 2419–2429.

Oro, A. E., and Scott, M. P. (1998). Splitting hairs: Dissecting roles of signaling systems in epidermal development. *Cell* **95,** 575–578.

Oro, A. E., Higgins, K. M., Hu, Z., Bonifas, J. M., Epstein, E. H., Jr., and Scott, M. P. (1997). Basal cell carcinomas in mice overexpressing sonic hedgehog. *Science* **276,** 817–821.

Oshima, H., Rochat, A., Kedzia, C., Kobayashi, K., and Barrandon, Y. (2001). Morphogenesis and renewal of hair follicles from adult multipotent stem cells. *Cell* **104,** 233–245.

Outram, S. V., Varas, A., Pepicelli, C. V., and Crompton, T. (2000). Hedgehog signaling regulates differentiation from double-negative to double-positive thymocyte. *Immunity* **13,** 187–197.

Pagan, S., Ros, M. A., Tabin, C., and Fallon, J. F. (1996). Surgical removal of limb bud Sonic hedgehog results in posterior skeletal defects. *Dev. Biol.* **180,** 35–40.

Pagan-Westphal, S. M., and Tabin, C. J. (1998). The transfer of left–right positional information during chick embryogenesis. *Cell* **93,** 25–35.

Palis, J., Robertson, S., Kennedy, M., Wall, C., and Keller, G. (1999). Development of erythroid and myeloid progenitors in the yolk sac and embryo proper of the mouse. *Development* **126,** 5073–5084.

Pankratz, M. J., and Hoch, M. (1995). Control of epithelial morphogenesis by cell signaling and integrin molecules in the *Drosophila* foregut. *Development* **121,** 1885–1898.

Paria, B. C., Ma, W., Tan, J., Raja, S., Das, S. K., *et al.* (2001). Cellular and molecular responses of the uterus to embryo implantation can be elicited by locally applied growth factors. *Proc. Natl. Acad. Sci. USA* **98,** 1047–1052.

Park, H. L., Bai, C., Platt, K. A., Matise, M. P., Beeghly, A., *et al.* (2000). Mouse Gli1 mutants are viable but have defects in SHH signaling in combination with a Gli2 mutation. *Development* **127,** 1593–1605.

Parmantier, E., Lynn, B., Lawson, D., Turmaine, M., Namini, S. S., *et al.* (1999). Schwann cell-derived Desert hedgehog controls the development of peripheral nerve sheaths. *Neuron* **23,** 713–724.

Parr, B. A., and McMahon, A. P. (1995). Dorsalizing signal Wnt-7a required for normal polarity of D–V and A–P axes of mouse limb. *Nature (London)* **374,** 350–353.

Pateder, D. B., Rosier, R. N., Schwarz, E. M., Reynolds, P. R., Puzas, J. E., *et al.* (2000). PTHrP expression in chondrocytes, regulation by TGF-β, and interactions between epiphyseal and growth plate chondrocytes. *Exp. Cell Res.* **256,** 555–562.

Pathi, S., Rutenberg, J. B., Johnson, R. L., and Vortkamp, A. (1999). Interaction of Ihh and BMP/Noggin signaling during cartilage differentiation. *Dev. Biol* **209,** 239–253.

Patten, I., and Placzek, M. (2002). Opponent activities of Shh and BMP signaling during floor plate induction in vivo. *Curr. Biol.* **12,** 47–52.

Paulson, R. B., Hayes, T. G., and Sucheston, M. E. (1985). Scanning electron microscope study of tongue development in the CD-1 mouse fetus. *J. Craniofac. Genet. Dev. Biol.* **5,** 59–73.

Pautou, M. P. (1977). Establissement de l'axe dorso-ventral dans le pied de l'embryon de poulet. *J. Embryol. Exp. Morphol.* **42,** 177–194.

Pearse, R. V., II, Vogan, K. J., and Tabin, C. J. (2001). Ptc1 and Ptc2 transcripts provide distinct readouts of Hedgehog signaling activity during chick embryogenesis. *Dev. Biol.* **239,** 15–29.

Pepicelli, C. V., Lewis, P. M., and McMahon, A. P. (1998). Sonic hedgehog regulates branching morphogenesis in the mammalian lung. *Curr. Biol.* **8,** 1083–1086.

Pera, E. M., and Kessel, M. (1997). Patterning of the chick forebrain anlage by the prechordal plate. *Development* **124,** 4153–4162.

Peters, K., Werner, S., Liao, X., Wert, S., Whitsett, J., and Williams, L. (1994). Targeted expression of a dominant negative FGF receptor blocks branching morphogenesis and epithelial differentiation of the mouse lung. *EMBO J.* **13,** 3296–3301.

Petrij, F., Giles, R. H., Dauwerse, H. G., Saris, J. J., Hennekam, R. C., *et al.* (1995). Rubinstein–Taybi syndrome caused by mutations in the transcriptional co-activator CBP. *Nature (London)* **376,** 348–351.

Piepenburg, O., Vorbruggen, G., and Jackle, H. (2000). *Drosophila* segment borders result from unilateral repression of hedgehog activity by wingless signaling. *Mol. Cell* **6,** 203–209.

Pierani, A., Brenner-Morton, S., Chiang, C., and Jessell, T. M. (1999). A sonic hedgehog-independent, retinoid-activated pathway of neurogenesis in the ventral spinal cord. *Cell* **97,** 903–915.

Pierucci-Alves, F., Clark, A. M., and Russell, L. D. (2001). A developmental study of the Desert hedgehog-null mouse testis. *Biol. Reprod.* **65,** 1392–1402.

Pietsch, T., Waha, A., Koch, A., Kraus, J., Albrecht, S., *et al.* (1997). Medulloblastomas of the desmoplastic variant carry mutations of the human homologue of *Drosophila* patched. *Cancer Res.* **57,** 2085–2088.

Pignoni, F., and Zipursky, S. (1997). Induction of *Drosophila* eye development by decapentaplegic. *Development* **124,** 271–278.

Placzek, M. (1995). The role of the notochord and floor plate in inductive interactions. *Curr. Opin. Genet. Dev.* **5,** 499–506.

Platt, K. A., Michaud, J., and Joyner, A. L. (1997). Expression of the mouse Gli and Ptc genes is adjacent to embryonic sources of hedgehog signals suggesting a conservation of pathways between flies and mice. *Mech. Dev.* **62,** 121–135.

Podlasek, C. A., Duboule, D., and Bushman, W. (1997). Male accessory sex organ morphogenesis is altered by loss of function of Hoxd-13. *Dev. Dyn.* **208,** 454–465.

Podlasek, C. A., Barnett, D. H., Clemens, J. Q., Bak, P. M., and Bushman, W. (1999). Prostate development requires Sonic hedgehog expressed by the urogenital sinus epithelium. *Dev. Biol.* **209,** 28–39.

Pola, R., Ling, L. E., Silver, M., Corbley, M. J., Kearney, M., *et al.* (2001). The morphogen Sonic hedgehog is an indirect angiogenic agent upregulating two families of angiogenic growth factors. *Nat. Med.* **7,** 706–711.

Poncet, C., Soula, C., Trousse, F., Kan, P., Hirsinger, E., Pourquie, O., Duprat, A. M., and Cochard, P. (1996). Induction of oligodendrocyte progenitors in the trunk neural tube by ventralizing signals: effects of notochord and floor plate grafts, and of sonic hedgehog. *Mech of Dev.* **60,** 13–22.

Pons, S., and Marti, E. (2000). Sonic hedgehog synergizes with the extracellular matrix protein vitronectin to induce spinal motor neuron differentiation. *Development* **127,** 333–342.

Porter, J. A., von Kessler, D. P., Ekker, S. C., Young, K. E., Lee, J. J., *et al.* (1995). The product of hedgehog autoproteolytic cleavage active in local and long-range signalling. *Nature (London)* **374,** 363–366.

Porter, J. A., Ekker, S. C., Park, W. J., von Kessler, D. P., Young, K. E., et al. (1996). Hedgehog patterning activity: Role of a lipophilic modification mediated by the carboxy-terminal autoprocessing domain. *Cell* **86**, 21–34.

Pourquie, O., Coltey, M., Teillet, M. A., Ordahl, C., and Le Douarin, N. M. (1993). Control of dorsoventral patterning of somitic derivatives by notochord and floor plate. *Proc. Natl. Acad. Sci. USA* **90**, 5242–5246.

Pownall, M. E., Strunk, K. E., and Emerson, C. J. (1996). Notochord signals control the transcriptional cascade of myogenic bHLH genes in somites of quail embryos. *Development* **122**, 1475–1488.

Pringle, N. P., Wei-Ping, Y., Guthrie, S., Roelink, H., Lumsden, A., Peterson, A. C., and Richardson, W. D. (1996). Determination of neuroepithelial cell fate: induction of the oligodendrocyte lineage by ventral midline cells and sonic hedgehog. *Dev. Biol.* **177**, 30–42.

Qu, S., Niswender, K. D., Ji, Q., van der Meer, R., Keeny, D., et al. (1997). Polydactyly and ectopic ZPA formation in Alx-4 mutant mice. *Development* **124**, 3999–4008.

Qu, S., Tucker, S. C., Ehrlich, J. S., Levorse, J. M., Flaherty, L. A., et al. (1998). Mutations in mouse Aristaless-like4 cause Strong's luxoid polydactyly. *Development* **125**, 2711–2721.

Radhakrishna, U., Blouin, J. L., Solanki, J. V., Dhorlani, G. M., and Antonarakis, S. E. (1996). An autosomal dominant triphalangeal thumb: Polysyndactyly syndrome with variable expression in a large Indian family maps to 7q36. *Am. J. Med. Genet.* **66**, 209–215.

Radhakrishna, U., Wild, A., Grzeschik, K. H., and Antonarakis, S. E. (1997). Mutation in GLI3 in postaxial polydactyly type, A. *Nat. Genet.* **17**, 269–271.

Radhakrishna, U., Bornholdt, D., Scott, H. S., Patel, U. C., Rossier, C., et al. (1999). The phenotypic spectrum of GLI3 morphopathies includes autosomal dominant preaxial polydactyly type-IV and postaxial polydactyly type-A/B; no phenotype prediction from the position of GLI3 mutations. *Am. J. Hum. Genet.* **65**, 645–655.

Raffel, C., Jenkins, R. B., Frederick, L., Hebrink, D., Alderete, B., et al. (1997). Sporadic medulloblastomas contain PTCH mutations. *Cancer Res.* **57**, 842–845.

Ramalho-Santos, M., Melton, D. A., and McMahon, A. P. (2000). Hedgehog signals regulate multiple aspects of gastrointestinal development. *Development* **127**, 2763–2772.

Rankin, C. T., Bunton, T., Lawler, A. M., and Lee, S. J. (2000). Regulation of left–right patterning in mice by growth/differentiation factor-1. *Nat. Genet.* **24**, 262–265.

Raskind, W. H., Conrad, E. U., Chansky, H., and Matsushita, M. (1995). Loss of heterozygosity in chondrosarcomas for markers linked to hereditary multiple exostoses loci on chromosomes 8 and 11. *Am. J. Hum. Genet.* **56**, 1132–1139.

Reddy, S., Andl, T., Bagasra, A., Lu, M. M., Epstein, D. J., et al. (2001). Characterization of Wnt gene expression in developing and postnatal hair follicles and identification of Wnt5a as a target of Sonic hedgehog in hair follicle morphogenesis. *Mech. Dev.* **107**, 69–82.

Reifenberger, J., Woler, M., Weber, R. G., Megahe, M., Razuicka, T., et al. (1998). Missense mutations in SMOH in sporadic basal cell carcinomas of the skin and primitive neuroectodermal tumors of the central nervous system. *Cancer Res.* **58**, 1798–1803.

Reifers, F., Bohli, H., Walsh, E. C., Crossley, P. H., Stanier, D. Y., and Brand, M. (1998). Fgf8 is mutated in zebrafish acerebellar (ace) mutants and is required for maintenance of midbrain–hindbrain boundary development and somitogenesis. *Development* **125**, 2381–2395.

Reppeto, M., Maziere, J. C., Citadelle, D., Dupuis, R., Meier, M., et al. (1990). Teratogenic effect of the cholesterol synthesis inhibitor AY9944 on rat embryos in vitro. *Teratology* **42**, 611–618.

Riddle, R. D., Johnson, R. L., Laufer, E., and Tabin, C. (1993). Sonic hedgehog mediates the polarizing activity of the ZPA. *Cell* **75**, 1401–1416.

Riddle, R. D., Ensini, M., Nelson, C. E., Tsuchida, T., Jessell, T. M., and Tabin, C. (1995). Induction of the LIM homeobox gene Lmx1 by WNT7a establishes dorsoventral pattern in the vertebrate limb. *Cell* **83**, 631–640.

Riechmann, V., Rehorn, K. P., Reuter, R., and Leptin, M. (1998). The genetic control of the distinction between fat body and gonadal mesoderm in *Drosophila*. *Development* **125**, 713–723.

Roach, E., Demyer, W., Conneally, P. M., Palmer, C., and Merritt, A. D. (1975). Holoprosencephaly: Birth data, benetic and demographic analyses of 30 families. *Birth Defects Orig. Artic. Ser.* **11**, 294–313.

Roberts, D. J., Johnson, R. L., Burke, A. C., Nelson, C. E., Morgan, B. A., and Tabin, C. (1995). Sonic hedgehog is an endodermal signal inducing Bmp-4 and Hox genes during induction and regionalization of the chick hindgut. *Development* **121**, 3163–3174.

Roberts, D. J., Smith, D. M., Goff, D. J., and Tabin, C. J. (1998). Epithelial–mesenchymal signaling during the regionalization of the chick gut. *Development* **125**, 2791–2801.

Roberts, W. M., Douglass, E. D., Peiper, S. C., Houghton, P. J., and Look, A. T. (1989). Amplification of the gli gene in childhood sarcoma. *Cancer Res.* **49**, 5407–5413.

Rochat, A., Kobayashi, K., and Barrandon, Y. (1994). Location of stem cells of human hair follicles by clonal analysis. *Cell* **76**, 1063–1073.

Rodriguez, I., and Basler, K. (1997). Control of compartmental affinity boundaries by hedgehog. *Nature (London)* **389**, 614–618.

Rodriguez Esteban, C., Capdevila, J., Economides, A. N., Pascual, J., Ortiz, A., and Izpisua Belmonte, J. C. (1999). The novel Cer-like protein Caronte mediates the establishment of embryonic left–right asymmetry. *Nature (London)* **401**, 243–251.

Roelink, H., Augsburger, A., Heemskerk, J., Korzh, V., Norlin, S., *et al.* (1994). Floor plate and motor neuron induction by vhh-1, a vertebrate homolog of hedgehog expressed by the notochord. *Cell* **76**, 761–775.

Roelink, H., Porter, J. A., Chiang, C., Tanabe, Y., Chang, D. T., *et al.* (1995). Floor plate and motor neuron induction by different concentrations of the amino-terminal cleavage product of sonic hedgehog autoproteolysis. *Cell* **81**, 445–455.

Roessler, E., and Muenke, M. (2001). Midline and laterality defects: Left and right meet in the middle. *Bioessays* **23**, 888–900.

Roessler, E., Belloni, E., Gaudenz, K., Jay, P., Berta, P., *et al.* (1996). Mutations in the human Sonic Hedgehog gene cause holoprosencephaly. *Nat. Genet.* **14**, 357–360.

Roessler, E., Ward, D. E., Gaudenz, K., Belloni, E., Scherer, S. W., *et al.* (1997). Cytogenetic rearrangements involving the loss of the Sonic Hedgehog gene at 7q36 cause holoprosencephaly. *Hum. Genet.* **100**, 172–181.

Rohr, K. B., Schulte-Merker, S., and Tautz, D. (1999). Zebrafish zic1 expression in brain and somites is affected by BMP and hedgehog signalling. *Mech. Dev.* **85**, 147–159.

Rong, P. M., Teillet, M. A., Ziller, C., and Le Douarin, N. M. (1992). The neural tube/notochord complex is necessary for vertebral but not limb and body wall striated muscle differentiation. *Development* **115**, 657–672.

Ros, M. A., Lopez-Martin, A., Simandl, B. K., Rodrigues, C., Izpisua Belmonte, J. C., *et al.* (1996). The limb field mesoderm determines initial limb bud anterior–posterior asymmetry and budding independent of sonic hedgehog or apical ectodermal gene expression. *Development* **122**, 2319–2330.

Roux, C., Horvath, C., and Dupuis, R. (1979). Teratogenic action and embryo lethality of AY9944: Prevention of a hypercholesterolemia-provoking diet. *Teratology* **19**, 35–38.

Rowe, D. A., and Fallon, J. F. (1982). The proximodistal determination of skeletal parts in the developing chick leg. *J. Embryol. Exp. Morphol.* **68**, 1–7.

Rowitch, D. H., St-Jacques, B., Lee, S. M., Flax, J. D., Snyder, E. Y., and McMahon, A. P. (1999). Sonic hedgehog regulates proliferation and inhibits differentiation of CNS precursor cells. *J. Neurosci.* **19**, 8954–8965.

Roy, S., Gardiner, D. M., and Bryant, S. V. (2000). Vaccinia as a tool for functional analysis in regenerating limbs: Ectopic expression of Shh. *Dev. Biol.* **218**, 199–205.

Saldanha, G., Jones, L., Shaw, J., Pringle, H., Walker, R., and Fletcher, A. (1998). Breast cancer cell line invasion and hedgehog pathway dysregulation. *J. Pathol.* **186**, A23.

Sander, M., Paydar, S., Ericson, J., Briscoe, J., Berber, E., *et al.* (2000). Ventral neural patterning by Nkx homeobox genes: Nkx6.1 controls somatic motor neuron and ventral interneuron fates. *Genes Dev.* **14,** 2134–2139.

Sanz-Ezquerro, J. J., and Tickle, C. (2000). Autoregulation of Shh expression and Shh induction of cell death suggest a mechanism for modulating polarising activity during chick limb development. *Development* **127,** 4811–4823.

Sarkar, L., Cobourne, M., Naylor, S., Smalley, M., Dale, T., and Sharpe, P. T. (2000). Wnt/Shh interactions regulate ectodermal boundary formation during mammalian tooth development. *Proc. Natl. Acad. Sci. USA* **97,** 4520–4524.

Sasaki, H., Hui, C., Nakafuku, M., and Kondoh, H. (1997). A binding site for Gli proteins is essential for HNF-3β floor plate enhancer activity in transgenics and can respond to Shh in vitro. *Development* **124,** 1313–1322.

Sato, N., Leopold, P. L., and Crystal, R. G. (1999). Induction of the hair growth phase in postnatal mice by localized transient expression of Sonic hedgehog. *J. Clin. Invest.* **104,** 855–864.

Saunders, J. W., Jr., and Fallon, J. F. (1948). The proximo-distal sequence of origin of the parts of the chick wing and the role of the ectoderm. *J. Exp. Zool.* **108,** 363–403.

Saunders, J. W., Jr., and Gasseling, M. T. (1968). "Ectoderm–Mesenchymal Interaction in the Origins of Wing Symmetry," pp. 78–97. Williams & Wilkins, Baltimore, MD.

Saunders, J. W., Jr., and Gasseling, M. T. (1983). New insights into the problem of pattern regulation in the limb bud of the chick embryo. *Prog. Clin. Biol. Res.* **110,** 67–76.

Schauerte, H. E., van Eeden, F. J., Fricke, C., Odenthal, J., Strahle, U., and Haffter, P. (1998). Sonic hedgehog is not required for the induction of medial floor plate cells in the zebrafish. *Development* **125,** 2983–2993.

Schilling, T. F., Concordet, J. P., and Ingham, P. W. (1999). Regulation of left–right asymmetries in the zebrafish by Shh and BMP4. *Dev. Biol.* **210,** 277–287.

Schipani, E., Lanske, B., Hunzelman, J., Luz, A., Kovacs, C. S., *et al.* (1997). Targeted expression of constitutively active receptors for parathyroid hormone and parathyroid hormone-related peptide delays endochondral bone formation and rescues mice that lack parathyroid hormone-related peptide. *Proc. Natl. Acad. Sci. USA* **94,** 13689–13694.

Schneider, A., Brand, T., Zweigerdt, R., and Arnold, H. (2000). Targeted disruption of the Nkx3.1 gene in mice results in morphogenetic defects of minor salivary glands: Parallels to glandular duct morphogenesis in prostate. *Mech. Dev.* **95,** 163–174.

Schneider, R. A., Hu, D., Rubenstein, J. L., Maden, M., and Helms, J. A. (2001). Local retinoid signaling coordinates forebrain and facial morphogenesis by maintaining FGF8 and SHH. *Development* **128,** 2755–2767.

Sekine, K., Ohuchi, H., Fujiwara, M., Yamasaki, M., Yoshizawa, T., *et al.* (1999). Fgf10 is essential for limb and lung formation. *Nat. Genet.* **21,** 138–141.

Selleck, M. A., Garcia-Castro, M. I., Artinger, K. B., and Bronner-Fraser, M. (1998). Effects of Shh and Noggin on neural crest formation demonstrate that BMP is required in the neural tube but not ectoderm. *Development* **125,** 4919–4930.

Sengel, P. (1975). Feather pattern development. *Ciba Found. Symp.* **0,** 51–70.

Sesgin, M. Z., and Stark, R. B. (1961). The incidence of congenital defects. *Plast. Reconst. Surg.* **27,** 261–266.

Sharpe, J., Lettice, L., Hecksher-Sorenson, J., Fox, M., Hill, R., and Krumlauf, R. (1999). Identification of sonic hedgehog as a candidate gene responsible for the polydactylous mouse mutant Sasquatch. *Curr. Biol.* **9,** 97–100.

Shimamura, K., and Rubenstein, J. L. (1997). Inductive interactions direct early regionalization of the mouse forebrain. *Development* **124,** 2709–2718.

Shimeld, S. M. (1999). The evolution of the hedgehog gene family in chordates: Insights from Amphioxus hedgehog. *Dev. Genes Evol.* **209,** 40–47.

Sive, H., and Bradley, L. (1996). A sticky problem: The *Xenopus* cement gland as a paradigm for anteroposterior patterning. *Dev. Dyn.* **205,** 265–280.

Small, K. M., and Potter, S. S. (1993). Homeotic transformations and limb defects in Hox A11 mutant mice. *Genes Dev.* **7,** 2318–2328.

Smith, D. M., Nielsen, C., Tabin, C. J., and Roberts, D. J. (2000). Roles of BMP signaling and Nkx2.5 in patterning at the chick midgut–foregut boundary. *Development* **127,** 3671–3681.

Solloway, M. J., Dudley, A. T., Bikoff, E. K., Lyons, K. M., Hogan, B. L., and Robertson, E. J. (1998). Mice lacking Bmp6 function. *Dev. Genet.* **22,** 321–339.

Sordino, P., van der Hoeven, F., and Duboule, D. (1995). Hox gene expression in teleost fins and the origin of vertebrate digits. *Nature (London)* **375,** 678–681.

Soula, C., Danesin, C., Kan, P., Grob, M., Poncet, C., and Cochard, P. (2001). Distinct sites of origin of oligodendrocytes and somatic motoneurons in the chick spinal cord: Oligodendrocytes arise from Nkx2.2-expressing progenitors by a Shh-dependent mechanism. *Development* **128,** 1369–1379.

Spinella-Jaegle, S., Rawadi, G., Kawai, S., Gallea, S., Faucheu, C., *et al.* (2001). Sonic hedgehog increases the commitment of pluripotent mesenchymal cells into the osteoblastic lineage and abolishes adipocytic differentiation. *J. Cell Sci.* **114,** 2085–2094.

Stark, D. R., Gates, P. B., Brockes, J. P., and Ferretti, P. (1998). Hedgehog family member is expressed throughout regenerating and developing limbs. *Dev. Dyn.* **212,** 352–363.

Stein, U., Eder, C., Karsten, U., Haensch, W., Walther, W., and Schlag, P. M. (1999). GLI gene expression in bone and soft tissue sarcomas of adult patients correlates with tumor grade. *Cancer Res.* **59,** 1890–1895.

Stenkamp, D., Frey, R., Prabhudesai, S., and Raymond, P. (2000). Function for Hedgehog genes in zebrafish retinal development. *Dev. Biol.* **220,** 238–252.

Stenn, K. S., and Paus, R. (2001). Controls of hair follicle cycling. *Physiol. Rev.* **81,** 449–494.

Stern, H. M., Brown, A. M., and Hauschka, S. D. (1995). Myogenesis in paraxial mesoderm: Preferential induction by dorsal neural tube and by cells expressing Wnt-1. *Development* **121,** 3675–3686.

Stickens, D., Clines, G. A., Burbee, D., Ramos, P., Thomas, S., *et al.* (1996). The EXT2 multiple exotoses gene defines a family of putative tumour suppressor genes. *Nat. Genet.* **14,** 25–32.

Stickens, D., Brown, D., and Evans, G. A. (2000). EXT genes are differentially expressed in bone and cartilage during mouse embryogenesis. *Dev. Dyn.* **218,** 452–464.

St-Jacques, B., Dassule, H. R., Karavanova, I., Botchkarev, V. A., Li, J., *et al.* (1998). Sonic hedgehog signaling is essential for hair development. *Curr. Biol.* **8,** 1058–1068.

St-Jacques, B., Hammerschmidt, M., and McMahon, A. P. (1999). Indian hedgehog signaling regulates proliferation and differentiation of chondrocytes and is essential for bone formation. *Genes Dev.* **13,** 2072–2086.

Stocker, K. M., and Carlson, B. M. (1990). Hensen's node, but not other biological signallers, can induce supernumerary digits in the developing chick limb bud. *Roux Arch. Dev. Biol.* **198,** 371–381.

Stratford, T., Horton, C., and Maden, M. (1996). Retinoic acid is required for the initiation of outgrowth in the chick limb bud. *Curr. Biol.* **6,** 1124–1133.

Strigini, M., and Cohen, S. (1997). A hedgehog activity gradient contributes to ap axial patterning of the *Drosophila* wing. *Development* **124,** 4697–4705.

Struhl, G., Barbash, D., and Lawrence, P. (1997a). Hedgehog organizes the pattern and polarity of epidermal-cells in the *Drosophila* abdomen. *Development* **124,** 2143–2154.

Struhl, G., Barbash, D., and Lawrence, P. (1997b). Hedgehog acts by distinct gradient and signal relay mechanisms to organize cell-type and cell polarity in the *Drosophila* abdomen. *Development* **124,** 2155–2165.

Strutt, D., and Mlodzik, M. (1997). Hedgehog is an indirect regulator of morphogenetic furrow progression in the *Drosophila* eye disc. *Development* **124,** 3233–3240.

Sukegawa, A., Narita, T., Kameda, T., Saitoh, K., Nohno, T., *et al.* (2000). The concentric structure of the developing gut is regulated by Sonic hedgehog derived from endodermal epithelium. *Development* **127,** 1971–1980.

Summerbell, D. (1974). A quantitative analysis of the effect of excision of the AER from the chick limb-bud. *J. Embryol. Exp. Morphol.* **32,** 651–660.

Summerbell, D. (1983). The effect of local application of retinoic acid to the anterior margin of the developing chick limb. *J. Embryol. Exp. Morphol.* **78,** 269–289.

Sun, T., Echelard, Y., Lu, R., Yuk, D., Kaing, S., *et al.* (2001). Olig bHLH proteins interact with homeodomain proteins to regulate cell fate acquisition in progenitors of the ventral neural tube. *Curr. Biol.* **11,** 1413–1420.

Sun, X., Lewandoski, M., Meyers, E. N., Liu, Y. H., Maxson, R. E., Jr., and Martin, G. R. (2000). Conditional inactivation of Fgf4 reveals complexity of signalling during limb bud development. *Nat. Genet.* **25,** 83–86.

Sussel, L., Marin, O., Kimura, S., and Rubenstein, J. L. (1999). Loss of Nkx2.1 homeobox gene function results in a ventral to dorsal molecular respecification within the basal telencephalon: Evidence for a transformation of the pallidum into the striatum. *Development* **126,** 3359–3370.

Sussman, C. R., Dyer, K. L., Marchionni, M., and Miller, R. H. (2000). Local control of oligodendrocyte development in isolated dorsal mouse spinal cord. *J. Neurosci. Res.* **59,** 413–420.

Suzuki, H. R., Sakamoto, H., Yoshida, T., Sugimura, T., Terada, M., and Solursh, M. (1992). Localization of Hstl transcripts to the apical ectodermal ridge in the mouse embryo. *Dev. Biol.* **150,** 219–222.

Suzuki, T., and Saigo, K. (2000). Related transcriptional regulation of *atonal* required for *Drosophila* larval eye development by concerted action of eyes absent, sine oculis and hedgehog signaling independent of fused kinase and cubitus interruptus. *Development* **127,** 1531–1540.

Tabata, T., and Kornberg, T. (1994). Hedgehog is a signalling protein with a key role in patterning *drosophila* imaginal discs. *Cell* **76,** 89–102.

Tabata, T., Eaton, S., and Kornberg, T. B. (1992). The *Drosophila hedgehog* gene is expressed specifically in posterior compartment cells and is a target of *engrailed* regulation. *Genes Dev.* **6,** 2635–2645.

Tajbakhsh, S., Borello, U., Vivarelli, E., Kelly, R., Papkoff, J., *et al.* (1998). Differential activation of Myf5 and MyoD by different Wnts in explants of mouse paraxial mesoderm and the later activation of myogenesis in the absence of Myf5. *Development* **125,** 4155–4162.

Takabatake, T., Ogawa, M., Takahashi, T., Mizuno, M., Okamoto, M., and Takeshima, K. (1997). *Hedgehog* and *patched* gene expression in adult ocular tissues. *FEBS Lett.* **410,** 485–489.

Takahashi, M., Tamura, K., Buescher, D., Masuya, H., Yonei-Tamura, S., *et al.* (1998). The role of Alx-4 in the establishment of anterioposterior polarity during vertebrate limb development. *Development* **125,** 4417–4425.

Takashima, S., and Murakami, R. (2001). Regulation of pattern formation in the *Drosophila* hindgut by wg, hh, dpp, and en. *Mech. Dev.* **101,** 79–90.

Takata, Y., Webster, N. J., and Olefsky, J. M. (1992). Intracellular signaling by a mutant human insulin receptor lacking the carboxyl-terminal tyrosine autophosphorylation sites. *J. Biol. Chem.* **267,** 9065–9070.

Talbot, W. S., Trevarrow, B., Halpern, M. E., Melby, A. E., Farr, G., and *et al.* (1995). A homeobox gene essential for zebrafish notochord development. *Nature (London)* **378,** 150–157.

Tanabe, Y., William, C., and Jessell, T. M. (1998). Specification of motor neuron identity by the MNR2 homeodomain protein. *Cell* **95,** 67–80.

Tanaka, M., Cohn, M. J., Ashby, P., Davey, M., Martin, P., and Tickle, C. (2000). Distribution of polarizing activity and potential for limb formation in mouse and chick embryos and possible relationships to polydactyly. *Development* **127,** 4011–4021.

Tashiro, S., Michiue, T., Higashijima, S., Zenno, S., Ishimaru, S., *et al.* (1993). Structure and expression of hedgehog, a *Drosophila* segment-polarity gene required for cell–cell communication. *Gene* **124**, 183–189.

Taylor, G., Lehrer, M. S., Jensen, P. J., Sun, T. T., and Lavker, R. M. (2000). Involvement of follicular stem cells in forming not only the follicle but also the epidermis. *Cell* **102**, 451–461.

Teillot, M., and Le Douarin, N. M. (1983). Consequences of neural tube and notochord excision on the development of the peripheral nervous system in the chick embryo. *Dev. Biol.* **98**, 192–211.

Teillet, M., Watanabe, Y., Jeffs, P., Duprez, D., Lapointe, F., and Le Douarin, N. (1998a). Sonic hedgehog is required for survival of both myogenic and chondrogenic somitic lineages. *Development* **125**, 2019–2030.

Teillet, M., Lapointe, F., and Le Douarin, N. M. (1998b). The relationships between notochord and floor plate in vertebrate development revisited. *Proc. Natl. Acad. Sci. USA* **95**, 11733–11738.

Teillet, M., Ziller, C., and Le Douarin, N. M. (1999). Quail–chick chimeras. *Methods Mol. Biol.* **97**, 305–318.

Tekki-Kessaris, N., Woodruff, R., Hall, A. C., Gaffield, W., Kimura, S., *et al.* (2001). Hedgehog-dependent oligodendrocyte lineage specification in the telencephalon. *Development* **128**, 2545–2554.

ten Berge, D., Brouwer, A., Korving, J., Reijnen, M. J., van Raaij, E. J., *et al.* (2001). Prx1 and Prx2 are upstream regulators of sonic hedgehog and control cell proliferation during mandibular arch morphogenesis. *Development* **128**, 2929–2938.

Testaz, S., Jarov, A., Williams, K. P., Ling, L. E., Koteliansky, V. E., *et al.* (2001). Sonic hedgehog restricts adhesion and migration of neural crest cells independently of the Patched–Smoothened–Gli signaling pathway. *Proc. Natl. Acad. Sci. USA* **98**, 12521–12526.

Theil, T., Kaesler, S., Grotewold, L., Bose, J., and Ruther, U. (1999). Gli genes and limb development. *Cell Tissue Res.* **296**, 75–83.

Thesleff, I., and Jernvall, J. (1997). The enamel knot: A putative signaling center regulating tooth development. *Cold Spring Harbor Symp. Quant. Biol.* **62**, 257–267.

Thesleff, I., and Sharpe, P. (1997). Signalling networks regulating dental development. *Mech. Dev.* **67**, 111–123.

Thesleff, I., Vaahtokari, A., and Partanen, A. M. (1995). Regulation of organogenesis: Common molecular mechanisms regulating the development of teeth and other organs. *Int. J. Dev. Biol.* **39**, 35–50.

Thomas, M. K., Rastalsky, N., Lee, J. H., and Habener, J. F. (2000). Hedgehog signaling regulation of insulin production by pancreatic *β*-cells. *Diabetes* **49**, 2039–2047.

Thomas, M. K., Lee, J. H., Rastalsky, N., and Habener, J. F. (2001). Hedgehog signaling regulation of homeodomain protein islet duodenum homeobox-1 expression in pancreatic beta-cells. *Endocrinology* **142**, 1033–1040.

Tickle, C. (1981a). Limb regeneration. *Am. Sci.* **69**, 639–646.

Tickle, C. (1981b). The number of polarizing region cells required to specify additional digits in the developing chick wing. *Nature (London)* **289**, 295–298.

Tickle, C., Summerbell, D., and Wolpert, L. (1975). Positional signaling and specification of digits in the chick limb morphogenesis. *Nature (London)* **254**, 199–202.

Ting-Berreth, S. A., and Chuong, C. M. (1996). Sonic Hedgehog in feather morphogenesis: Induction of mesenchymal condensation and association with cell death. *Dev. Dyn.* **207**, 157–170.

Tint, G. S., Irons, M., Elias, E. R., Batta, A. K., Frieden, R., *et al.* (1994). Defective cholesterol biosynthesis associated with the Smith-Lemli-Opitz syndrome. *N. Engl. J. Med.* **330**, 107–113.

Tomlinson, A., and Ready, D. (1987). Cell fate in the *Drosophila* ommatidium. *Dev. Biol.* **120**, 366–376.

Torok, M. A., Gardiner, D. M., Izpisua Belmonte, J. C., and Bryant, S. V. (1999). Sonic hedgehog (shh) expression in developing and regenerating axolotl limbs. *J. Exp. Zool.* **284**, 197–206.

Traiffort, E., Charytoniuk, D., Watroba, L., Faure, H., Sales, N., and Ruat, M. (1999). Discrete localizations of hedgehog signalling components in the developing and adult rat nervous system. *Eur. J. Neurosci.* **11,** 3199–3214.

Traiffort, E., Moya, K. L., Faure, H., Hassig, R., and Ruat, M. (2001). High expression and anterograde axonal transport of aminoterminal sonic hedgehog in the adult hamster brain. *Eur. J. Neurosci.* **14,** 839–850.

Treier, M., O'Connell, S., Gleiberman, A., Price, J., Szeto, D. P., *et al.* (2001). Hedgehog signaling is required for pituitary gland development. *Development* **128,** 377–386.

Trousse, F., Marti, E., Gruss, P., Torres, M., and Bovolenta, P. (2001). Control of retinal ganglion cell axon growth: A new role for Sonic hedgehog. *Development* **128,** 3927–3936.

Tsukui, T., Capdevila, J., Tamura, K., Ruiz-Lozano, P., Rodriguez-Esteban, C., *et al.* (1999). Multiple left–right asymmetry defects in Shh$^{-/-}$ mutant mice unveil a convergence of the shh and retinoic acid pathways in the control of Lefty-1. *Proc. Natl. Acad. Sci. USA* **96,** 11376–11381.

Tsukurov, O., Boehmer, A., Flynn, J., Nicolai, J. P., Hamel, B. C., *et al.* (1994). A complex bilateral polysyndactyly disease locus maps to chromosome 7q36. *Nat. Genet.* **6,** 282–286.

Unden, A. B., Zaphiropoulos, P. G., Bruce, K., Toftgard, R., and Stahle-Backdahl, M. (1997). Human patched (PTCH) mRNA is overexpressed consistently in tumor cells of both familial and sporadic basal cell carcinoma. *Cancer Res.* **57,** 2336–2340.

Valentini, R. P., Brookhiser, W. T., Park, J., Yang, T., Briggs, J., *et al.* (1997). Post-translational processing and renal expression of mouse Indian hedgehog. *J. Biol. Chem.* **272,** 8466–8473.

van den Akker, E., Reijen, M., Korving, J., Brouwer, A., Meijlink, F., and Deschamps, J. (1999). Targeted inactivation of Hoxb8 affects survival of a spinal ganglion and causes aberrant limb reflexes. *Mech. Dev.* **89,** 103–114.

van Den Brink, G. R., de Santa Barbara, P., and Roberts, D. J. (2001). Epithelial cell differentiation—a Mather of choice. *Science* **294,** 2115–2116.

Van Hui, W., Wuyts, W., Hendricks, J., Speleman, F., Wauters, J., *et al.* (1998). Identification of a third EXT-like gene (EXTL3) belonging to the EXT gene family. *Genomics* **47,** 230–237.

Varga, Z., Amores, A., Lewis, K. E., Yan, Y. L., Postlethwait, J. H., *et al.* (2001). Zebrafish smoothened functions in ventral neural tube specification and axon tract formation. *Development* **128,** 3497–3509.

Vervoort, M., Crozatier, M., Valle, D., and Vincent, A. (1999). The COE transcription factor Collier is a mediator of short-range Hedgehog-induced patterning of the *Drosophila* wing. *Curr. Biol.* **9,** 632–639.

Vied, C., and Horabin, J. I. (2001). The sex determination master switch, Sex-lethal, responds to Hedgehog signaling in the *Drosophila* germline. *Development* **128,** 2649–2660.

Vogel, A., and Tickle, C. (1993). FGF-4 maintains polarizing activity of posterior limb bud cells in vivo and in vitro. *Development* **119,** 199–206.

Vogel, A., Rodriguez, C., Warnken, W., and Izpisua Belmonte, J. C. (1995). Dorsal cell fate specified by chick Lmx1 during vertebrate limb development. *Nature (London)* **378,** 716–720.

Vogel, A., Rodriguez, C., and Izpisua Belmonte, J. C. (1996). Involvement of Fgf-8 in initiation, outgrowth and patterning of the vertebrate limb. *Development* **122,** 1737–1750.

Vorechovsky, I., Tingby, O., Hartman, M., Stromberg, B., Nister, M., *et al.* (1997). Somatic mutations in the human homologue of *Drosophila* patched in primitive neuroectodermal tumours. *Oncogene* **15,** 361–366.

Vortkamp, A., Gessler, M., and Grzeschik, K. H. (1991). GLI3 zinc-finger gene interrupted by translocations in Greig syndrome families. *Nature (London)* **352,** 539–540.

Vortkamp, A., Lee, K., Lanske, B., Segre, G. V., Kronenberg, H. M., and Tabin, C. J. (1996). Regulation of rate of cartilage differentiation by Indian hedgehog and PTH-related protein. *Science* **273,** 613–622.

Vortkamp, A., Pathi, S., Peretti, G. M., Caruso, E. M., Zaleske, D. J., and Tabin, C. J. (1998). Recapitulation of signals regulating embryonic bone formation during postnatal growth and in fracture repair. *Mech. Dev.* **71,** 65–76.

Wada, T., Kagawa, T., Ivanova, A., Zalc, B., Shirasaki, R., *et al.* (2000). Dorsal spinal cord inhibits oligodendrocyte development. *Dev. Biol.* **227**, 42–55.

Wagner, M., Thaller, C., Jessell, T. M., and Eichele, G. (1990). Polarizing activity and retinoid synthesis in the floor plate of the neural tube. *Nature (London)* **3245**, 819–822.

Wallace, V. A. (1999). Purkinje-cell-derived Sonic hedgehog regulates granule neuron precursor cell proliferation in the developing mouse cerebellum. *Curr. Biol.* **9**, 445–448.

Wallace, V. A., and Raff, M. (1999). A role for Sonic hedgehog in axon-to-astrocyte signalling in the rodent optic nerve. *Development* **126**, 2901–2909.

Wallin, J., Wilting, J., Koseki, H., Fritsch, R., Christ, B., and Balling, R. (1994). The role of Pax-1 in axial skeleton development. *Development* **120**, 1109–1121.

Wanek, N., Gardiner, D. M., Muneoka, K., and Bryant, S. V. (1991). Conversion by retinoic acid of anterior cells into ZPA cells in the chick wing bud. *Nature (London)* **350**, 81–83.

Wang, L. C., Liu, Z. Y., Gambardella, L., Delacour, A., Shapiro, R., *et al.* (2000). Conditional disruption of hedgehog signaling pathway defines its critical role in hair development and regeneration. *J. Invest. Dermatol.* **114**, 901–908.

Wang, M. Z., Jin, P., Bumcrot, D. A., Marigo, V., McMahon, A. P., *et al.* (1995). Induction of dopaminergic neuron phenotype in the midbrain by Sonic hedgehog protein. *Nat. Med.* **1**, 1184–1188.

Warburton, D., Schwarz, M., Tefft, D., Flores-Delgado, G., Anderson, K. D., and Cardoso, W. V. (2000). The molecular basis of lung morphogenesis. *Mech. Dev.* **92**, 55–81.

Wassif, C. A., Maslen, C., Kachilele-Linjewile, S., Lin, D., Linck, L. M., *et al.* (1998). Mutations in the human sterol Δ^7-reductase gene at 11q12–13 cause Smith-Lemli-Opitz syndrome. *Am. J. Hum. Genet.* **63**, 55–62.

Watanabe, Y., and Nakamura, H. (2000). Control of chick tectum territory along dorsoventral axis by Sonic hedgehog. *Development* **127**, 1131–1140.

Waterham, H. R., Wijburg, F. A., Hennekam, R. C., Vreken, P., Poll-The, B. T., *et al.* (1998). Smith-Lemli-Opitz syndrome is caused by mutations in the 7-dehydrocholesterol reductase gene. *Am. J. Hum. Genet.* **63**, 329–338.

Watt, F. M. (1998). Epidermal stem cells: Markers, patterning and the control of stem cell fate. *Philos. Trans. R. Soc. Lond. B Biol. Sci.* **353**, 831–837.

Weaver, M., Yingling, J. M., Dunn, N. R., Bellusci, S., and Hogan, B. L. (1999). Bmp signaling regulates proximal–distal differentiation of endoderm in mouse lung development. *Development* **126**, 4005–4015.

Wechsler-Reya, R. J., and Scott, M. P. (1999). Control of neuronal precursor proliferation in the cerebellum by Sonic Hedgehog. *Neuron* **22**, 103–114.

Weinberg, E. S., Allende, M. L., Kelly, C. S., Abdelhamid, A., Murakami, T., *et al.* (1996). Developmental regulation of zebrafish *MyoD* in wild type, *no tail* and *spadetail* embryos. *Development* **122**, 271–280.

Weir, E. C., Philbrick, W. M., Amling, M., Neff, L. A., Baron, R., and Broadus, A. E. (1996). Targeted overexpression of parathyroid hormone-related peptide in chondrocytes causes chondrodysplasia and delayed endochondral bone formation. *Proc. Natl. Acad. Sci. USA* **93**, 10240–10245.

Werner, C. A., Dohner, H., Joos, S., Trumper, L. H., Baudis, M., *et al.* (1997). High-level DNA amplifications are common genetic aberrations in B-cell neoplasms. *Am. J. Pathol.* **151**, 335–342.

Wessells, N. K. (1965). Morphology and proliferation during early feather development. *Dev. Biol.* **12**, 131–153.

Wessells, R., Grumbling, G., Donaldson, T., Wang, S., and Simcox, A. (1999). Tissue-specific regulation of vein/EGF receptor signaling in *Drosophila*. *Dev. Biol.* **216**, 243–259.

Wessels, N. K. (1970). Mammalian lung development: interactions in formation and morphogenesis of tracheal buds. *J. Exp. Zool.* **175**, 455–466.

Wild, A., Kalff-Suske, M., Vortkamp, A., Bornholdt, D., Konig, R., and Grzeschik, K. H. (1997). Point mutations in human GL13 cause Greig syndrome. *Hum. Mol. Genet.* **6**, 1979–1984.

Williams, Z., Tse, V., Hou, L., Xu, L., and Silverberg, G. D. (2000). Sonic hedgehog promotes proliferation and tyrosine hydroxylase induction of postnatal sympathetic cells in vitro. *Neuroreport* **11**, 3315–3319.

Wise, C. A., Clines, G. A., Massa, H., Trask, B. J., and Lovett, M. (1997). Identification and localization of the gene for EXTL, third member of the multiple exostoses gene family. *Genome Res.* **7**, 10–16.

Wolpert, L. (1969). Positional information and the spatial pattern of cellular differentiation. *J. Theor. Biol.* **25**, 1–47.

Wolpert, L. (1994). The evolutionary origin of development: Cycles, patterning, privilege and continuity. *Dev. Suppl.* 79–84.

Wolter, M., Reifenberger, J., Sommer, C., Ruzicka, T., and Reifenberger, G. (1997). Mutations in the human homologue of the *Drosophila* segment polarity gene patched (PTCH) in sporadic basal cell carcinomas of the skin and primitive neuroectodermal tumors of the central nervous system. *Cancer Res.* **57**, 2581–2585.

Wu, Q., Zhang, Y., and Chen, Q. (2001). Indian hedgehog is an essential component of mechanotransduction complex to stimulate chondrocyte proliferation. *J. Biol. Chem.* **276**, 35290–35296.

Wuyts, W., Van Hul, W., Hendrickx, J., Speleman, F., Wauters, J., *et al.* (1997). Identification and characterization of a novel member of the EXT gene family, EXTL2. *Eur. J. Hum. Genet.* **5**, 382–389.

Xie, J., Johnson, R. L., Zhang, X., Bare, J. W., Waldman, F. M., *et al.* (1997). Mutations of the PATCHED gene in several types of sporadic extracutaneous tumors. *Cancer Res.* **57**, 2369–2372.

Xie, J., Murone, M., Luoh, S. M., Ryan, A., Gu, O., *et al.* (1998). Activating Smoothened mutations in sporadic basal-cell carcinoma. *Nature (London)* **39**, 90–92.

Yang, Y., and Niswander, L. (1995). Interaction between the signaling molecules WNT7a and SHH during vertebrate limb development: Dorsal signals regulate anterior–posterior patterning. *Cell* **80**, 939–947.

Yang, Y., Drossopoulou, G., Chuang, P. T., Duprez, D., Marti, E., *et al.* (1997). Relationship between dose, distance and time in Sonic Hedgehog mediated regulation of anterioposterior polarity in the chick limb. *Development* **124**, 4393–4404.

Yasugi, S. (1993). The role of epithelial–mesenchymal interactions in differentiation of epithelium of vertebrate digestive organs. *Dev. Growth Differ.* **35**, 1–9.

Ye, W., Shimamura, K., Rubenstein, J. L., Hynes, M. A., and Rosenthal, A. (1998). FGF and Shh signals control dopaminergic and serotonergic cell fate in the anterior neural plate. *Cell* **93**, 755–766.

Yelon, D., Ticho, B. S., Halpern, M. E., Ruvinsky, I., Ho, R. K., *et al.* (2000). The bHLH transcription factor hand2 plays parallel roles in zebrafish heart and pectoral fin development. *Development* **127**, 2573–2582.

Yokouchi, Y., Sakiyama, J., and Kuroiwa, A. (1995). Coordinated expression of Abd-B subfamily genes of the HoxA cluster in the developing digestive tract of chick embryo. *Dev. Biol.* **169**, 76–89.

Yokouchi, Y., Vogan, K. J., Pearse, R. V., II, and Tabin, C. J. (1999). Antagonistic signaling by Caronte, a novel Cerberus-related gene, establishes left–right asymmetric gene expression. *Cell* **98**, 573–583.

Yonei, S., Tamura, K., Ohsugi, K., and Ide, H. (1995). MRC-5 cells induce the AER prior to the duplicated pattern formation in chick limb bud. *Dev. Biol.* **170**, 542–552.

Yoshida, E., Noshiro, M., Kawamoto, T., Tsutsumi, S., Kuruta, Y., and Kato, Y. (2001). Direct inhibition of Indian hedgehog expression by parathyroid hormone (PTH)/PTH-related peptide and up-regulation by retinoic acid in growth plate chondrocyte cultures. *Exp. Cell Res.* **265**, 64–72.

Zecca, M., Basler, K., and Struhl, G. (1995). Sequential organizing activities of *engrailed, hedgehog* and *decapentaplegic* in the *Drosophila* wing. *Development* **121**, 2265–2278.

Zeng, X., Goetz, J. A., Suber, L. M., Scott, W. J., Jr., Schreiner, C. M., and Robbins, D. J. (2001). A freely diffusible form of Sonic hedgehog mediates long-range signaling. *Nature (London)* **411,** 716–720.

Zguricas, J., Heus, H., Morales-Peralta, E., Breedveld, G., Kuyt, B., *et al.* (1999). Clinical and genetic studies on 12 preaxial polydactyly families and refinement of the localization of the gene responsible to a 1.9 cM region on chromosome 7q36. *J. Med. Genet.* **36,** 32–40.

Zhang, J., Rosenthal, A., de Sauvage, F. J., and Shivdasani, R. A. (2001). Downregulation of Hedgehog signaling is required for organogenesis of the small intestine in *Xenopus. Dev. Biol.* **229,** 188–202.

Zhang, X., and Yang, X. (2001). Temporal and spatial effects of Sonic hedgehog signaling in chick eye morphogenesis. *Dev. Biol.* **233,** 271–290.

Zhang, X., Ramalho-Santos, M., and McMahon, A. (2001). Smoothened mutants reveal redundant roles for Shh and Ihh signaling including regulation of L/R asymmetry by the mouse node. *Cell* **105,** 781–792.

Zhang, Y., and Kalderon, D. (2000). Regulation of cell proliferation and patterning in *Drosophila* oogenesis by Hedgehog signaling. *Development* **127,** 2165–2176.

Zhang, Y., and Kalderon, D. (2001). Hedgehog acts as a somatic stem cell factor in the *Drosophila* ovary. *Nature (London)* **410,** 599–604.

Zhang, Y., Zhao, X., Hu, Y., St Amand, T., Zhang, M., *et al.* (1999). Msx1 is required for the induction of Patched by Sonic hedgehog in the mammalian tooth germ. *Dev. Dyn.* **215,** 45–53.

Zhang, Y., Zhang, Z., Zhao, X., Yu, X., Hu, Y., *et al.* (2000). A new function of BMP4: Dual role for BMP4 in regulation of Sonic hedgehog expression in the mouse tooth germ. *Development* **127,** 1431–1443.

Zhao, X., Zhang, Z., Song, Y., Zhang, X., Zhang, Y., *et al.* (2000). Transgenically ectopic expression of Bmp4 to the Msx1 mutant dental mesenchyme restores downstream gene expression but represses Shh and Bmp2 in the enamel knot of wild type tooth germ. *Mech. Dev.* **99,** 29–38.

Zhou, Q., Choi, G., and Anderson, D. J. (2001). The bHLH transcription factor Olig2 promotes oligodendrocyte differentiation in collaboration with Nkx2.2. *Neuron* **31,** 791–807.

Zhu, L., Marvin, M. J., Gardiner, A., Lassar, A. B., Mercola, M., *et al.* (1999). Cerberus regulates left–right asymmetry of the embryonic head and heart. *Curr. Biol.* **9,** 931–938.

Ziegle, A., Leffell, D. J., Kunala, S., Sharma, H. W., Gailani, M., *et al.* (1993). Mutation hotspots due to sunlight in the p53 gene of nonmelanoma skin cancers. *Proc. Natl. Acad. Sci. USA* **90,** 4216–4220.

Zou, H., Wieser, R., Massague, J., and Niswander, L. (1997). Distinct roles of type I bone morphogenetic protein receptors in the formation and differentiation of cartilage. *Genes Dev.* **11,** 2191–2203.

Zuniga, A., Haramis, A. P., McMahon, A. P., and Zeller, R. (1999). Signal relay by BMP antagonism controls the SHH/FGF4 feedback loop in vertebrate limb buds. *Nature (London)* **401,** 598–602.

2

Genomic Imprinting: Could the Chromatin Structure Be the Driving Force?

Andras Paldi
Institut Jacques Monod, CNRS, École Pratique des Hautes Études
75005 Paris, France

Genomic imprinting is traditionally defined as an epigenetic process leading to parental origin-dependent monoallelic expression of some genes. The current paradigm considers this unusual expression mode as the biological *raison d être* of imprinting. The present chapter proposes a critical review of our ideas about genomic imprinting in light of more recent investigatory progress. Many observations are difficult to explain on the basis of the current paradigm. Studies of allelic expression of many imprinted genes and other characteristics of chromatin domains containing clustered imprinted genes, such as replication and chromatin structure, revealed an unexpectedly complex situation that challenged the role of genomic imprinting as a mechanism of transcriptional regulation. The emerging picture is that parental imprinting is a feature of large chromatin domains with their own domain-wide characteristics. The primary biological function of imprinting may reside in the differential chromatin structure of the parental chromosomal regions and not in the monoallelic expression of some of the genes contained within them. © 2003, Elsevier Science (USA).

I. Introduction

The intriguing phenomenon of genomic imprinting attracts more and more attention in the scientific community of geneticists. Reports on the discovery of new imprinted genes or new features of imprinted genes and reviews summarizing the

Current Topics in Developmental Biology, Vol. 53

most recent progress have become a regular topic in the most prestigious scientific journals. In spite of the tremendous amount of experimental data accumulated, the fundamental questions about imprinting remain unanswered. We still do not know the biological significance of imprinting, how and why it has evolved, and the underlying mechanism. Why this relative lack of success? Our knowledge is certainly insufficient for a full understanding of the phenomenon at the present stage. In addition, we cannot exclude the possibility that our way of approaching the problem is not optimal and that the conceptual framework in current use may also need to be amended. Such a conceptual framework, also called a paradigm, is a set of ideas based on initial observations and generally accepted by the scientific community. The paradigm determines the way a problem is addressed and how the observations and experimental data are interpreted (Kuhn, 1996). Observations that do not fit the paradigm are considered exceptions or just put aside. A new conceptual framework is needed if the number of uninterpretable observations increases to an untolerably high level. Many examples in the history of science illustrate how some observations became easily interpretable once the way of asking the question has changed. One of the best examples in genetics is the discovery and rediscovery of Mendel's laws.

The present article proposes a critical review of our ideas on genomic imprinting in light of the most recent progress. The current paradigm of genomic imprinting was defined more than 10 years ago, before the first imprinted gene was discovered. Our knowledge has progressed considerably since then. A large number of unexpected new observations have now accumulated that cannot be explained on the basis of the generally accepted paradigm, and are hence most frequently overlooked. The inability of the current paradigm to successfully explain the majority of observations may stem from the erroneous definition of imprinting. The aim of this review is to understand how the current definition was formulated and to highlight major contradictions, existing between the current paradigm and more recent experimental observations, that should initiate a critical reexamination of our current conceptual framework and should help to formulate alternative hypotheses. One possible alternative is that the chromatin structural differences between homologs per se may have a biological function selected and maintained during evolution.

II. How Imprinting Was Discovered

To better understand how the current conceptual framework of the parental imprinting phenomenon has developed, it is interesting to undertake a brief historical overview.

The inequality of some reciprocal crosses, and the existence of mutations that caused different phenotypes depending on their parental origin, were known for a long time but they received little attention at the molecular level. Two types of observations made almost simultaneously helped to reach the "critical mass" necessary to recognize the existence of parental origin-dependent heritable effects as a separate group of phenomena. First, pronuclear transplantation studies performed on mouse zygotes showed the absolute necessity for a biparental genetic contribution for normal development (McGrath and Solter, 1984; Surani *et al.,* 1984). The developmental abnormalities of gynogenetic and androgenetic embryos suggested that the paternal and maternal genomes play complementary roles during fetal development. Second, systematic genetic studies of mice with Robertsonian translocations confirmed that some, but not all, chromosomal regions must be inherited from both parents for normal embryonal development (Cattanach and Kirk, 1985). These latter pioneering studies led to the construction of a low-resolution chromosomal imprinting map of the mouse genome. This map is frequently used as a reference for the identification of new imprinted genes.

The conclusion drawn from these experiments was clear: the paternal and maternal genomes are not functionally equivalent and the presence of both parental genetic contributions is an absolute requirement for normal development. On the basis of this conclusion the following well-known hypothesis was proposed: some genes that play a key role during mammalian development might be expressed monoallelically. For some of these gense only the paternal, for others only the maternal, allele might be active. The active or inactive nature of an allele was supposed to be determined by an epigenetic mark acquired in the germ line of the parents and transmitted to the zygote. The imprint is supposed to be maintained on the allele in somatic cells and to determine its activity, but it is erased in the germ line, where a new imprint is subsequently established according to the sex of the individual. This was clearly an "ad hoc" hypothesis, because its aim was only to explain the experimental data. However, the hypothesis was quickly extended and lost its ad hoc nature as it became the theoretical framework of the newly emerging field of "genomic," "parental," or "gametic" imprinting. The new theory found strong support in the subsequent discovery of imprinted genes.

Why did the scientific community believe the functional nonequivalence of the parental genomes needed a special theoretical explanation? The phenomenon was identified on the basis of phenotypic effects of genetic events (full parental genome duplication or chromosomal translocations). These phenotypes do not follow the mendelian rules of inheritance. Because we live in the era of molecular biology, a molecular explanation was immediately proposed: the genes responsible for these phenotypes are expressed monoallelically in a parental origin-dependent way. It is assumed that both alleles of a gene are usually expressed in all cells. Although the generality of this rule has never been carefully verified at the molecular level, it is generally considered self-evident. From this viewpoint, imprinted genes appeared as an exception to the rule. Therefore, genes in the mammalian genome were classified in two clearly defined categories. The overhelming majority of the genes are thought to be nonimprinted and expressed biallelically. Imprinted genes expressed monoallelically are supposed to represent only a minority. Imprinted genes were thought to have special function during embryonic development as judged on the basis of dramatic phenotypes observed. A special molecular mechanism specific for these genes was assumed to regulate their expression. This mechanism is called genomic, parental, or gametic imprinting. The low number of naturally occurring mutations with parental origin-dependent phenotype was explained by the exception-al nature of the imprinted genes. Identification of these genes, elucidation of special biological functions of imprinting, and description of the specific molec-ular mechanism(s) involved are defined by the paradigm as the final goal of the research. (General aspects of imprinting has been widely discussed by various authors in a number of reviews. Here I propose a selection of them: Solter, 1988; Monk and Surani, 1990; Efstratiadis, 1994; Barlow, 1995; Tilghman, 1999; Pfeifer, 2000; Pardo-Manuel de Villena *et al.*, 2000; Reik and Walter, 2001.)

It is worth mentioning that many other exceptions to the mendelian rules were known for a long time before the advent of molecular genetics. Genetic phenom-ena such as penetrance, expressivity, and epistasis were successfully integrated in the classic genetic theory without considering them a violation of mendelian rules, in spite of the fact that no satisfactory molecular explanation exists for them. By contrast, imprinting was discovered after the revolution of molecular genetics that introduced new experimental methodology. Although the initial hy-pothesis emphasized the nonmendelian inheritance of phenotypes associated with imprinted genes, most attention was focused on the study of molecular mecha-nisms of gene expression. This subtle shift of focus was initially a consequence of the molecular experimental approach, but rapidly resulted in a modification of the way the phenomenon of imprinting is considered.

III. Identification of Imprinted Genes

As indicated above, the phenomenon of imprinting was identified on the basis of the phenotype of monoparental embryos or mice with monoparental inheritance of some chromosomal regions. In the same way, the parental origin-dependent phenotype of heterozygous mutants helped identify the first two imprinted genes: *Igf2r* and *Igf2*.

The imprinting of *Igf2r* was discovered by a candidate gene approach (Barlow *et al.*, 1991). This gene was mapped to the region of Chr17 deleted in T^{hp} mutant mice. T^{hp} is one of the rare natural mutations characterized by a parental origin-dependent phenotype (Johnson, 1974). Heterozygous mice die if they inherit the deletion from the mother but are viable if the deletion is transmitted by the father. In agreement with the inheritance of the phenotype, the *Igf2r* gene was found to be expressed from the maternal allele (Barlow *et al.*, 1991).

The imprinting of *Igf2* was recognized when the parental origin-dependent phe-notype of the heterozygous deletion of the gene created by homologous recombi-nation was observed (DeChiara *et al.*, 1991). Heterozygous mutants grew smaller when they inherited the mutation from their fathers. The phenotypes of homozygous mutant animals were identical. However, when heterozygous small females were mated to wild-type males, the offspring were of normal size. Therefore, the phenotype followed precisely the nonmendelian inheritance predicted for an imprinted gene. Indeed, no mRNA derived from the paternal allele could be found in the tissues of these mice. Interestingly, one more imprinted gene, *Mash2*, was also discovered in the same fortuitous way on the basis of the phenotype observed in mice deleted for this gene by homologous recombination (Guillemot *et al.*, 1995). The third gene to be identified was *H19*. This gene is one of the favorite genes of the imprinting comunity (Bartolomei *et al.*, 1991).

H19 mRNA is abundant in the postimplantation embryo. Perhaps this is why it was independently cloned by several laboratories (Davis *et al.*, 1987; Pachnis *et al.*, 1988; Poirier *et al.*, 1991) using independent experimental approaches. Despite considerable efforts, its function is still unknown. It is not clear from the initial publication why the imprinted status of the gene had been tested (Bartolomei *et al.*, 1991). In any case, only mRNAs transcribed from the maternal allele were found. *H19* is the first example of imprinted genes identified solely on the basis of their monoallelic expression and without any associated phenotype.

IV. Ambiguous Definition

The discovery of the first few imprinted genes had confirmed the initial hypothesis and led to the widely accepted consensus that a gene is considered imprinted if it is expressed monoallelically depending on the parental origin of the allele. Monoallelic expression became the only criterion in the identification of imprinted genes. An associated parental origin-dependent phenotype is no longer needed for a gene to be considered imprinted. All current definitions place monoallelic expression at the center, as the final biological significance of imprinting. Some examples: "An epigenetic imprinting mechanism . . . restricts expression of a subset of mammalian genes to one parental chromosome" (Sleutels *et al.*, 2000); "a subset of genes in the mammalian genome are imprinted—they are expressed from only one of the two alleles, maternal or paternal" (Ben-Porath and Cedar, 2000); "genomic imprinting refers to the differential expression of the two alleles of a gene that is dependent on their parent of origin" (Schmidt *et al.*, 2000); "to date autosomal imprinting, as defined in this review as the differential expression of the two parental alleles of a gene, has been demonstrated only in eutherian (i.e., placental, nonmarsupial) mammals" (Tilghman, 1999). These examples illustrate well that monoallelic expression is believed to play the central role in the phenomenon of imprinting. The shift in emphasis from the parental origin-dependent phenotype to monoallelic gene expression was the logical consequence of the molecular genetic approach, which focuses almost exclusively on the expression of genes and leaves out the structural and topological aspects of the genome.

Experimental strategies for the identification and study of imprinted genes, such as cDNA libraries prepared by subtraction hybridization of monoparental and wild-type embryonic RNAs (Kanedo-Ishino *et al.*, 1995), differential display of gene expression of reciprocal crosses of well-defined genotypes (Hayashizaki *et al.*, 1994), or techniques that directly compare the expression of parental alleles of individual candidate genes (Bartolomei *et al.*, 1991), are all based on this consensus. More than 50 imprinted genes are known and most of them were identified in this way (a comprehensive database of imprinted genes can be found on the Web: www.mgu.har.mrc.ac.uk/research/imprinted/imprin.html).

In addition to restricting the experimental strategies employed to the study of imprinting, the shift of the definition of imprinting from the phenotype to gene expression seriously narrowed the field of investigations. All other parental origin-dependent genetic phenomena were excluded from the field of genomic imprinting. For example, imprinting was considered a potential cause of many human genetic disorders that do not follow mendelian inheritance, such as myotonic dystrophy or Huntington disease (Reik, 1989). These phenomena are now considered unrelated to imprinting.

The restriction of the definition to monoallelic expression is the source of serious conceptual contradictions. The expression of imprinted genes has turned out to be more complex than initially suspected. Now it is clear that the oversimplified view that imprinted genes are expressed exclusively from one parental allele is not only incorrect, but has also biased investigations and flawed the biological interpretation of the phenomenon.

It was found that imprinted genes are frequently expressed biallelically in normal tissues. *Igf2*, for example, is expressed biallelically in the choroid plexus of the brain (DeChiara *et al.*, 1991). *Kvlqt1* is monoallelically expressed in all tissues except the heart muscle (Lee *et al.*, 1997). Interestingly, it is in the heart muscle that this gene plays an essential biological function. Individuals with homozygous mutations in the gene have a heart disease called "long QT syndrome" (Lee *et al.*, 1997), but heterozygous individuals are normal. For some imprinted genes mono-allelic expression is almost an exception. For example, the candidate gene of the Angelman syndrome, *UB3A*, is biallelically expressed in all tissues of the organism, except some parts of the brain (Rougeulle *et al.*, 1997; Vu and Hoffman, 1997). It was frequently observed that the allelic usage of imprinted genes can change during embryogenesis. The paternal allele of the *H19* gene, for example, is gradually silenced during the first stages of postimplantation development (Svensson *et al.*, 1998). The mouse *Ins2* gene is biallelically expressed in the visceral yolk sac during the first half of gestation, but gradually shifts to monoallelic paternal expression at midgestation (Deltour *et al.*, 1995). In the pancreas, biallelic expression of *Ins2* is maintained at all stages (Deltour *et al.*, 1995; Giddings *et al.*, 1994). Several new genes have been described that display preferential expression of one parental allele. The *Ipl* and *Impt1* genes (Qian *et al.*, 1997) were shown to be expressed predominantly, but not exclusively, from the maternal allele in most tissues examined. These genes were classified as imprinted (Dao *et al.*, 1998; Qian *et al.*, 1997). The human *TSSC4* and *TSSC6* genes were found to display a slight allelic preference in some fetal tissues (Lee *et al.*, 1999). The authors concluded that these two genes "are not substantially imprinted when compared with other genes in this region" (Lee *et al.*, 1999). *ORCTL2* is another human gene in the same chromosomal region (Cooper *et al.*, 1998). It also shows preferential expression of the maternal allele in several fetal tissues. According to the authors this gene is imprinted, "albeit with varying degrees of "leakiness," ... such that one allele (paternal) is expressed at 20–30% the level of the preferentially expressed allele (maternal)." This way of considering imprinted genes differs from that suggested by the definitions cited above. It implies that there might be intermediate levels of imprinting, that is, that the imprinting status of a gene can be relative. The sharp distinction between imprinted and nonimprinted genes is not possible anymore.

In addition, the expression of the same gene can be simultaneously mono- and biallelic in the same tissue at a given stage. The human *IGF2* gene is transcribed biallelically from the most upstream promoter (P1), whereas the three other promoters (P2–P4) function predominantly on the paternal allele (Ekström *et al.*, 1995; Vu and Hoffman, 1994). Interestingly, all four transcripts encode the same protein.

The example of the *Gnas* gene is the most unusual and illustrates perfectly that the problem of gene regulation by imprinting cannot be reduced simply to a question of whether an allele is transcribed or not. Both the mouse *Gnas* locus and human *GNAS1* locus are transcribed from three different promoters (Hayward *et al.*, 1998; Peters *et al.*, 1999). Transcripts from P1, which encode the Nesp protein, are derived from the maternal chromosome only. Transcripts from P2 encode the Xlαs protein and are derived from the paternal chromosome, whereas transcripts from P3 encode the α subunit of the stimulatory G protein and are derived from both alleles. In addition, an antisense transcript that starts upstream of the P2 exon and spans the P1 region is derived from the paternal allele in most, but not all, tissues. It is unclear whether the *Gnas* gene, or rather the *Gnas* locus, must be considered as imprinted or nonimprinted because transcription occurs on both parental chromosomes and biallelic transcripts overlap with the full length of the monoallelic transcripts. This example does not fit the current definition, according to which an imprinted gene is transcribed only from one parental chromosome, because the locus is transcriptionally competent on both chromosomes. One way to interpret the observation is to consider that the imprint is restricted to the promoter of a gene and not to the whole gene or locus, as usually thought. In this case the primary transcript starting from the P1 promoter on the maternal chromosome spans the P2 promoter and includes the full length of the Xlαs transcript normally derived from the maternal chromosome. Nevertheless, splicing between the Xlαs exon and the second *Gnas* exon on the P1 primary transcript has never been observed, indicating that the parental origin determines not only the activity of the promoters but also the way the splicing apparatus functions on the primary transcripts transcribed from those promoters.

In addition to intraindividual variations, interindividual variations of imprinting in human and mouse are also well documented. The human *IGF2R* gene, for example, is biallelically expressed in most tested cases and monoallelic expression occurs only in some individuals (Smrzka *et al.*, 1995; Xu *et al.*, 1993). Therefore, mono- or biallelic expression of a gene may also depend on the genetic background in addition to the parentally transmitted epigenetic marks. Genetic background-dependent polymorphism of allelic expression of several genes has also been shown both in laboratory and wild mice (Jiang *et al.*, 1998; Vrana *et al.*, 1998). This strange feature has not attracted too much attention so far, although it suggests that parental origin-dependent monoallelic expression is a highly variable trait capable of continued evolutionary modification, and also that both mono- and biallelic expression of these genes is fully compatible with normal development and life.

In summary, allelic expression analysis of imprinted genes has unraveled an unexpectedly complex situation that challenges the role of parental imprinting as a mechanism of transcriptional regulation. In contrast with the usual definition, imprinted genes are not always expressed in a parental origin-dependent way, as it was originally thought; they can go through different expression states from strictly monoallelic to strictly biallelic in different tissues at different stages of life. In fact, all known imprinted genes are expressed biallelically in at least one tissue and in at least one developmental stage. As a consequence, the difference between imprinted and nonimprinted genes has become "fuzzy." The only way to determine whether a gene is not imprinted on the basis of its expression pattern is to analyze its allelic expression in all tissues and at all developmental stages. This is obviously impossible and most genes are therefore described as nonimprinted. Paradoxically, it is easier to show that a gene is imprinted than to demonstrate that it is not! Therefore, the inability to define unambiguously what an imprinted gene is has created a considerable conceptual difficulty. Defining parental origin-dependent monoallelic gene expression as a basic criterion for the definition of an imprinted gene made the frontier between imprinted and nonimprinted genes disappear and introduced ambiguity in the biological interpretations of the phenomenon.

V. Mechanism

Research on the molecular mechanisms of imprinting represents a major scientific effort (Pfeifer, 2000). Because parental imprinting is considered a mechanism of transcriptional regulation, most of the work is oriented toward molecular aspects that have something to do with gene regulation. It is generally accepted that "any parent of origin-specific mark must be epigenetic in nature" (Pfeifer, 2000). The term "epigenetic" means in this context that the imprint must not change the primary sequence of the genome. Methylation became the major candidate for such a mechanism long before the discovery of the first imprinted gene. Methylation profiles of several transgenes were found to be dependent on parental origin

(Reik *et al.*, 1987; Sapienza *et al.*, 1989; Swain *et al.*, 1987), even in the absence of expression. Later, regions with differential methylation (DMRs) on the paternal and maternal chromosomes around many imprinted genes were identified: *Igf2r*, *H19*, *Igf2*, *SNRPN*, *U2af1-rs1*, *Gnas*, and *GTL2*. These DMRs are usually relatively short, CpG-rich DNA segments located in the flanking regions but also in the coding or intronic sequences of these genes. Several detailed reviews on DMRs have been published (Constancia *et al.*, 1998; Neumann and Barlow, 1996; Reik and Walter, 2001).

Although methylation is clearly part of the epigenetic differences between the parental alleles, its role is unlikely to be essential, because not all imprinted genes are associated with differentially methylated CpG islands. Frequently cited experimental evidence in favor of the key role of CpG methylation in the regulation of imprinted genes comes from the analysis of mice mutant for the *Dnmt-1* gene (Li *et al.*, 1993). This enzyme catalyzes the faithful copying of methylation patterns on the newly sythesized DNA strand during replication and, thus, is responsible for the maintenace of methylation patterns. The lack of this activity is lethal; homozygous mutant embryos die before midgestation (Li *et al.*, 1992). The gradual loss of methylation on the DMRs is correlated with the loss of difference in expression between the two parental alleles of *H19*, *Igf2*, and *Igf2r* (Li *et al.*, 1992). It was concluded that methylation plays a key role in imprinting. However, the loss of methylation induces widespread, major chromosomal abnormalities and disorganization of the chromatin (Viegas-Pequignot, personal communication), frequent mutations, and deregulated gene expression (Walsh *et al.*, 1998). In this context it is difficult to ascribe the changes in the expression of imprinted genes to a gene-specific effect. In addition, the *Mash2* gene escapes this phenomenon, because its expression remains monoallelic in the mutant embryos (Caspary *et al.*, 1998; Tanaka *et al.*, 1999).

Many DMRs are already differentially methylated at the moment of fertilization and conserve their allele-specific methylation pattern during preimplantation development. Some authors used to say that these genes "escape" or "resist" demethylation, or that they are "protected" from demethylation (Reik and Walter, 2001). The use of this term is misleading, because the maintenance of the methylated state is an active process. To conserve the methylation of a DNA element in a cell where the whole genome loses its methylation by passive dilution (Rougier *et al.*, 1998) or, perhaps, by a nonspecific active demethylating mechanism (Mayer *et al.*, 2000a; Oswald *et al.*, 2000), the methyltransferase enzyme responsible for the maintenance must be targeted to the DMR of only one parental chromosome. This would be impossible if the methylation was the only mark that allows the cell to differentiate between the two alleles. The maintenance of the methylated state is therefore an active, continuous process taking place throughout preimplantation development. It is likely that methylation is just one important component of a more complex molecular system associated with imprinted regions that allows the cell to recognize and distinguish the two parental alleles. This molecular system is likely to involve other components of the chromatin structure that cooperate with methylation to maintain allelic differences. Functional, but also chromatin structural, studies support this view.

When investigation is extended to species other than the laboratory mouse, the general role of DNA methylation becomes uncertain even in the case of the prototype of imprinted genes, such as *Igf2r*. As mentioned above, *IGF2R* is biallelically expressed in most humans. Yet, the parental origin-dependent allelic methylation of the intronic CpG island is conserved even in these individuals (Smrzka *et al.,* 1995). In contrast, in the opposum, *Igf2r* was found to be expressed monoallelically from the maternal allele despite the total absence of differentially methylated regions (Killian *et al.,* 2000).

Studies of chromatin structure revealed differences in nucleosome positioning, histone acetylation, or nuclease sensitivity between the two parental alleles of imprinted genes. In the case of the *SNRPN* gene (small nuclear ribonucleoprotein N-coding gene), one of the best studied imprinted genes (Nicholls *et al.,* 1998), the silencing of one parental allele is correlated with the presence of a differential methylated region (DMR) in the promoter region. The CpG island corresponding to the promoter of this gene is unmethylated on the paternally transmitted active allele and fully methylated on the inactive maternal allele (Nicholls *et al.,* 1998). The methylated DMR of the maternal allele is specifically associated with hypoacetylated histones, whereas the gene body and the entire paternal allele, including the promoter, are associated with acetylated histones. Treatment of the cells with the DNA methyltransferase inhibitor 5-azadeoxycytidine induced demethylation of the CpG island and restored gene expression on the maternal allele. The reactivation was associated with increased H4 acetylation (Saitoh and Wada, 2000).

Epigenetic modifications such as DNA methylation or acetylation of histones are commonly presented as the components of "chromatin structure" in the literature on regulation of gene expression. Therefore, differences between the two homologous alleles of an imprinted gene can be presented as differences in chromatin structure. These features are not specific to imprinted genes; rather, they are related to the active or inactive state of the alleles. Similar differences in chromatin structural features can be detected between any expressed and silent nonallelic genes in the genome. The description of various aspects of the chromatin structure of imprinted genes differentiating the two parental chromosomes can certainly help to understand how these genes are regulated, in the same way as comparative studies of inactive and active gene copies can help in understanding gene regulation. However, such a molecular description, limited to gene regulatory sequences, sheds little light on the mechanism of genomic imprinting itself. As a result, most discussions about the mechanisms of imprinting are, in fact, about mechanisms of gene regulation rather than mechanisms of imprinting. Imprinted genes are expressed differently on homologous chromosomes because they are imprinted and not the opposite! A description of structural differences between parental homologs is not an explanation of their function.

Expressed genes and their regulatory sequences are not the only components of the genome that have chromatin structure. Genes are surrounded by noncoding sequences that represent the overhelming majority of the genome. Noncoding and repetitive sequences can also be methylated or demethylated, and associated with acetylated histones, and so on. In some cases they can even be transcribed. In other words, every part of the genome has a characteristic chromatin structure. Although this is obvious, most frequently it is neglected for the simple reason that geneticists are more interested in gene expression than in the study of noncoding parts of the genome, often considered nonfunctional or even "junk." Although observations suggesting that imprinting is a feature of large genomic blocks (see below), almost nothing is known about the chromatin structure of imprinted genomic regions outside the regulatory sequences of some imprinted genes.

To explain the biallelic expression of imprinted genes in some tissues, some developmental stages, or both, other authors proposed the existence of a reading mechanism (Efstratiadis, 1994; Barlow, 1995; Reik and Walter, 2001). The function of this hypothetical mechanism would be to recognize specifically the imprint in order to protect it from erasure and execute its transcriptional function (Efstratiadis, 1994). It is postulated that in the tissues where this mechanism is not functional, the imprint is "not read" and the expression of the imprinted gene is biallelic. The hypothesis of a specific reading mechanism raises conceptual and practical problems. Such a specific, sophisticated, and complex mechanism should have appeared simultaneously and should have coevolved with the mechanism of imprinting. It is unclear what makes the difference between the parental alleles: is it the parental imprint or the reading mechanism? If a protein, for example, binds one parental allele to regulate its expression, but does not bind the other allele, what criteria will determine whether it is part of the imprinting mechanism or the reading mechanism? Are there as many reading mechanisms as imprinted genes? If such a hypothetical reading mechanism existed, it should be gene specific with nonoverlapping tissue specificity. It should be able to recognize the different imprinted genes independently of each other and produce a unique mono- and biallelic expression pattern in a tissue-specific way for each gene individually. For example, the *Ins2* gene is expressed biallelically in the pancreas, whereas its nearest neighbor on the same chromosome, *Igf2*, is expressed only from the paternal allele in this tissue (Deltour *et al.*, 1995). Both of these genes are expressed biallelically in the brain and paternally in the yolk sac (DeChiara *et al.*, 1991; Deltour *et al.*, 1995; Giddings *et al.*, 1994) during the second half of gestation. No direct proof for the existence of such a hypothetical reading mechanism specific for imprinted genes has been found so far. As mentioned above, all the mechanisms found to regulate imprinted genes are the same as those found for any other gene in the genome. The question is, how do general gene regulatory mechanisms lead to imprinted expression in some tissues but to biallelic expression in others? It is possible to resolve this problem without postulating specific reading mechanisms.

Observations of the *Igf2* and *H19* genes, using primary transcript analysis of the nuclei of individual fetal liver cells, revealed that transcription of these genes is frequently biallelic (Jouvenot *et al.,* 1999). Nevertheless, transcripts from the second parental allele do not accumulate in the cytoplasm. The data suggest that the transcriptional silencing of the second parental allele is incomplete, and the transcripts derived from it are less stable than the transcripts of the highly transcribed allele. Although the possible involvement of posttranscriptional processes in the monoallelic expression of an imprinted gene appears surprising at first glance, it can be explained in light of the stochastic nature of the gene expression process (Misteli, 2001). Gene expression is a multistep process involving transcription, RNA processing, transport from the nucleus, and translation. Initiation of transcription is mostly a stochastic process (Hume, 2000; Misteli, 2001). The probability of successful initiation of the process determines whether a gene is transcribed efficiently. If the probability of transcription is equally low on the two chromosomes, transcription of the two alleles in the same cell will be a rare event. It is likely that the parental alleles of an imprinted gene have different probabilities of transcription because of differences in chromatin structure. Therefore, one parental allele is less efficiently transcribed than the other, resulting in a large proportion of cells with monoallelic transcription, and a small proportion of cells with biallelic transcription. However, both transcription and primary transcript processing take place in the same nuclear structures in the same nuclear compartment and are likely to be performed by components of the same multiproteinic complex (Lemon and Tjian, 2000). A small difference in the efficiency at each step of mature mRNA production from the two parental alleles in an individual cell may result in a significant overall difference in abundance between the two populations of mRNAs in the whole tissue. This overall difference will be more pronounced for genes with a low transcription rate. Therefore, it is not necessary to invoke a special reading mechanism to understand why imprinted genes are sometimes expressed biallelically. It is suffucent to suppose that imprinted genes have low transcription rates in those tissues, where they are expressed monoallelically, high transcriptional rates when they expressed biallelically, and intermediate rates when preferential expression of one allele is detected. (However, this explanation is difficult to reconcile with the hypothetical role of imprinting as a mechanism of monoallelic gene regulation.) On the basis of the above-depicted considerations, we have hypothesized that imprinted gene clusters evolved from clusters of genes with low transcription rates (Ohlsson *et al.,* 2001). In fact, random monoallelic expression of genes with low rates of transcription also occurs in clusters (Ohlsson *et al.,* 2001).

Whether a gene is considered highly expressed, that is, the steady-state mRNA level is high, depends mainly on the stability of the RNA in the cytoplasm rather than on the transcriptional activity (Hume, 2000). In fact, the gene expression level is more efficiently regulated at the level of mRNA stability than at the level of transcription. This point is well illustrated by the imprinted *H19* gene, which is one of the most abundantly expressed genes in the mouse embroy. Studies have shown that upregulation of *H19* expression in differentiating myoblasts is due to increased stability of the RNA resulting in increased steady-state mRNA levels, while the transcription rate remains unchanged (Milligan *et al.,* 2000).

As was suggested by Constancia *et al.* (1998), "it could be postulated that the 'hard work' of distinguishing the two parental alleles has already been accomplished by the imposition of the differential methylation pattern. The next step is the 'blind' transcriptional response. . . ."

Evolution did not need to devise a new specific molecular mechanism for differential epigenetic marking of some genes in the two parental genomes. The existence of epigenetic differences between the parental chromosomes can be explained by the action of already existing mechanisms. Although the paternal and maternal genomes have identical gene composition, their capacity to express these genes is radically different at the moment of fertilization. The male and female germ cells are two highly specialized cell types. As indicated by the dramatically different recombination frequency during female and male meiosis, the chromatin structure of imprinted domains is certainly different in the two cell types (Paldi *et al.*, 1995; Robinson and Lalande, 1995). The epigenetic state of the two parental genomes in a zygote reflects the chromatin state in the sperm cell and oocyte that fuse. For example, both the level and the distribution of methylated CpG dinucleotides are different in the two genomes (Monk *et al.*, 1987; Rougier *et al.*, 1998). The paternal genome is associated mainly with protamines whereas the maternal genome contains histones. The acetylation pattern of histones that replace the protamines during the first cell cycle is also different compared with the maternal genome. Therefore, genome-wide epigenetic differences between the parental chromosomes at the moment of fertilization is the consequence of the different morphology and physiology of the male and female gametes. These epigenetic characteristics are gradually erased during the early stages of development, after the fusion of the two gametes (fertilization). However, if the erasure of the original epigenetic state in a chromosomal region is incomplete or partial, the two homologous alleles or regions will retain some epigenetic differences. Therefore, epigenetic differences between the parental chromosomes can appear spontaneously without the action of specific molecular mechanisms. The critical question concerns why some differences are maintained. Incomplete erasure of the previous chromatin state is facilitated by the spatial segregation of the two parental genomes in the nucleus observed during the first stages of development (Mayer *et al.*, 2000b). Those epigenetic differences between the parental homologs that survive the critical period of preimplantation development are then conserved at later stages by the usual mechanisms of cellular memory responsible for the maintenance of epigenetic states in the genome.

VI. Unexplained Features of Imprinting

Some features of imprinting could not be predicted and are impossible to explain on the basis of the current paradigm. The most important of these is the organization of imprinted genes in clusters. Imprinted genes are not randomly distributed throughout the genome; instead, they are concentrated in discrete chromosomal regions. The size of these clusters can be large, up to several hundred kilobases or even several megabases. They may contain many genes with a spectrum of allelic expression specificity going from predominantly paternal, through equilibrated biallelic to predominantly maternal expression. The question of how imprinting evolved cannot be successfully addressed without considering the clustered arrangement of imprinted genes.

Clustering of several imprinted genes in subchromosomal domains suggests the existence of coordinated regulation of imprinting in these regions. This idea received strong support from the discovery of so-called imprinting mutations in the human 15q13 region (Buiting *et al.,* 1995; Sutcliffe *et al.,* 1994). Small deletions that removed the 5' end of the *SNRPN* gene on chromosome 15q13 of some patients with Prader-Willi syndrome (PWS) or Angelman syndrome (AS) had a dramatic effect on the whole (2-Mb-long) imprinted domain. The paternal chromosome of these PWS patients carried an inappropriate maternal methylation profile and all imprinted genes normally active on the paternal chromosme were silent. The opposite was found in the AS patients: their maternal chromosome had a paternal methylation profile (Buiting *et al.,* 1995; Sutcliffe *et al.,* 1994). These observations suggest that the deletions removed a chromosomal element required for the establishment of the proper imprint over the entire domain in the germ line. It was proposed that this element, named, the "imprinting center" (IC), regulates the imprint switch process in the female and male germ lines (Saitoh *et al.,* 1996). The position of the IC and its role seem to be conserved between mouse and human, because a targeted deletion of the putative IC region in the mouse genome produced an effect similar to that observed in humans (Yang *et al.,* 1998). On the basis of the shortest region of deletion overlap (SRO) detected in PWS and AS patients, two distinct regulatory elements were identified within the IC region (Ohta *et al.,* 1999a,b). The PWS-SRO was estimated to be only 4.3 kb long, and it coincides with the promoter of the differentially methylated *SNRPN* gene. It is believed to contain the element involved in the initiation of the paternal imprint in the male germ line. The AS-SRO is only about 1 kb long, located 30 kb away (Ohta *et al.,* 1999a), and seems to control the establishment of the maternal imprint in the female germ line (Ohta *et al.,* 1999b). Therefore, the establishment of the parental imprint over the whole 2-Mb chromosomal region in the male and female germ lines seems to be under the control of two closely located, small, but clearly distinct elements.

Interestingly, studies of humans and mice show that deletion of the IC element at a later, postzygotic stage also affects the imprinting of the entire cluster (Bielinska *et al.,* 2000). In somatic tissues the mutant chromosome had lost its paternal epigenetic identity and adopted a maternal methylation profile at all analyzed loci of the cluster. The main conclusion from these observations is clear: if a somatic mutation can result in the complete loss of paternal epigenotype, then the IC region is not only required for the establishment of the paternal imprint in the male germ line, but also for the postzygotic maintenance of the imprint. This is an important observation, because it indicates that both the establishment of the parental imprint in the germ line and its maintenance in somatic cells depend on the same factors. These two phases had previously been regarded as two independent and separate processes (Reik and Walter, 2001). In light of the data on somatic IC deletions, it has become impossible to determine precisely when establishment of the imprint ends and maintenance starts.

Another unexplained feature of imprinted genomic regions is their asynchronous replication. It was found that the paternal and maternal homologs of these regions follow different replication kinetics during the S phase of the cell cycle (Kitsberg *et al.*, 1993; Knoll *et al.*, 1994). It is usually the paternal copy that replicates earlier (Kitsberg *et al.*, 1993). The difficulty in interpreting these observations stems from the fact that asynchronous replication is not a structural feature, but a cellular process with a time dimension. As eukaryotic replication is influenced by chromatin structure, it is not surprising that regions of the genome with known structural differences have allelic differences in replication timing. From this point of view, replication asynchrony can be considered an indicator of differential chromatin structure of the homologous chromosomal regions.

Two aspects of imprinted chromatin domain replication might be relevant for the understanding of imprinting. First, replication asynchrony characterizes large genomic regions that contain imprinted genes. Paternally, maternally, or biallelically expressed genes as well as intergenic segments all follow the same replication kinetics within the same region. Thus, rather than being gene specific, determination of replication timing is region specific and depends only on the parental origin. Second, replication asynchrony, in addition to being independent of gene expression within the cluster, is independent of the differentiation state of the cell. Imprinted domains replicate asynchronously in a variety of cell types, for example, pluripotential embryonic stem cells, lymphocytes, and fibroblasts (Gunaratne *et al.*, 1995; Kitsberg *et al.*, 1993; Knoll *et al.*, 1994), irrespective of gene expression in the cluster. Analysis of mice has shown that asynchronous replication is first detectable during the first cell cycle after fertilization and maintained even in the germ line until the beginning of meiosis (Simon *et al.*, 1999). These observations clearly indicate that recognition of the parental origin of homologous chromosomes operates in all normal cells of the organism. No other known characteristic of imprinted genes, imprinted gene clusters, or both is maintained in all cells of the organism. Even DMRs that are differentially methylated in the gametes and conserve this difference during development and cell differentiation become demethylated in primordial germ cells (Brandeis *et al.*, 1993; Kerjean *et al.*, 2000; Tada *et al.*, 1998).

Asynchronous replication depends on the presence of paternal and maternal chromosomes in the cell. Imprinted regions in parthenogenetic embryos and embryonic stem cells replicate synchronously (Simon *et al.*, 1999). In cells of Prader-Willi syndrome patients with maternal uniparental disomies (UDPs), both copies of chromosome 15q13 replicate at the same time (LaSalle and Lalande, 1995).

However, the timing of replication is significantly delayed compared with the replication timing of the maternal chromosome in a normal cell. Conversely, the replication timing of the two paternal chromosomes in the cells of Angelman syndrome patients with paternal UDPs is also shifted during the S phase compared with the paternal chromosome in normal cells. These observations suggest that the two chromosomes are able to recognize each other and that the replication timing of imprinted chromosomal domains on each chromosome is an intrinsic property that depends exclusively on the parental origin.

Asynchronous replication may represent an interesting link between imprinted genes and genes with random monoallelic expression. The latter category of genes can also be organized in clusters (Kelly and Locksley, 2000) and replicate asynchronously (Hollander, 1999; Simon et al., 1999). A potential evolutionary link between these two groups has been hypothesized (Ohlsson et al., 2001).

In light of the above-cited data it is difficult to defend the idea that imprinting is a mechanism of gene regulation by which only one of the parental copies of a gene is expressed. Evolutionary conservation of imprinting indicates that it is indeed important, but it does not seem to be a mechanism to regulate gene expression levels even if it can influence it. In any case, it is difficult to define whether a gene is imprinted solely on the basis of its allelic expression. Imprinting is a feature of large chromatin domains with their own domain-wide characteristics. These domains may contain many genes with variable allelic expression patterns. Although parental origin-dependent monoallelic expression is an important feature of imprinting, it is possible that this is a secondary effect of a phenomenon whose biological function remains to be determined.

VII. Biological Significance: Toward a Paradigm Shift

The biological significance and evolutionary reasons for imprinting are unclear. Whatever is the primary biological function of imprinting, by silencing one allele of an autosomal gene mammals lose the main advantage of diploidy over haploidy, that is, masking of deleterious mutations. The mutational load in a population associated with an imprinted locus is equal to the load in the case of dominant mutations at a nonimprinted locus (Spencer, 1997). Therefore, there is a high cost to pay in maintaining imprinting. To survive this selective pressure on the evolutionary time scale, advantages of having imprinting must be greater than the drawbacks associated with the haploid expression of the genes concerned. What could be the biological function that outweighs the increased mutational load?

According to the current paradigm, genomic imprinting has evolved to regulate the level of gene expression through the inactivation of one parental allele and it has been selected during evolution because of the need to maintain haploid levels of expression at key loci at key developmental stages. As argued above, imprinting does not seem to be a mechanism to regulate gene expression levels. In addition, if imprinting had evolved as a mechanism of gene dosage control, higher than haploid levels of expression at imprinted loci would have a strong effect. Yet, many imprinted genes can be overexpressed in transgenic animals without any effect (Ainscough et al., 1997; Bucchini et al., 1986; Hatada et al., 1997; Pfeifer et al., 1996). Although higher than diploid level overexpression of Igf2 and Igf2r was shown to be deleterious (Sun et al., 1997; Zaina et al., 1998), the statement that imprinting in general is a process to regulate gene dosage is unjustified.

More than 14 theories have been proposed to account for the existence of imprinting, all of them based on the assumption that imprinting regulates gene dosage (Haig and Trivers, 1995; Hurst, 1997). The most popular hypothesis is the well-known parental conflict model (Haig and Graham, 1991; Moore and Haig, 1991). According to it, imprinting has evolved because of the conflicting interests of the paternal and maternal genomes. This hypothesis is based on several assumptions: (1) there is a conflict of interest between the paternal and maternal genomes over maternal resources because of unequal parental investment between the sexes; (2) this conflict exists only in populations with polygamous reproduction; and (3) imprinting is the consequence of the "arms race" between the parental genomes because imprinted genes control embryonic growth and resource transfer from the mother to the fetus. The hypothesis has been presented in detail several times (Hurst, 1997; Hurst and McVean, 1997); hence, I draw attention to only a few crucial points.

According to the conflict model, imprinting has a function only in polygamous populations (Moore and Haig, 1991) and it is predicted to be "relaxed" in monogamous species. However, imprinting has also been observed in the monogamous mouse Peromyscus polionotus (Vrana et al., 1998). To reconcile this observation with the conflict model, it was proposed that the costs of removing imprinting may be higher than the costs of maintaining it (Moore and Mills, 1999). Interestingly, this argument is a modern reformulation of an old, nineteenth century argument, advanced by the fixists to oppose Darwin, aiming to exclude the possibility of gradual evolutionary change.

The conflict model, as well as most other models, is based on the idea that imprinting is specific to the biological function of the genes of interest, that is, growth regulation. However, imprinted genes do not share the same or even similar biological roles and it is highly unlikely that they serve a common purpose. If the primary cause of the imprinting of a gene were related to its biological function, there should be as many independent causes as imprinted genes. Although this possibility cannot theoretically be excluded, it raises more questions than it answers. Most importantly, the clustering of imprinted genes remains to be explained. In addition, as described in the previous section, many observations suggest that a diploid or higher level of expression of imprinted genes is usually compatible with normal development and life, suggesting that the requirement for functional haploidy—if any—is not strict.

In light of the contradictions between the assumption of a requirement for functional haploidy as the basis of parental imprinting and the experimental data, it is necessary to examine alternative possibilities. The reason for the evolution of imprinting can be far removed from the coding capacity of the nucleotide sequences contained in imprinted genes. Some biological functions of imprinting other than gene dosage regulation could be favored by natural selection. These functions may be mediated by the differential chromatin structures of homologous chromosomal regions. Maintaining these structural differences might be important for some cellular processes, ranging in occurrence from the fertilized zygote to the meiotic cell.

Taking into account the fact that asynchronous replication of imprinted domains is the only characteristic shared between all imprinted domains and all loci within them and at all stages (Simon *et al.,* 1999), it seems reasonable to postulate that imprinting has something to do with the coordinated replication of the genome. Each chromosome of a diploid cell must undergo one and only one round of replication during the cell cycle. The existence of relatively large blocks with structural differences between the homologs provides a mark that might serve to identify each chromosome. Such identification may be necessary for their coordinated replication and surveillance of normal chromosomal number. Identification may be mediated by interactions between the homologs (Paldi and Jouvenot, 1999). This aspect has remained unexplored so far. Only sporadic observations of the existence of interactions between homologous imprinted domains have been reported in human and mouse somatic cells (Duvillié *et al.,* 1998; LaSalle and Lalande, 1996).

C. Sapienza and colleagues have proposed a more elaborate, new hypothesis based on similar logic (Pardo-Manuel de Villena *et al.,* 2000). According to this hypothesis, structural differences between the parental chromosomes are necessary for "homologous pairing of chromosomes at meiosis and the associated processes of DNA repair and recombination during both meiosis and mitosis" (Pardo-Manuel de Villena *et al.,* 2000). Such interactions are important to promote recombination between homologs, rather than sister chromatids, during meiosis and repair of DNA double-strand breaks and any associated mitotic recombination between sister chromatids rather than between homologous chromosomes during mitosis. Conversely, imprinted chromatin domains could serve as templates by which homologous chromosomes might identify each other. A number of observations unexplained on the basis of the current paradigm support the hypothesis that imprinting is a general mechanism for distinguishing homologous chromosomes from each other. Most important, there are observations that suggest a direct interaction between the parental homologs of imprinted chromosomal regions (LaSalle and Lalande, 1996; Duvillié *et al.,* 1998).

Because no direct research has been done so far to test these particular hypotheses, their validity remains to be established. Nevertheless, the basic idea that the chromatin structural differences between homologs per se may have a biological function selected and maintained during evolution represents an alternative to the current paradigm based on the role of the function of the genes contained in imprinted regions. There are well-known precedents for structural features of the genome that are maintained because of a function other than coding capacity. The best known examples are centromeres and telomeres, but regions with the capacity to bind structural elements in the nucleus also could be listed in this category.

On occasional, as a by-product, these structural differences can result in parental origin-dependent monoallelic expression. In the majority of cases, it represents a burden with which the cell must cope. This constitutes a constraint for which region of the genome may or may not be imprinted. Regions with genes encoding essential functions are unlikely to become imprinted because of the high mutational load of the associated functional haploidy. On the other hand, once a monoallelic expression pattern has been established, it can be recruited for novel tasks. However, in these cases the biological function of imprinted expression is likely to vary from gene to gene.

Postulating that parentally imprinted differential chromatin structure has a biological function implies that imprinting must be widespread among eukaryotes. Although unusual phenomena of parental origin-dependent heredity are frequent in nature, the possibility that imprinting could exist in nonmammalian species has never been seriously examined. A major challenge for future research will be the exploration of various aspects of genomic imprinting beyond monoallelic expression and in a wide variety of species.

Acknowledgments

I thank Denise Barlow, Mounia Guenatri, Jacques Jami, Rolf Ohlsson, Françoise Poirier, Carmen Sapienza, and Pascale Schaerly for helpful discussions, comments, and suggestions on drafts of the manuscript.

References

Ainscough, J., Kolde, T., Tada, M., Barton, S., and Surani, M. (1997). Imprinting of *Igf2* and *H19* from a 130 kb YAC transgene. *Development* **124**, 3621–3632.

Barlow, D. (1995). Gametic imprinting in mammals. *Science* **270**, 1610–1613.

Barlow, D., Stöger, R., Herrmann, B., Saito, K., and Schweifer, N. (1991). The mouse insulin-like growth factor type-2 receptor is imprinted and closely linked to the Tme locus. *Nature (London)* **349**, 84–87.

Bartolomei, M., Zemel, S., and Tilghman, S. (1991). Parental imprinting of the mouse H19 gene. *Nature (London)* **351**, 153–155.

Ben-Porath, I., and Cedar, H. (2000). Imprinting: Focusing on the center. *Curr. Opin. Genet. Dev.* **10**, 550–554.

Bielinska, B., Blaydes, S., Buiting, K., Yang, T., Krajewska-Walasek, M., Horsthemke, B., and Brannan, C. (2000). De novo deletions of *SNRPN* exon 1 in early human and mouse embryos result in a paternal to maternal imprint switch. *Nat. Genet.* **25**, 74–78.

Brandeis, M., Kafri, T., Ariel, M., Chaillet, J. R., McCarrey, J., Razin, A., and Cedar, H. (1993). The ontogeny of allele-specific methylation associated with imprinted genes in the mouse. *EMBO J.* **12,** 3669–3677.

Bucchini, D., Ripoche, M., Stinnakre, M., Debois, P., Lores, P., Monthioux, E., Absil, E., Lepesant, J., Pictet, R., and Jami, J. (1986). Pancreatic expression of human insulin gene in transgenic mice. *Proc. Natl. Acad. Sci. USA* **83,** 2511–2515.

Buiting, K., Saitoh, S., Gross, S., Dittrich, B., Schwartz, S., Nicholls, R., and Horsthemke, B. (1995). Inherited microdeletions in the Angelman and Prader-Willi syndromes define an imprinting centre on human chromosome 15. *Nat. Genet.* **9,** 395–400.

Caspary, T., Cleary, M. A., Baker, C. C., Guan, X. J., and Tilghman, S. M. (1998). Multiple mechanisms regulate imprinting of the mouse distal chromosome 7 gene cluster. *Mol. Cell. Biol.* **18,** 3466–3474.

Cattanach, B. M., and Kirk, M. (1985). Differential activity of maternally and paternally derived chromosome regions in mice. *Nature (London)* **315,** 496–498.

Constancia, M., Pickard, B., Kelsey, G., and Reik, W. (1998). Imprinting mechanisms. *Genome Res.* **8,** 881–900.

Cooper, P., Smilinich, N., Day, C., Nowak, N., Reid, L., Pearsall, R., Reece, M., Prawitt, D., Landers, J., Housman, D., Winterpacht, A., Zabel, B., Pelletier, J., Weisman, B., Shows, T., and Higgins, M. (1998). Divergently transcribed overlapping genes expressed in liver and kidney in the 11p15 imprinted domain. *Genomics* **49,** 38–51.

Dao, D., Frank, D., Qian, N., O'Keefe, D., Vosatka, R., Walsh, C., and Tycko, B. (1998). *IMPT1,* an imprinted gene similar to polyspecific transporter and multi-drug resistance genes. *Hum. Mol. Genet.* **7,** 597–608.

Davis, R., Weintraub, H., and Lassar, A. (1987). Expression of single transfected cDNA converts fibroblasts to myoblasts. *Cell* **24,** 987–1000.

DeChiara, T., Robertson, E., and Efstratiadis, A. (1991). Parental imprinting of the mouse insulin-like growth factor ll gene. *Cell* **64,** 849–859.

Deltour, L., Montagutelli, X., Guenet, J., Jami, J., and Paldi, A. (1995). Tissue- and developmental stage-specific imprinting of the mouse proinsulin gene, *Ins2. Dev. Biol.* **168,** 686–688.

Duvillié, B., Bucchini, D., Tang, T., Jami, J., and Pàldi, A. (1998). Imprinting at the mouse *Ins2* locus: Evidence for *cis-* and *trans-*allelic interactions. *Genomics* **47,** 52–57.

Efstratiadis, A. (1994). Parental imprinting of autosomal mammalian genes. *Curr. Opin. Genet. Dev.* **4,** 265–280.

Ekström, T. J., Cui, H., Li, X., and Ohlsson, R. (1995). Promoter-specific *IGF2* imprinting status and its plasticity during human liver development. *Development* **121,** 309–316.

Giddings, S. J., King, C. D., Harman, K. W., Flood, J. F., and Carnaghi, L. R. (1994). Allele specific inactivation of insulin 1 and 2, in the mouse yolk sac, indicates imprinting. *Nat. Genet.* **6,** 310–313.

Guillemot, F., Caspary, T., Tilghman, S., Gilbert, D. J., Jenkins, N. A., Anderson, D. J., Joyner, A. L., Rossant, J., and Nagy, A. (1995). Genomic imprinting of *Mash-2,* a mouse gene required for trophoblast development. *Nat. Genet.* **9,** 235–241.

Gunaratne, P., Nakao, M., Ledbetter, D., Sutcliffe, J., and Chinault, A. (1995). Tissue-specific and allele-specific replication timing control in the imprinted human Prader-Willi syndrome region. *Genes Dev.* **9,** 808–820.

Haig, D., and Graham, C. (1991). Genomic imprinting and the strange case of the insulin-like growth factor II receptor. *Cell* **64,** 1045–1046.

Haig, D., and Trivers, R. (1995). The evolution of parental imprinting: A review of hypotheses. *In* "Genomic Imprinting: Causes and Consequences" (R. Ohlsson, K. Hall, and M. Ritzen, eds.), pp. 17–28. Cambridge University Press, Cambridge.

Hatada, I., Nabetani, A., Arai, Y., Ohishi, S., Suzuki, M., Miyabara, S., Nishimune, Y., and Mukai, T. (1997). Aberrant methylation of an imprinted gene U2af1-rs1(SP2) caused by its own transgene. *J. Biol. Chem.* **272,** 9120–9122.

Hayashizaki, Y., Shibata, H., Hirotsune, S., Sugino, H., Okazaki, Y., Sasaki, N., Hirose, K., Imoto, H., Okuizumi, H., Maramatzu, M., Komatsubara, H., Shiroshi, T., Moriwaki, K., Katsuki, M., Hatano, N., Sasaki, H., Ueda, T., Takagi, N., Plass, C., and Chapman, V. M. (1994). Identification of an imprinted U2af binding protein related sequence on mouse chromosome 11 using the RLGS method. *Nat. Genet.* **6,** 33–40.

Hayward, B., Moran, V., Strain, L., and Bonthorn, D. (1998). Bidirectional imprinting of a single gene: GNAS1 encodes maternally, paternally, and biallelically derived proteins. *Proc. Natl. Acad. Sci. USA* **95,** 15475–15480.

Hollander, G. (1999). On the stochastic regulation of interlukin-2 transcription. *Semin. Immunol.* **11,** 357–367.

Hume, D. (2000). Probability in transcriptional regulation and its implication for leukocyte differentiation and inducible gene expression. *Blood* **96,** 2323–2328.

Hurst, L. (1997). Evolutionary theories of genomic imprinting. In "Frontiers in Molecular Biology: Genomic Imprinting" (W. Reik and A. Surani, eds.), pp. 211–239. Oxford University Press, Oxford.

Hurst, L., and McVean, G. (1997). Growth effects of uniparental disomies and the conflict theory of genomic imprinting. *Trends Genet.* **13,** 436–442.

Jiang, S., Hemann, M. A., Lee, M. P., and Feinberg, A. P. (1998). Strain-dependent developmental relaxation of imprinting of an endogenous mouse gene, KvLQT1. *Genomics* **53,** 395–399.

Johnson, D. R. (1974). Hairpin-tail: A case of post-reductional gene action in the mouse egg. *Genetics* **76,** 795–805.

Jouvenot, Y., Poirier, F., Jami, J., and Paldi, A. (1999). Biallelic transcription of Igf2 and H19 in individual cells suggests a post-transcriptional contribution to genomic imprinting. *Curr. Biol.* **9,** 1199–1202.

Kanedo-Ishino, T., Kuroiwa, Y., Miyoshi, N., Kohda, T., Suzuki, R., Yokoyama, M., Viville, S., Barton, S. C., Ishino, F., and Surani, M. A. (1995). Peg1/Mest imprinted gene on chromosome 6 identified by cDNA subtraction hybridization. *Nat. Genet.* **11,** 52–59.

Kelly, B., and Locksley, R. (2000). Coordinate regulation of the IL-4, IL-13, and IL-5 cytokine clusters in Th2 clones revealed by allelic expression patterns. *J. Immunol.* **165,** 2982–2986.

Kerjean, A., Dupont, J., Vasseur, C., LeTessier, D., Cuisset, L., Paldi, A., Jouannet, P., and Jeanpierre, M. (2000). Establishment of the parental methylation imprint of the human H19 and MEST/PEG1 genes during spermatogenesis. *Hum. Mol. Genet.* **9,** 2183–2187.

Killian, J., Byrd, J., Jirtle, J., Munday, B., Stoskopf, M., MacDonald, R., and Jirtle, R. (2000). M6P/IGF2R imprinting evolution in mammals. *Mol. Cell* **5,** 707–716.

Kitsberg, D., Selig, S., Brandeis, M., Simon, I., Keshet, I., Driscoll, D. J., Nicholls, R. D., and Cedar, H. (1993). Allele-specific replication timing of imprinted gene regions. *Nature (London)* **364,** 459–463.

Knoll, J. H. M., Cheng, S.-D., and LaLande, M. (1994). Allele-specificity of DNA replication timing in the Angelman/Prader Willi syndrome imprinted chromosomal region. *Nat. Genet.* **6,** 41–46.

Kuhn, T. (1996). "The Structure of Scientific Revolutions." University of Chicago Press, Chicago.

LaSalle, J., and Lalande, M. (1995). Domain organization of allele-specific replication within the GABRB3 gene cluster requires a biparental 15q11-13 contribution. *Nat. Genet.* **9,** 386–394.

LaSalle, J. M., and Lalande, M. (1996). Homologous association of oppositely imprinted chromosomal domains. *Science* **272,** 725–728.

Lee, M., Brandenbourg, S., Landes, G., Adams, M., Miller, G., and Feinberg, A. (1999). Two novel genes in the center of the 11p15 imprinted domain escape genomic imprinting. *Hum. Mol. Genet.* **8,** 683–690.

Lee, M. P., Hu, R. J., Johnson, L. A., and Feinberg, A. P. (1997). Human KVLQT1 gene shows tissue-specific imprinting and encompasses Beckwith-Wiedemann syndrome chromosomal rearrangements. *Nat. Genet.* **15,** 181–185.

Lemon, B., and Tjian, R. (2000). Orchestrated response: A symphony of transcription factors for gene control. *Genes Dev.* **14,** 2551–2569.

Li, E., Bestor, T., and Jaenisch, R. (1992). Targeted mutation of the DNA methyltransferase gene results in embryonic lethality. *Cell* **69,** 915–926.

Li, E., Beard, C., and Jaenisch, R. (1993). Role for DNA methylation in genomic imprinting. *Nature (London)* **366,** 362–365.

Mayer, W., Nivelau, A., Walter, J., Fundele, R., and Haaf, T. (2000a). Demethylation of the zygotic paternal genome. *Nature (London)* **403,** 501–502.

Mayer, W., Smith, A., Fundele, R., and Haaf, T. (2000b). Spatial separation of parental genomes in preimplantation mouse embryos. *J. Cell Biol.* **148,** 629–634.

McGrath, J., and Solter, D. (1984). Completion of mouse embryogenesis requires both the maternal and paternal genomes. *Cell* **37,** 179–183.

Milligan, L., Bisbal, A., Brunel, W., Forne, T., and Cathala, G. (2000). H19 gene expression is up-regulated exclusively by stabilization of the RNA during muscle cell differentiation. *Oncogene* **19,** 5810–5816.

Misteli, T. (2001). Protein dynamics: Implication for nuclear architecture and gene expression. *Science* **291,** 843–847.

Monk, M., Bolibelik, M., and Lehnert, S. (1987). Temporal and regional changes in DNA methylation in the embryonic, extra-embryonic and germ cell lineages during mouse embryo development. *Development* **99,** 371–382.

Monk, M., and Surani, M., eds. (1990). "Genomic Imprinting: Papers Presented at a Meeting of the British Society for Developmental Biology at the University of Manchester, April 1990." *Development* (Suppl.) Vol. 89. Company of Biologists, Cambridge.

Moore, T., and Haig, D. (1991). Genomic imprinting in mammalian development: A parental tug-of-war. *Trends Genet.* **7,** 45–49.

Moore, T., and Mills, W. (1999). Imprinting and monogamy. *Nat. Genet.* **22,** 130–131.

Neumann, B., and Barlow, D. (1996). Multiple roles for DNA methylation in gametic imprinting. *Curr. Opin. Genet. Dev.* **6,** 159–163.

Nicholls, R., Saitoh, S., and Horsthemke, B. (1998). Imprinting in Prader-Willi and Angelman syndromes. *Trends Genet.* **14,** 194–200.

Ohlsson, R., Paldi, A., and Graves, J. (2001). Did genomic imprinting and X chromosome inactivation arise from stochastic expression? *Trends Genet.* **17,** 136–141.

Ohta, T., Buiting, K., Kokkonenen, H., McCandless, S., Heeger, S., Leisti, H., Driscoll, D., Cassidy S., Horsthemke, B., and Nicholls, R. (1999a). Molecular mechanism of Angelman syndrome in two large families involves an imprinting mutation. *Am. J. Hum. Genet.* **64,** 385–396.

Ohta, T., Gray, T., Rogan, P., Buiting, K., Gabriel, J., Saitoh, S., Mularidhar, B., Bilienska, B., Krajewska-Walasek, M., Driscoll, D., Horsthemke, B., Butler, M., and Nicholls, R. (1999b). Imprinting-mutation mechanisms in Prader-Willi syndrome. *Am. J. Hum. Genet.* **64,** 397–413.

Oswald, J., Engemann, S., Lane, N., Mayer, W., Olek, A., Fundele, R., Dean, W., Reik, W., and Walter, J. (2000). Active demethylation of the paternal genome in the mouse zygote. *Curr. Biol.* **10,** 475–478.

Pachnis, V., Brannan, C. I., and Tilghman, S. M. (1988). The structure and expression of a novel gene activated in early mouse embryogenesis. *EMBO J.* **3,** 673–681.

Paldi, A., and Jouvenot, Y. (1999). Allelic trans-sensing and imprinting. *Results Probl. Cell Differ.* **25,** 271–282.

Paldi, A., Gyapay, G., and Jami, J. (1995). Imprinted chromosomal regions of the human genome display sex-specific meiotic recombination frequencies. *Curr. Biol.* **5,** 1030–1035.

Pardo-Manuel de Villena, F. D., Casa-Esperon, E. D. L., and Sapienza, C. (2000). Natural selection and the function of genome imprinting: Beyond the silenced minority. *Trends Genet.* **16,** 573–579.

Peters, J., Wroe, S., Wells, C., Miller, H., Bodle, D., Beechey, C., Williamson, C., and Kelsey, G. (1999). A cluster of oppositely imprinted transcripts at the *Gnas* locus in the distal imprinting region of mouse chromosome 2. *Proc. Natl. Acad. Sci. USA* **96,** 3830–3835.

Pfeifer, K. (2000). Mechanisms of genomic imprinting. *Am. J. Hum. Genet.* **67,** 777–787.

Pfeifer, K., Leighton, P., and Tilghman, S. M. (1996). The structural H19 gene is required for transgene imprinting. *Proc. Natl. Acad. Sci. USA* **93**, 13876–13883.

Poirier, F., Chan, C.-T., Timmons, P., Robertson, E., Evans, M., and Rigby, P. (1991). The murine *H19* gene is activated during embryonic stem cell differentiation *in vitro* and at the time of implantation in the developing embryo. *Development* **113**, 1105–1114.

Qian, N., Frank, D., Keefe, D., Dao, D., Zhao, L., Yuan, L., Wang, Q., Keating, M., Walsh, C., and Tycko, B. (1997). The IPL gene on chromosome 11p15.5 is imprinted in humans and mice and is similar to TDAG51, implicated in Fas expression and apoptosis. *Hum. Mol. Genet.* **6**, 2021–2029.

Reik, W. (1989). Genomic imprinting and genetic disorders in man. *Trends Genet.* **5**, 331–336.

Reik, W., and Walter, J. (2001). Genomic imprinting: Parental influence on the genome. *Nat. Rev. Genet.* **2**, 21–32.

Reik, W., Collick, A., Norris, M., Barton, S., and Surani, M. (1987). Genomic imprinting determines methylation of parental alleles in transgenic mice. *Nature (London)* **328**, 248–251.

Robinson, W. P., and Lalande, M. (1995). Sex-specific meiotic recombination in the Prader-Willi/Angelman syndrome imprinted region. *Hum. Mol. Genet.* **4**, 801–806.

Rougeulle, C., Glatt, H., and Lalande, M. (1997). The Angelman syndrome candidate gene, *UBE3A/E6-AP*, is imprinted in the brain. *Nat. Genet.* **17**, 14–15.

Rougier, N., Bourc'his, D., Gomes, D., Niveleau, A., Plachot, M., Paldi, A., and Viegas-Pequignot, E. (1998). Chromosome methylation patterns during mammalian preimplantation development. *Genes Dev.* **12**, 2108–2113.

Saitoh, S., and Wada, T. (2000). Parent-of-origin specific histone acetylation and reactivation of a key imprinted gene locus in Prader-Willi syndrome. *Am. J. Hum. Genet.* **66**, 1958–1962.

Saitoh, S., Buiting, K., Rogan, P., Buxton, J., Driscoll, D., Arnemann, J., König, R., Malcolm, S., Horsthemke, B., and Nicholls, R. (1996). Minimal definition of the imprinting center and fixation of a chromosome 15q11-13 epigenotype by imprinting mutations. *Proc. Natl. Acad. Sci. USA* **93**, 7811–7815.

Sapienza, C., Paquette, J., Tran, T. H., and Peterson, A. (1989). Epigenetic and genetic factors affect transgene methylation imprinting. *Development* **107**, 165–168.

Schmidt, J., Matteson, P., Jones, B., Guan, X., and Tilghman, S. (2000). The Dlk1 and Gtl2 genes are linked and reciprocally imprinted. *Genes Dev.* **14**, 1997–2002.

Simon, I., Tenzen, T., Reubinoff, B. E., Hillman, D., McCarrey, J. R., and Cedar, H. (1999). Asynchronous replication of imprinted genes is established in the gametes and maintained during development. *Nature (London)* **401**, 929–932.

Sleutels, F., Barlow, D. P., and Lyle, R. (2000). The uniqueness of the imprinting mechanism. *Curr. Opin. Genet. Dev.* **10**, 229–233.

Smrzka, O., Fae, I., Stoger, R., Kurzbauer, R., Fischer, G., Henn, T., Weith, A., and Barlow, D. (1995). Conservation of a maternal-specific methylation signal at the human IGF2R locus. *Hum. Mol. Genet.* **4**, 1945–1952.

Solter, D. (1988). Differential imprinting and expression of maternal and paternal genomes. *Annu. Rev. Genet.* **22**, 127–146.

Spencer, H. (1997). Mutation-selection balance under genomic imprinting at an autosomal locus. *Genetics* **147**, 281–287.

Sun, F., Dean, W., Kelsey, G., Allen, N., and Reik, W. (1997). Transactivation of *Igf2* in a mouse model of Beckwith-Wiedemann syndrome. *Nature (London)* **389**, 809–815.

Surani, M., Barton, S., and Norris, M. (1984). Development of reconstituted mouse eggs suggests imprinting of the genome during gametogenesis. *Nature (London)* **308**, 548–550.

Sutcliffe, J. S., Nakao, M., Christian, S., Örstavik, K. H., Tommerup, N., Ledbetter, D. H., and Beaudet, A. L. (1994). Deletions of a differentially methylated CpG island at the SNRPN gene define a putative imprinting control region. *Nat. Genet.* **8**, 52–58.

Svensson, K., Mattsson, R., James, T., Wentzel, P., Pilartz, M., MacLaughlin, J., Miller, S., Olsson, T., Eriksson, U., and Ohlsson, R. (1998). The paternal allele of the *H19* gene is progressively silenced during early mouse development: The acetylation status of histones may be involved in the generation of variegated expression patterns. *Development* **125,** 61–69.

Swain, J. L., Stewart, T. A., and Leder, P. (1987). Parental legacy determines methylation and expression of an autosomal transgene: A molecular mechanism for parental imprinting. *Cell* **50,** 719–727.

Tada, T., Tada, M., Hilton, K., Barton, S., Sado, T., Takagi, N., and Surani, M. (1998). Epigenotype switching of imprintable loci in embryonic germ cells. *Dev. Genes Evol.* **207,** 551–562.

Tanaka, M., Puchyr, M., Gertsenstein, M., Harpal, K., Jaenisch, R., Rossant, J., and Nagy, A. (1999). Parental origin-specific expression of Mash2 is established at the time of implantation with its imprinting mechanism highly resistant to genome-wide demethylation. *Mech. Dev.* **87,** 129–142.

Tilghman, S. M. (1999). The sins of the fathers and mothers: Genomic imprinting in mammalian development. *Cell* **96,** 185–193.

Vrana, P., Guan, X., Ingram, R., and Tilghman, S. (1998). Genomic imprinting is disrupted in interspecific *Peromyscus* hybrids. *Nat. Genet.* **20,** 362–365.

Vu, T., and Hoffman, A. R. (1994). Promoter-specific imprinting of the human insulin-like growth factor-II gene. *Nature (London)* **371,** 714–717.

Vu, T., and Hoffman, A. (1997). Imprinting of the Angelman syndrome gene, *UBE3A*, is restricted to brain. *Nat. Genet.* **17,** 12–13.

Walsh, C., Chaillet, J., and Bestor, T. (1998). Transcription of IAP endogenous retroviruses is constrained by methylation. *Nat. Genet.* **20,** 116–117.

Xu, Y., Goodyer, C. G., Deal, C., and Polychronakos, C. (1993). Functional polymorphism in the parental imprinting of the human *IGF2R* gene. *Biochem. Biophys. Res. Commun.* **197,** 747–754.

Yang, T., Adamson, T., Resnick, J., Leff, S., Wevrick, R., Franke, U., Jenkins, N., Copeland, N., and Brannan, C. (1998). A mouse model for Prader-Willi syndrome imprinting-centre mutations. *Nat. Genet.* **19,** 25–31.

Zaina, S., Newton, R. V., Paul, M. R., and Graham, C. F. (1998). Local reduction of organ size in transgenic mice expressing a soluble insulin-like growth factor II/mannose-6-phosphate receptor. *Endocrinology* **139,** 3886–3895.

3

Ontogeny of Hematopoiesis: Examining the Emergence of Hematopoietic Cells in the Vertebrate Embryo

*Jenna L. Galloway and Leonard I. Zon**
Division of Hematology/Oncology
Harvard Medical School and Howard Hughes Medical Institute
Children's Hospital
Boston, Massachusetts 02115

Hematopoietic stem cells (HSCs) are responsible for generating all the lineages of the blood. During vertebrate development, waves of hematopoietic activity can be found in distinct anatomical sites, and they contribute to both embryonic and adult hematopoiesis. The origin of the HSCs that ultimately give rise to all the adult blood lineages has been a controversial issue in the field of hematopoiesis. Studies of amniotes have linked HSC activity to the aorta–gonad–mesonephros (AGM) region, whereas others suggest that the yolk sac is the true source of HSCs. This review describes both primitive and definitive hematopoiesis in mice, humans, chicks,

*To whom correspondence should be addressed. E-mail: zon@enders.tch.harvard.edu

Current Topics in Developmental Biology, Vol. 53

frogs, and zebrafish and examines the current debate over the embryonic origins of
HSCs.   © 2003, Elsevier Science (USA).

I. Introduction

Hematopoiesis is a dynamic process that requires the continuous coordination of
many cellular events throughout the life span of the organism. It is defined as
the differentiation of multipotent, self-renewing stem cells into all lineages of the
blood. The ability to maintain a continuous high rate of proliferation and differentiation of hematopoietic cells permits approximately 1 billion red blood cells and
100 million white blood cells to be produced per hour in the adult human (Sieff
and Nathan, 1993). This rate of differentiation and proliferation is unparalleled in
most other adult tissues, and the ultimate source of this activity is the hematopoietic stem cell (HSC). The HSC lies at the top of the hematopoietic hierarchy both
because of its ability to create progeny of all blood cell lineages and its apparently
limitless capacity to self-renew (Fleischman *et al.,* 1982; Harrison *et al.,* 1988;
Spangrude *et al.,* 1988; Jordan and Lemischka, 1990; Chaddah *et al.,*1996; Osawa
et al., 1996). This review focuses on the emergence of HSC activity in the early
embryo and the sites of hematopoiesis among different vertebrates.

Vertebrate hematopoiesis is thought to occur in two successive waves, primitive
and definitive, that differ in cell types produced and anatomic location. Whereas the
primitive wave occurs transiently, generating only nucleated erythrocytes expressing genes encoding embryonic globins and macrophages, the definitive program
lasts the life of the organism and produces HSCs capable of giving rise to all blood
lineages. The differentiation of HSCs into the various blood cell types is believed
to occur through a series of committed progenitors that maintain great proliferative potential but that lack significant self-renewal ability (Morrison *et al.,* 1997).
Oligopotent definitive progenitors, which have been identified in the mouse, give
rise to either the lymphoid or myeloerythroid lineage (Kondo *et al.,* 1997; Akashi
et al., 2000). Common lymphoid progenitors give rise to T, B, and natural killer
cells, whereas common myeloid progenitors produce monocytes, granulocytes,
megakaryocytes, and erythrocytes. Specific assays have been developed in order
to distinguish HSCs from committed progenitors. The most stringent assay for
defining HSCs is the murine long-term multilineage repopulating (LTR) assay, in
which donor cells are transplanted into a host mouse that has been lethally irradiated to remove endogenous HSC activity. Donor cells are considered HSCs if they
are able to reconstitute all lineages of the blood indefinitely.

A. History

Studies examining the origins of HSCs have traced the location of stem cell activity in different vertebrates during embryogenesis and early adulthood. In most

vertebrates, the primitive wave originates on the extraembryonic yolk sac (YS) whereas the definitive wave occurs at an intraembryonic site later in development. Over the past few decades, a controversy has emerged concerning whether definitive HSCs ultimately derive from an extraembryonic or intraembryonic location. The original hypothesis put forth by Moore and Owen (1965, 1967) stated that all HSCs originate from the YS. This was established by experiments in which host chick bodies were grafted onto a donor YS. After several days of incubation, donor cells from the YS were found in the hematopoietic organs of the host embryo, demonstrating the ability of the YS to contribute to definitive hematopoiesis. Further evidence supporting this hypothesis came from experiments performed in mice in which day 7 postcoitum (dpc) embryos grown *in vitro* without a YS developed normally, but lacked hematopoietic cells (Moore and Metcalf, 1970). However, the original avian grafting experiments of Moore and Owen were performed after circulation had begun, and possible contamination from circulating HSCs may have occurred. Therefore, this hypothesis was reconsidered, and the experiments were repeated with embryos and YSs grafted before the onset of circulation. When precirculation embryos were used, host cells were the sole contributors to definitive hematopoiesis, suggesting that definitive HSCs are derived from an intraembryonic location in the avian system (Dieterlen-Liévre, 1975; Lassila *et al.*, 1978, 1982).

An intraembryonic hematopoietic region closely associated with the aortic endothelium had been found to contain erythropoietic foci in the chick (Dieterlen-Liévre and Martin, 1981). Later studies further characterized the hematopoietic capacity of the avian aortic region and also explored the hematopoietic potential of the murine AGM. The mouse initially appeared to mimic the chick in that the AGM possessed hematopoietic activity capable of LTR of irradiated adult mice whereas the murine YS was restricted to primitive hematopoiesis (Medvinsky *et al.*, 1993; Müller *et al.*, 1994). These results in mice, in addition to the earlier chick chimera experiments, indicated that the AGM region has the potential to be the sole source of intraembryonic HSCs. However, further examination by Yoder and colleagues (1997b) demonstrated that the murine YS does in fact contain HSCs, and their ability to repopulate a host is dependent on a supportive environment determined by the age of the recipient (Yoder and Hiatt, 1997). More recently, studies with AGM-derived stromal lines in mice have supported a dual model in which cells from both the YS and AGM isolated before the onset of circulation are capable of contributing to definitive hematopoiesis (Matsuoka *et al.*, 2001).

B. Models

Currently three general models describing the origin of HSCs have been put forth. The first model asserts that the YS and AGM HSCs arise independently, whereupon they both colonize the fetal liver and produce definitive hematopoietic cells. In the second model, all hematopoietic stem cells are derived from the YS, and the AGM cells provide the correct environmental cues to enable YS cells to mature into HSCs and contribute to adult hematopoiesis. The third model states that the AGM is the sole source of HSCs, and any definitive activity seen in the YS is derived from AGM cells that travel through the circulation to the YS. Different vertebrate systems have been used to study the origins of HSCs and to examine the evidence pertaining to each of these models.

II. Mammalian Hematopoiesis

A. Mouse Hematopoiesis

1. Yolk Sac

Hematopoietic cells are derived from the mesodermal germ layer, which is formed from cells migrating through the primitive streak between the presumptive ectoderm and the endoderm. On the basis of the location of ingression along the anterior–posterior axis of the primitive streak, the cells are fated to become either dorsal (notochord, somites) or ventral (blood, mesenchyme) mesoderm (Tam and Behringer, 1997). The same layer of mesoderm extends both into the body of the embryo and extraembryonically, into the YS. At stage 7–7.5 dpc, the extraembryonic mesodermal cells in association with the visceral endoderm form the blood islands, the outer cells of which later differentiate into endothelial cells, whereas the inner cells become primitive erythroblasts (Haar and Ackerman, 1971). It is at this time that the first primitive erythroid and macrophage precursors can be observed morphologically, and the expression of hematopoietic genes can be detected (Palis *et al.*, 1999, 2001).

Expression of the hematopoietic genes *stem cell leukemia* (*scl*) and *lmo2* marks the appearance of presumptive HSCs in the yolk sac at 7 dpc (Palis *et al.*, 2001). Targeted disruption of either gene results in a loss of YS hematopoiesis *in vivo* and an inability to produce definitive colonies *in vitro* (Begley *et al.*, 1989; Warren *et al.*, 1994; Shivdasani *et al.*, 1995; Robb *et al.*, 1995; Yamada *et al.*, 1998). Subsequent to their expression, GATA1, a zinc finger DNA-binding protein, is detected in the YS (Pevny *et al.*, 1995; Palis *et al.*, 2001). The *GATA1* gene is expressed in cells of the myeloerythroid lineage, but it is only required for the maturation of erythroid progenitors beyond the proerythroblast stage (Pevny *et al.*, 1991, 1995; Weiss *et al.*, 1994). The nucleated primitive red blood cells then enter circulation between 8.5 and 9 dpc, and by 9 dpc the primitive erythroid potential of the YS has disappeared (Palis *et al.*, 1999).

In addition to primitive erythropoiesis, definitive hematopoietic activity has been attributed to the YS. Palis and colleagues (1999) discovered small enucleated cells expressing adult *globin,* termed burst-forming unit-erythroid (BFU-E) cells, at 8 dpc in the YS before circulation begins. By 9 dpc, BFU-E cells are found in the bloodstream, and at about 11 dpc BFU-E cells are found in the liver, suggesting this activity originates from the YS (Palis *et al.*, 1999). The YS also harbors multipotent cells capable of generating myeloid and B-lymphoid cells at 8.5–9 dpc and other progenitors that give rise to erythroid cells, macrophages, granulocytes, mast cells, and megakaryocytes at 9 dpc (Huang *et al.*, 1994; Godin *et al.*, 1995; Yoder *et al.*, 1997a; Palis *et al.*, 1999). YS cells taken at 9–10 dpc can reconstitute the blood of conditioned newborn recipients for up to 1 year, yet are incapable of LTR of conditioned adult recipients (Yoder and Hiatt, 1997; Yoder *et al.*, 1997b). Matsuoka and colleagues have shown that YS cells isolated at 8 dpc have LTR-HSC activity in adult recipients when cultured for 4 days with AGM-derived stromal cells. Therefore, it appears that the AGM cells are supportive for definitive HSCs from the YS and that the YS possesses definitive HSC potential (Matsuoka *et al.*, 2001).

2. Aorta–Gonad–Mesonephros

The mesodermal germ layer is divided into the extraembryonic and intraembryonic regions, and the intraembryonic lateral plate mesoderm is subdivided by a coelomic cavity into the somatopleura, which contacts the ectoderm above, and the splanchnopleura, which contacts the endoderm below. After 9.5 dpc, the splanchnopleura will first give rise to the aorta, followed by the mesonephros, mesentery, and gonads (Cumano and Godin, 2001). Hematopoietic progenitors have been discovered in the paraaortic splanchnopleure (P-Sp)/AGM region as early as 9 dpc in the mouse (Medvinsky *et al.*, 1993). At 10 dpc, the AGM region is capable of long-term multilineage repopulation of primary and serial adult recipients (Müller *et al.*, 1994). After 13 dpc, the hematopoietic progenitor activity observed in the AGM can no longer be detected (Sánchez *et al.*, 1996).

Morphologically, cells have been observed protruding into the lumen of the dorsal aorta between 9.5 and 11 dpc, and these cells are thought to represent HSCs exiting the AGM region to seed the fetal liver (Garcia-Porrero *et al.*, 1995; North *et al.*, 1999). The hematopoietic cells within the AGM region express genes encoding transcriptional regulators associated with HSC formation, such as *scl*, *aml1*, c-*myb*, and *lmo2*, as well as genes encoding the cell surface markers CD34 and c-Kit (Delassus *et al.*, 1999). Targeted disruption of the *aml1* gene completely eliminates the definitive program and the formation of the intraaortic cell clusters within the AGM (Okuda *et al.*, 1996; Wang *et al.*, 1996; North *et al.*, 1999). The proto-oncogene c-*myb* selectively affects definitive hematopoiesis and the hematopoietic cells in the P-Sp/AGM but does not affect the primitive program (Mucenski *et al.*, 1991; Mukouyama *et al.*, 1999). Because c-*myb*$^{-/-}$ mice have some hematopoietic activity, it has been suggested that c-*myb* may have more of a role in proliferation rather than differentiation of HSCs (Gewirtz and Calabretta, 1988; Mucenski *et al.*, 1991). The transmembrane proteins CD34 and c-Kit have both been shown to be important stem cell markers in the bone marrow (Krause *et al.*, 1994; Geissler *et al.*, 1988; Tan *et al.*, 1990). These markers have been used to characterize the AGM and other hematopoietic tissues in mice and also to identify the corresponding sites of hematopoiesis in vertebrates.

3. Liver

The progression of hematopoietic activity proceeds sequentially from the YS/AGM to the liver, spleen, and finally to the bone marrow. The fetal liver does not produce hematopoietic cells *de novo,* but is instead colonized by cells originating from the YS and /or the AGM, although no studies documenting this migration have been performed in mice (Johnson and Moore, 1975; Houssaint, 1981).

The liver primordium is first observed in mice at 10 dpc as an endodermal outgrowth from the foregut beneath the pericardial cavity. The three-dimensional arrangement of prehepatocytes serves as a framework in the liver that supports the hematopoietic cells (Rifkind *et al.,* 1969; Medlock and Haar, 1983). Colonization of the fetal liver occurs after 10 dpc as indicated by *aml1* expression at 10.5 dpc and LTR activity between 11 and 12 dpc (North *et al.,* 1999; Müller *et al.,* 1994). By 12 dpc, nonnucleated erythroid cells are found in circulation and the population of primitive erythrocytes declines, signaling the start of definitive hematopoiesis (Rifkind *et al.,* 1969). Erythropoiesis is the main activity of the fetal liver despite the presence of myeloid and lymphoid progenitors (Delassus and Cumano, 1996; Mebius and Akashi, 2000; Mebius *et al.,* 2001). At 14 dpc, maturing erythroid cells form the erythroblastic islands, and as hematopoietic activity declines, their size decreases accordingly (Sasaki and Iwatsuki, 1997; Sasaki and Sonoda, 2000). After birth, definitive hematopoiesis in the liver subsides, and the erythroblastic islands disappear.

4. Spleen and Bone Marrow

As a predominantly erythropoietic organ at the end of fetal life, the spleen aids in the transition from fetal liver to bone marrow hematopoiesis. The splenic rudiment forms as a mesenchymal thickening in the dorsal mesogastrium. It is colonized by HSCs at 12 dpc; however, it is not hematopoietic until 16 dpc, when immature erythroid cells and small lymphocytes can be morphologically observed (Sasaki and Matsumura, 1988; Godin *et al.,* 1999).

Although HSCs are not detected in the bone marrow until 18 dpc, it is believed that colonization by hematopoietic cells has already occurred (Godin *et al.,* 1999). This is demonstrated by the appearance of B cell precursors as early as 15 dpc, and lymphopoiesis by 17 dpc within the bone marrow, which remains the site of hematopoiesis throughout adult life (Delassus and Cumano, 1996) (Figure 1A).

B. Human Hematopoiesis

Examination of hematopoietic stem cells in developing human embryos has relied on immunohistochemistry with known surface antigens and *in vitro* colony

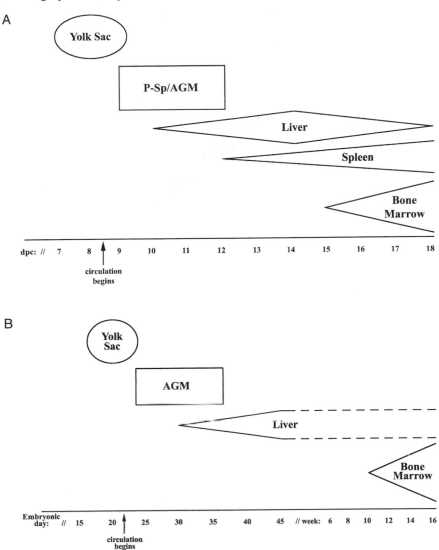

Figure 1 Approximate time of appearance of hematopoietic activity in different tissues of mice (A) and humans (B) during embryogenesis (based on current knowledge).

assays in which the potential of clonogenic hematopoietic progenitors in the YS and intraembryonic tissues can be determined (Huyhn *et al.,* 1995). The CD34 glycoprotein marks both vascular endothelial cells and the earliest multipotential hematopoietic stem cells (Andrews *et al.,* 1989; Terstappen *et al.,* 1991; Fina *et al.,* 1990). The pan-leukocyte surface molecule CD45 can be used to discriminate between the two populations of CD34+ cells, as it recognizes only hematopoietic cells (Shah *et al.,* 1988). Although technical limitations restrict the study of human hematopoiesis, it currently appears that the human program closely parallels the mouse program in both the sequential progression of hematopoietic activity from one site to another and the embryonic tissues involved.

1. Yolk Sac

Primitive erythroblasts and $CD34^+$ hematopoietic cells can be distinguished after 18.5 days of development in the YS, indicating hematopoiesis has begun (Bloom and Bartelmez, 1940). The heart begins beating at day 22 (four somites), after which erythroid cells are found in the circulation and within the embryo (Tavian *et al.,* 1999). After day 24 of development, YS mesoderm is no longer differentiating, and the intravascular blood islands are not present (Tavian *et al.,* 1999).

2. Aorta–Gonad–Mesonephros

Hematopoietic cells with both lymphoid and myeloid potential have been detected in the P-Sp and the aorta between 24 and 34 days, but this potential has never been observed in the YS (Tavian *et al.,* 2001). Tavian and colleagues (1996, 1999) have shown that the preumbilical aortic region can give rise to progenitor cells of both the erythroid and granulocytic lineages. The first $CD34^+$ hematopoietic cells are detected in the mesenchyme surrounding the dorsal aorta on day 26 (24 somites) whereas $CD34^+$ hematopoietic cells are present in the endothelium of the dorsal aorta and the ventral cell clusters between days 27 and 36. Morphological analysis on day 35 shows unorganized endothelial cells near the ventral clusters in the dorsal aorta, suggesting that hematopoietic cells have budded from the subaortic mesoderm (Tavain *et al.,* 1999).

3. Liver

On day 23, the liver rudiment begins developing, after which it is colonized by hematopoietic cells (Tavian *et al.,* 1999). $CD45^+CD34^+$ hematopoietic progenitors are first detected in the liver on day 30, and the $CD34^+$ cell population increases in number until 42 days, whereupon the liver is the main hematopoietic organ (Tavian *et al.,* 1999). At this time, erythroid cells are the major cell type found in the embryonic liver and are located extravascularly (Fukuda, 1974). The liver essentially ceases to be hematopoietic by birth, although scattered erythroid precursors may be present in some neonates.

4. Bone Marrow

The bones of the human embryo are merely cartilaginous rudiments between weeks 6 and 8.5 of gestation (Charbord *et al.,* 1996). Between weeks 8.5 and 9, chondrolysis occurs, and osteoblasts, osteoclasts, and perichondral precursors invade the marrow cavity. After week 10.5, $CD45^+$ hematopoietic cells, $CD15^+$ cells, $CD68^+$ cells, and a low frequency of $CD34^+$ hematopoietic cells mark the appearance of hematopoiesis in the bone marrow. Bone formation is complete by week 16, and hematopoiesis continues in the bone marrow for the life span of the individual (Charbord et al., 1996) (Figure 1B).

III. Avian Hematopoiesis

A. Yolk Sac

As in mice, chick mesoderm forms during gastrulation from the invagination of epiblast cells through the primitive streak. The mesodermal layer spans from the central embryonic location of the area pellucida to the area vasculosa, and it is subdivided into the splanchnopleure and the somatopleure. The intraembryonic mesoderm is composed of the paraxial mesoderm, which forms the somites, and the lateral plate mesoderm, which forms the somatopleural and splanchnopleural sheets. The YS blood islands develop in the area vasculosa surrounding the embryo. The extraembryonic mesodermal cells differentiate into primitive red blood cells by day 1.5 (2–4 somites), and embryonic circulation begins by day 2 (16-17 somites) once the vascular network has developed (Evans, 1997).

The contribution of the YS to definitive hematopoiesis has been examined in multiple quail–chick or chick–chick chimera experiments, in which a host embryo is grafted onto a donor YS (Dieterlen-Liévre, 1975; Lassila et al., 1978, 1982; Martin et al., 1978; Dieterlen-Liévre and Martin, 1981). For the chick–chick chimeras, antigens expressed on the surface of red blood cells or sex chromosomal analysis has been used to allow for a distinction between donor and host. These experiments demonstrated that the donor YS contributes only to primitive hematopoiesis, and by embryonic day 17–18 (E17–18), the majority of blood cells in circulation are definitive and arise from the host (Lassila et al., 1982). Therefore, in the avian system the primitive blood cells are YS derived, whereas all definitive hematopoietic cells stem from an intraembryonic location.

B. Intraaortic Clusters/Paraaortic Foci

In chicks, two regions in the AGM contain hematopoietic activity: the intraaortic clusters and the paraaortic foci (Dieterlen-Liévre and Martin, 1981). Between E3 and E4 the intraaortic clusters are closely associated with the endothelium within the dorsal aorta and express genes encoding c-Myb and CD45 (Vandenbunder et al., 1989; Jaffredo et al., 1999). Grafting experiments between chick and quail have demonstrated the hematopoietic activity of the AGM. After transplanting an E3–E4 quail aorta into a permissive chick environment, hematopoietic and endothelial cells labeled with the quail antibody QH1 could be detected. In contrast, no QH1[+] hematopoietic cells were present when other blood vessels were grafted, suggesting that the hematopoietic potential is unique to the aorta (Dieterlen-Liévre *et al.*, 1988).

The paraaortic foci arise between E6 and E8 in the ventral wall of the aorta and display diffuse hematopoietic activity, which is indicated by hemoglobin and the expression of c-*myb* (Dieterlen-Liévre and Martin, 1981; Vandenbunder *et al.*, 1989; Jaffredo *et al.*, 1998). The hematopoietic activity corresponds temporally with the colonization of the splenic, thymic, and bursal rudiments (Dieterlen-Liévre, 1975). As in mammals, avian bone marrow is the site of definitive hemato-poiesis, yet the hematopoietic activity in the paraaortic foci is lost 2 days before it is seen in the bone marrow, suggesting there is potentially another source of definitive HSCs. These HSCs may derive from the allantois, as it is capable of producing hematopoietic cells that seed the bone marrow (Caprioli *et al.*, 1998).

C. Bone Marrow

The stroma of the bone marrow forms from osteogenic and perivascular cells, which are derived from the bone mesenchyme anlage. The hematopoietic cell populations of the marrow originate from an extrinsic source as in mice and humans, and these cells colonize the marrow after E10 (Le Douarin *et al.*, 1975; Dieterlen-Liévre and Martin, 1981). Before hatching, the bone marrow has become the primary source of hematopoietic cells, and it continues to produce blood cells throughout the life of the adult (Figure 2A).

IV. *Xenopus* Hematopoiesis

A. Ventral Blood Island

In *Xenopus*, signals emanating from the future endoderm induce the formation of the mesoderm in the overlying blastomeres during the early blastula stage. Fate-mapping experiments demonstrate that ventral mesoderm gives rise to embryonic blood and the dorsal mesoderm gives rise to notochord and somites (Dale and Slack, 1987; Moody, 1987). However, studies have indicated that dorsal mesoderm can in fact contribute to embryonic blood cells in the ventral blood island (VBI) (Lane and Smith, 1999; Mills *et al.*, 1999). The VBI in *Xenopus* is functionally equivalent to the mammalian and avian YS, and it forms at 24 hours postfertilization (hpf) from ventral mesoderm in association with presumptive endothelium and presump-tive hepatic endoderm (Mangia *et al.*, 1970). During neurula stages, *scl*, c-*myb*, and *GATA1* are expressed in the developing VBI (Turpen *et al.*, 1997). After the formation of the VBI, cells differentiate into primitive erythrocytes, which express

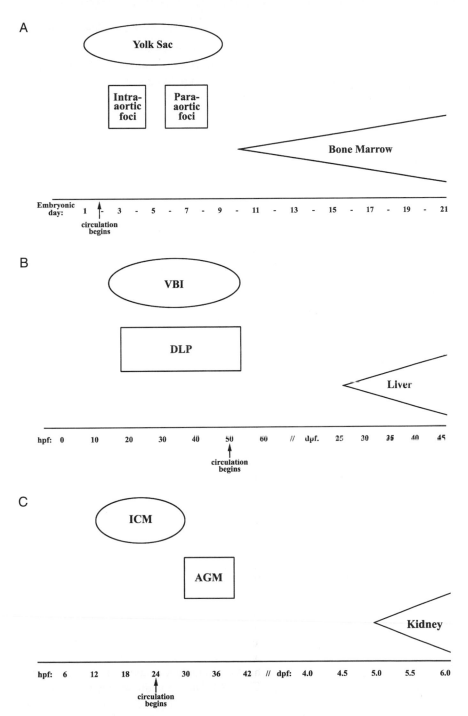

Figure 2 Approximate shift in hematopoietic activity temporally and spatially in chicks (A), *Xenopus* (B), and zebrafish (C) (based on current knowledge).

globin at 40 hpf, and by 50 hpf circulation is established (Mangia *et al.,* 1970). The hematopoietic cells of the VBI also contribute to definitive hematopoiesis, and they have been shown to colonize the liver and thymus, where they later give rise to definitive erythrocytes, thymocytes, and B lymphocyte precursors (Kau and Turpen, 1983; Smith *et al.,* 1989; Bechtold *et al.,* 1992).

B. Dorsal Lateral Plate

The dorsal lateral plate (DLP) is composed of the pronephros, postcardinal veins, and dorsal aorta, and it is equivalent to the mammalian and avian AGM (Kau and Turpen, 1983; Maéno *et al.,* 1985b). The c-*myb* gene is expressed in the DLP region as early as 18 hpf, after which the expression of *scl* at 26 hpf and *GATA1* at 36 hpf is detected in the DLP (Turpen *et al.,* 1997). Through transplantation experiments, it has been determined that the hematopoietic cells of the DLP mesoderm do not contribute to primitive erythropoiesis. Instead, these cells colonize the liver and thymus, and later produce the definitive hematopoietic populations during the larval period and throughout the life of the adult (Kau and Turpen, 1983; Maéno *et al.,* 1985a,b; Bechtold *et al.,* 1992; Chen and Turpen, 1995).

C. Liver

The liver rudiment develops from the foregut of the neurula embryo between the VBI and the heart (Nieuwkoop and Faber, 1967). By 2.5–3 dpf, the liver occupies the hepatic cavity, whereupon growth of the intestine shifts it toward an anterior and dorsal region of the intestine. In *Xenopus,* cells from the VBI and DLP colonize the liver, which will serve as the definitive site of hematopoiesis in both larval and adult stages. The contribution of the VBI and DLP to definitive hematopoiesis has been examined in *Xenopus* primarily because these locations are anatomically distinct and the system is amenable to transplantation experiments. By 26 dpf, definitive erythrocytes are the predominant cell type produced in the liver, with the majority of these originating from the DLP (Chen and Turpen, 1995). Most of the circulating erythrocytes at this time are of the primitive lineage and stem from the VBI. Over the next several hours of development, the relative contribution of VBI and DLP cells to erythropoiesis in the liver undergoes a dynamic change. During metamorphosis at 35 dpf, DLP-derived erythropoiesis decreases and is surpassed by VBI-derived erythropoiesis. In addition, definitive hematopoietic cells originating from the VBI undergo lymphopoiesis in the thymus, and this activity reaches a peak at 35 dpf (Bechtold *et al.,* 1992). It is not until 42 dpf that DLP-derived definitive red blood cells enter the circulation (Chen and Turpen, 1995). After metamorphosis, the contribution of VBI-derived cells to erythropoiesis in the liver and lymphopoiesis in the thymus declines, and it is replaced by DLP-derived cells (Bechtold *et al.,* 1992; Chen and Turpen, 1995) (Figure 2B).

V. Zebrafish Hematopoiesis

A. Intermediate Cell Mass

In the zebrafish embryo, mesoderm is induced by signals emanating from the underlying yolk cell and, as in other vertebrates, members of the transforming growth factor β (TGF-β) superfamily are involved in inducing ventral mesodermal fates (Kishimoto *et al.*, 1997; Dick *et al.*, 2000; Schmid *et al.*, 2000). After gastrulation, expression of *scl* and *lmo2* in the posterior lateral plate mesoderm likely signifies the presence of the first HSCs, and *GATA1* expression at about 12 hpf (five somites) signals the appearance of the first erythroid precursors (Thompson *et al.*, 1998). After detaching from the posterior lateral plate, the erythroid precursor cells migrate medially to form the ICM, and at 15 hpf they begin expressing embryonic *globin* (Al-Adhami and Kunz, 1977; Willett *et al.*, 1999). The ICM is believed to be the teleost equivalent of the YS blood islands in mammals, and it is composed of a mass of primitive erythroid cells enclosed within the trunk axial vein (Al-Adhami and Kunz, 1977). After 24 hpf, the heart begins to beat and round, nucleated erythroblasts enter the circulation, where they subsequently differentiate into primitive erythrocytes (Willett *et al.*, 1999).

Through the use of mutagenesis screens in zebrafish, several mutants with defects in hematopoiesis have been identified (Ransom *et al.*, 1996). The mutant *cloche*, which has severely reduced hematopoietic and endothelial cells, fails to express *scl* in the posterior lateral plate between 2 and 12 somites (Stainier *et al.*, 1995; Liao *et al.*, 1997). This mutant also lacks macrophages and definitive markers in the thymus, suggesting that the mutation may act at the HSC level (Thompson *et al.*, 1998; N. Trede, personal communication). In another zebrafish mutant, *moonshine*, there is no *GATA1* expression, and only the erythroid blood lineage is disrupted (Ransom *et al.*, 1996). Interestingly, the *bloodless* mutant has no primitive hematopoiesis, yet the definitive program appears undisturbed. This zebrafish mutant behaves unlike any of the other mouse mutants, in which there are defects in definitive or both primitive and definitive hematopoiesis. The gene responsible for this mutation is being cloned, and its discovery will provide insight into the molecular pathways required only for primitive hematopoiesis.

B. Aorta–Gonad–Mesonephros Equivalent

The expression of c-*myb* in the ventral wall of the dorsal aorta at approximately 31 hpf marks the prospective zebrafish equivalent of the mammalian AGM (Thompson *et al.*, 1998). The *aml1* (*runxa*) gene is also expressed in the dorsal aorta at this time (C. Erter, personal communication). Thompson and colleagues (1998) speculate that the c-*myb*-expressing cells are the definitive progenitors that ultimately colonize the kidney, the adult site of hematopoiesis in the zebrafish.

C. Kidney

Kidney differentiation is morphologically apparent by 48 hpf, when renal tubules and glomeruli appear in the pronephros (Drummond, 2000). By 5 dpf, definitive hematopoiesis is underway, erythrocytes and granulocytes are detected within the pronephros, and elliptical nucleated red blood cells expressing adult *globin* are found in the circulation (Willett *et al.,* 1999). Reticular cells lining the sinusoidal blood vessels in the pronephros likely serve as the proper support for the developing hematopoietic cells. Over the next few weeks, there is an increase in hematopoietic tissue, and after 3 weeks of development lymphoid cells can be observed in the pronephros (Willett *et al.,* 1999) (Figure 2C).

VI. Summary

Hematopoiesis is essential for the development of an organism, and many aspects of its ontogeny have been conserved. In most vertebrates, cells derived from the ventral mesoderm contribute to primitive hematopoiesis whereas cells from the dorsal lateral plate mesoderm are involved in the definitive program (Table I). All fetal and adult sites of blood formation, such as the liver and bone marrow in mice, are not intrinsically hematopoietic and are instead colonized by HSCs that provide the organism with a lifetime supply of blood. The origin of HSCs has been a major question among investigators, and several models regarding a YS or an AGM contribution to definitive hematopoiesis have been proposed.

The contribution of the VBI and DLP to definitive hematopoiesis has been undisputedly determined in *Xenopus,* and yet the role of the YS and the AGM in definitive hematopoiesis in mice has been at the root of an ongoing controversy. Currently, it appears that murine YS cells are capable of contributing to definitive hematopoiesis, but they require a supportive environment in order to attain HSC potential (Yoder *et al.,* 1997b; Matsuoka *et al.,* 2001). This could be interpreted

Table I   Comparison of Sites of Primitive and Definitive Hematopoiesis in Different Vertebrates

	Primitive	Intraembryonic	Fetal	Adult
Mouse	Yolk sac	AGM	Liver	Bone marrow
Human	Yolk sac	AGM	Liver	Bone marrow
Chick	Yolk sac	Intra /paraaortic foci, allantois	—	Bone marrow
Frog	VBI	DLP	Liver	Liver
Zebrafish	ICM	AGM	Kidney	Kidney

Abbreviations: AGM, Aorta–gonad–mesonephros; DLP, dorsal lateral plate; ICM, intermediate cell mass; VBI, ventral blood island.

as either a homing issue and/or a requirement for specific environmental cues. Because murine YS cells accomplish LTR of newborn but not adult recipients, YS cells may not be capable of seeding the marrow directly (Yoder and Hiatt, 1997; Yoder *et al.,* 1997b). Instead, they may need to follow the sequence of hematopoietic colonization from the YS to the liver and finally to the bone marrow. More recent work has shown that cells derived from the AGM provide instructive signals enabling YS cells to mature into HSCs capable of seeding adult marrow (Matsuoka *et al.,* 2001). In addition, the AGM has been shown to produce HSCs *de novo* before circulation (Cumano *et al.,* 2001). Taken together, the findings suggest a model in which YS cells receive specific instructive signals, possibly from supportive cells from the AGM, to mature into HSCs, and ultimately, the YS- and AGM-derived HSCs are responsible for hematopoiesis in the fetal liver and the bone marrow of the mouse. There are no data directly supporting this model in mammals; however, fate-mapping experiments tracking the movements of YS or AGM cells in the early embryo would resolve the debate over the source of HSCs.

In contrast to mice, chick chimera experiments demonstrate that definitive cells are entirely derived from an intraembryonic location (Dieterlen-Liévre, 1975; Lassila *et al.,* 1978, 1982). It has been proposed that because the chick chimera experiments were performed with unconditioned hosts in which the host HSCs were not ablated, the microenvironment could not permit the colonization by extrinsic cells (Yoder, 2001). However, the grafting experiments performed with the allantois, which demonstrated colonization of the chicks bone marrow by an exogenous source, would argue against this conclusion (Caprioli *et al.,* 1998). Additional experiments in chicks are needed to clarify this issue. Further insights into the developmental biology of HSCs may come from other vertebrate systems such as the zebrafish, where genetic screens and transgenic animals with promoter–reporter gene constructs can be designed to target specific aspects of hematopoiesis. Ultimately, the examination of multiple vertebrate systems will answer many of the questions regarding the origins of HSCs and allow for a more complete understanding of the molecular pathways of hematopoiesis.

Acknowledgments

We thank Noëlle Paffett-Lugassy, Alan Davidson, David Travor, Barry Paw, Caroline Burns, Kim Dooley, and José Rivera-Feliciano for helpful advice and critical reading of the manuscript.

References

Akashi, K., Traver, D., Miyamoto, T., and Weissman, I. L. (2000). A clonogenic common myeloid progenitor that gives rise to all myeloid lineages. *Nature (London)* **404,** 193–197.
Al-Adhami, M. A., and Kunz, Y. W. (1977). Ontogenesis of haematopoietic sites in *Brachydaniorerlo. Dev. Growth Differ.* **19,** 171–179.

Andrews, R. G., Singer, J. W., and Bernstein, I. D. (1989). Precursors of colony-forming cells in humans can be distinguished from colony-forming cells by expression of the CD33 and CD34 antigens and light scatter properties. *J. Exp. Med.* **169,** 1721–1731.

Bechtold, T. E., Smith, P. B., and Turpen, J. B. (1992). Differential stem cell contributions to thymocyte succession during development of *Xenopus laevis*. *J. Immunol.* **148,** 2975–2982.

Begley, C. G., Aplan, P. D., Denning, S. M., Haynes, B. F., Waldmann, T. A., and Kirsch, I. R. (1989). The gene SCL is expressed during early hematopoiesis and encodes a differentiation-related DNA-binding motif. *Proc. Natl. Acad. Sci. USA* **86,** 10128–10132.

Bloom, W., and Bartelmez, G. W. (1940). Hematopoiesis in young human embryos. *Am. J. Anat.* **67,** 21–53.

Caprioli, A., Jaffredo, T., Gautier, R., Dubourg, C., and Dieterlen-Liévre, F. (1998). Blood-borne seeding by hematopoietic and endothelial precursors from the allantois. *Proc. Natl. Acad. Sci. USA* **95,** 1641–1646.

Chaddah, M. R., Wu, D. D., and Phillips, R. A. (1996). Variable self-renewal of reconstituting stem cells in long-term bone marrow cultures. *Exp. Hematol.* **24,** 497–508.

Charbord, P., Tavian, M., Humeau, L., and Peault, B. (1996). Early ontogeny of the human marrow from long bones: An immunohistochemical study of hematopoiesis and its microenvironment. *Blood* **87,** 4109–4119.

Chen, X. D., and Turpen, J. B. (1995). Intraembryonic origin of hepatic hematopoiesis in *Xenopus laevis*. *J. Immunol.* **154,** 2557–2567.

Cumano, A., and Godin, I. (2001). Pluripotent hematopoietic stem cell development during embryogenesis. *Curr. Opin. Immunol.* **13,** 166–171.

Cumano, A., Ferraz, J. C., Klaine, M., Di Santo, J. P., and Godin, I. (2001). Intraembryonic, but not yolk sac hematopoietic precursors, isolated before circulation, provide long-term multilineage reconstitution. *Immunity* **15,** 477–485.

Dale, L., and Slack, J. M. (1987). Regional specification within the mesoderm of early embryos of *Xenopus laevis*. *Development* **100,** 279–295.

Delassus, S., and Cumano, A. (1996). Circulation of hematopoietic progenitors in the mouse embryo. *Immunity* **4,** 97–106.

Delassus, S., Titley, I., and Enver, T. (1999). Functional and molecular analysis of hematopoietic progenitors derived from the aorta–gonad–mesonephros region of the mouse embryo. *Blood* **94,** 1495–1503.

Dick, A., Hild, M., Bauer, H., Imai, Y., Maifeld, H., Schier, A. F., Talbot, W. S., Bouwmeester, T., and Hammerschmidt, M. (2000). Essential role of Bmp7 (snailhouse) and its prodomain in dorsoventral patterning of the zebrafish embryo. *Development* **127,** 343–354.

Dieterlen-Liévre, F. (1975). On the origin of haemopoietic stem cells in the avian embryo: An experimental approach. *J. Embryol. Exp. Morphol.* **33,** 607–619.

Dieterlen-Liévre, F., and Martin, C. (1981). Diffuse intraembryonic hemopoiesis in normal and chimeric avian development. *Dev. Biol.* **88,** 180–191.

Dieterlen-Liévre, F., Pardanaud, L., Yassine, F., and Cormier, F. (1988). Early haemopoietic stem cells in the avian embryo. *J. Cell. Sci. Suppl.* **10,** 29–44.

Drummond, I. A. (2000). The zebrafish pronephros: A genetic system for studies of kidney development. *Pediatr. Nephrol.* **14,** 428–435.

Evans, T. (1997). Developmental biology of hematopoiesis. *Aplastic Anemia and Stem Cell Biol.* **11,** 1115–1147.

Fina, L., Molgaard, H. V., Robertson, D., Bradley, N. J., Monaghan, P., Delia, D., Sutherland, D. R., Baker, M. A., and Greaves, M. F. (1990). Expression of the CD34 gene in vascular endothelial cells. *Blood* **75,** 2417–2426.

Fleischman, R. A., Custer, R. P., and Mintz, B. (1982). Totipotent hematopoietic stem cells: Normal self-renewal and differentiation after transplantation between mouse fetuses. *Cell* **30,** 351–359.

Fukuda, T. (1974). Fetal hemopoiesis. II. Electron microscopic studies on human hepatic hemopoiesis. *Virchows Arch. B Cell Pathol.* **16,** 249–270.

Garcia-Porrero, J. A., Godin, I. E., and Dieterlen-Liévre, F. (1995). Potential intraembryonic hemogenic sites at pre-liver stages in the mouse. *Anat. Embryol. (Berl.)* **192,** 425–435.

Geissler, E. N., Ryan, M. A., and Housman, D. E. (1988). The dominant-white spotting (W) locus of the mouse encodes the c-*kit* proto-oncogene. *Cell* **55,** 185–192.

Gewirtz, A. M., and Calabretta, B. (1988). A c-*myb* antisense oligodeoxynucleotide inhibits normal human hematopoiesis *in vitro. Science* **242,** 1303–1306.

Godin, I., Dieterlen-Liévre, F., and Cumano, A. (1995). Emergence of multipotent hemopoietic cells in the yolk sac and paraaortic splanchnopleura in mouse embryos, beginning at 8.5 days postcoitus. *Proc. Natl. Acad. Sci. USA* **92,** 773–777.

Godin, I., Garcia-Porrero, J. A., Dieterlen-Liévre, F., and Cumano, A. (1999). Stem cell emergence and hemopoietic activity are incompatible in mouse intraembryonic sites. *J. Exp. Med.* **190,** 43–52.

Haar, J. L., and Ackerman, G. A. (1971). A phase and electron microscopic study of vasculogenesis and erythropoiesis in the yolk sac of the mouse. *Anat. Rec.* **170,** 199–223.

Harrison, D. E., Astle, C. M., and Lerner, C. (1988). Number and continuous proliferative pattern of transplanted primitive immunohematopoietic stem cells. *Proc. Natl. Acad. Sci. USA* **85,** 822–826.

Houssaint, E. (1981). Differentiation of the mouse hepatic primordium. II. Extrinsic origin of the haemopoietic cell line. *Cell Differ.* **10,** 243–252.

Huang, H., Zettergren, L. D., and Auerbach, R. (1994). In vitro differentiation of B cells and myeloid cells from the early mouse embryo and its extraembryonic yolk sac. *Exp. Hematol.* **22,** 19–25.

Huyhn, A., Dommergues, M., Izac, B., Croisille, L., Katz, A., Vainchenker, W., and Coulombel, L. (1995). Characterization of hematopoietic progenitors from human yolk sacs and embryos. *Blood* **86,** 4474–4485.

Jaffredo, T., Gautier, R., Eichmann, A., and Dieterlen-Liévre, F. (1998). Intraaortic hemopoietic cells are derived from endothelial cells during ontogeny. *Development* **125,** 4575–4583.

Jaffredo, T., Gautier, R., Brajeul, V., and Dieterlen-Liévre, F. (1999). [Affiliation between endothelial and intra-embryonic hematopoietic stem cells]. *J. Soc. Biol.* **193,** 165–170.

Johnson, G. R., and Moore, M. A. (1975). Role of stem cell migration in initiation of mouse foetal liver haemopoiesis. *Nature (London)* **258,** 726–728.

Jordan, C. T., and Lemischka, I. R. (1990). Clonal and systemic analysis of long-term hematopoiesis in the mouse. *Genes Dev.* **4,** 220–232.

Kau, C. L., and Turpen, J. B. (1983). Dual contribution of embryonic ventral blood island and dorsal lateral plate mesoderm during ontogeny of hemopoietic cells in *Xenopus laevis. J. Immunol.* **131,** 2262–2266.

Kishimoto, Y., Lee, K. H., Zon, L., Hammerschmidt, M., and Schulte-Merker, S. (1997). The molecular nature of zebrafish swirl: BMP2 function is essential during early dorsoventral patterning. *Development* **124,** 4457–4466.

Kondo, M., Weissman, I. L., and Akashi, K. (1997). Identification of clonogenic common lymphoid progenitors in mouse bone marrow. *Cell* **91,** 661–672.

Krause, D. S., Ito, T., Fackler, M. J., Smith, O. M., Collector, M. I., Sharkis, S. J., and May, W. S. (1994). Characterization of murine CD34, a marker for hematopoietic progenitor and stem cells. *Blood* **84,** 691–701.

Lane, M. C., and Smith, W. C. (1999). The origins of primitive blood in *Xenopus:* Implications for axial patterning. *Development* **126,** 423–434.

Lassila, O., Eskola, J., Toivanen, P., Martin, C., and Dieterlen-Liévre, F. (1978). The origin of lymphoid stem cells studied in chick yolk sac–embryo chimaeras. *Nature (London)* **272,** 353–354.

Lassila, O., Martin, C., Toivanen, P., and Dieterlen-Liévre, F. (1982). Erythropoiesis and lymphopoiesis in the chick yolk-sac–embryo chimeras: Contribution of yolk sac and intraembryonic stem cells. *Blood* **59,** 377–381.

Le Douarin, N. M., Houssaint, E., Jotereau, F. V., and Belo, M. (1975). Origin of hematopoietic stem cells in embryonic bursa of Fabricius and bone marrow studied through interspecific chimeras. *Proc. Natl. Acad. Sci. USA* **72,** 2701–2705.

Liao, W., Bisgrove, B. W., Sawyer, H., Hug, B., Bell, B., Peters, K., Grunwald, D. J., and Stainier, D. Y. (1997). The zebrafish gene cloche acts upstream of a flk-1 homologue to regulate endothelial cell differentiation. *Development* **124,** 381–389.

Maéno, M., Tochinai, S., and Katagiri, C. (1985a). Differential participation of ventral and dorsolateral mesoderms in the hemopoiesis of *Xenopus,* as revealed in diploid–triploid or interspecific chimeras. *Dev. Biol.* **110,** 503–508.

Maéno, M., Todate, A., and Katagiri, C. (1985b). The localization of precursor cells for larval and adult hematopoietic cells in *Xenopus laevis* in two regions of embryos. *Dev. Growth Differ.* **27,** 137–148.

Mangia, F., Procicchiami, G., and Manelli, H. (1970). On the development of the blood island in *Xenopus laevis* embryos: Light and electron microscope study. *Acta Embryol. Exp.* **2,** 163–184.

Martin, C., Beaupain, D., and Dieterlen-Liévre, F. (1978). Developmental relationships between vitelline and intra-embryonic haemopoiesis studied in avian "yolk sac chimaeras." *Cell Differ.* **7,** 115–130.

Matsuoka, S., Tsuji, K., Hisakawa, H., Xu, M., Ebihara, Y., Ishii, T., Sugiyama, D., Manabe, A., Tanaka, R., Ikeda, Y., Asano, S., and Nakahata, T. (2001). Generation of definitive hematopoietic stem cells from murine early yolk sac and paraaortic splanchnopleures by aorta–gonad–mesonephros region-derived stromal cells. *Blood* **98,** 6–12.

Mebius, R., and Akashi, K. (2000). Precursors to neonatal lymph nodes: LT β^+CD45$^+$CD4$^+$CD3$^-$ cells are found in fetal liver. *Curr. Top. Microbiol. Immunol.* **251,** 197–201.

Mebius, R. E., Miyamoto, T., Christensen, J., Domen, J., Cupedo, T., Weissman, I. L., and Akashi, K. (2001). The fetal liver counterpart of adult common lymphoid progenitors gives rise to all lymphoid lineages, CD45$^+$CD4$^+$CD3$^-$ cells, as well as macrophages. *J. Immunol.* **166,** 6593–6601.

Medlock, E. S., and Haar, J. L. (1983). The liver hemopoietic environment. I. Developing hepatocytes and their role in fetal hemopoiesis. *Anat. Rec.* **207,** 31–41.

Medvinsky, A. L., Samoylina, N. L., Müller, A. M., and Dzierzak, E. A. (1993). An early pre-liver intraembryonic source of CFU-S in the developing mouse. *Nature (London)* **364,** 64–67.

Mills, K. R., Kruep, D., and Saha, M. S. (1999). Elucidating the origins of the vascular system: A fate map of the vascular endothelial and red blood cell lineages in *Xenopus laevis. Dev. Biol.* **209,** 352–368.

Moody, S. A. (1987). Fates of the blastomeres of the 16-cell stage *Xenopus* embryo. *Dev. Biol.* **119,** 560–578.

Moore, M. A., and Metcalf, D. (1970). Ontogeny of the haemopoietic system: Yolk sac origin of in vivo and in vitro colony forming cells in the developing mouse embryo. *Br. J. Haematol.* **18,** 279–296.

Moore, M. A., and Owen, J. J. (1965). Chromosome marker studies on the development of the haemopoietic system in the chick embryo. *Nature (London)* **208,** 956 passim.

Moore, M. A., and Owen, J. J. (1967). Chromosome marker studies in the irradiated chick embryo. *Nature (London)* **215,** 1081–1082.

Morrison, S. J., Wandycz, A. M., Hemmati, H. D., Wright, D. E., and Weissman, I. L. (1997). Identification of a lineage of multipotent hematopoietic progenitors. *Development* **124,** 1929–1939.

Mucenski, M. L., McLain, K., Kier, A. B., Swerdlow, S. H., Schreiner, C. M., Miller, T. A., Pietryga, D. W., Scott, W. J. Jr., and Potter, S. S. (1991). A functional c-*myb* gene is required for normal murine fetal hepatic hematopoiesis. *Cell* **65,** 677–689.

Mukouyama, Y., Chiba, N., Mucenski, M. L., Satake, M., Miyajima, A., Hara, T., and Watanabe, T. (1999). Hematopoietic cells in cultures of the murine embryonic aorta–gonad–mesonephros region are induced by c-Myb. *Curr. Biol.* **9,** 833–836.

Müller, A. M., Medvinsky, A., Strouboulis, J., Grosveld, F., and Dzierzak, E. (1994). Development of hematopoietic stem cell activity in the mouse embryo. *Immunity* **1,** 291–301.

Nieuwkoop, P., and Faber, J. (1967). Normal table of *Xenopus laevis* (Daudin). North Holland, Amsterdam.

North, T., Gu, T. L., Stacy, T., Wang, Q., Howard, L., Binder, M., Marin-Padilla, M., and Speck, N. A. (1999). Cbfa2 is required for the formation of intra-aortic hematopoietic clusters. *Development* **126,** 2563–2575.

Okuda, T., van Deursen, J., Hiebert, S. W., Grosveld, G., and Downing, J. R. (1996). AML1, the target of multiple chromosomal translocations in human leukemia, is essential for normal fetal liver hematopoiesis. *Cell* **84,** 321–330.

Osawa, M., Hanada, K., Hamada, H., and Nakauchi, H. (1996). Long-term lymphohematopoietic reconstitution by a single CD34-low/negative hematopoietic stem cell. *Science* **273,** 242–245.

Palis, J., Robertson, S., Kennedy, M., Wall, C., and Keller, G. (1999). Development of erythroid and myeloid progenitors in the yolk sac and embryo proper of the mouse. *Development* **126,** 5073–5084.

Palis, J., Chan, R. J., Koniski, A., Patel, R., Starr, M., and Yoder, M. C. (2001). Spatial and temporal emergence of high proliferative potential hematopoietic precursors during murine embryogenesis. *Proc. Natl. Acad. Sci. USA* **98,** 4528–4533.

Pevny, L., Simon, M. C., Robertson, E., Klein, W. H., Tsai, S. F., D'Agati, V., Orkin, S. H., and Costantini, F. (1991). Erythroid differentiation in chimaeric mice blocked by a targeted mutation in the gene for transcription factor GATA-1. *Nature (London)* **349,** 257–260.

Pevny, L., Lin, C. S., D'Agati, V., Simon, M. C., Orkin, S. H., and Costantini, F. (1995). Development of hematopoietic cells lacking transcription factor GATA-1. *Development* **121,** 163–172.

Ransom, D. G., Haffter, P., Odenthal, J., Brownlie, A., Vogelsang, E., Kelsh, R. N., Brand, M., van Eeden, F. J., Furutani-Seiki, M., Granato, M., Hammerschmidt, M., Heisenberg, C. P., Jiang, Y. J., Kane, D. A., Mullins, M. C., and Nusslein-Volhard, C. (1996). Characterization of zebrafish mutants with defects in embryonic hematopoiesis. *Development* **123,** 311–319.

Rifkind, R. A., Chui, D., and Epler, H. (1969). An ultrastructural study of early morphogenetic events during the establishment of fetal hepatic erythropoiesis. *J. Cell Biol.* **40,** 343–365.

Robb, L., Lyons, I., Li, R., Hartley, L., Kontgen, F., Harvey, R. P., Metcalf, D., and Begley, C. G. (1995). Absence of yolk sac hematopoiesis from mice with a targeted disruption of the *scl* gene. *Proc. Natl. Acad. Sci. USA* **92,** 7075–7079.

Sánchez, M. J., Holmes, A., Miles, C., and Dzierzak, E. (1996). Characterization of the first definitive heamtopoietic stem cells in the AGM and liver of the mouse embryo. *Immunity* **5,** 513–525.

Sasaki, K., and Iwatsuki, H. (1997). Origin and fate of the central macrophages of erythroblastic islands in the fetal and neonatal mouse liver. *Microsc. Res. Tech.* **39,** 398–405.

Sasaki, K., and Matsumura, G. (1988). Spleen lymphocytes and haemopoiesis in the mouse embryo. *J. Anat.* **160,** 27–37.

Sasaki, K., and Sonoda, Y. (2000). Histometrical and three-dimensional analyses of liver hematopoiesis in the mouse embryo. *Arch. Histol. Cytol.* **63,** 137–146.

Schmid, B., Furthauer, M., Connors, S. A., Trout, J., Thisse, B., Thisse, C., and Mullins, M. C. (2000). Equivalent genetic roles for bmp7/snailhouse and bmp2b/swirl in dorsoventral pattern formation. *Development* **127,** 957–967.

Shah, V. O., Civin, C. I., and Loken, M. R. (1988). Flow cytometric analysis of human bone marrow. IV. Differential quantitative expression of T-200 common leukocyte antigen during normal hemopoiesis. *J. Immunol.* **140,** 1861–1867.

Shivdasani, R. A., Mayer, E. L., and Orkin, S. H. (1995). Absence of blood formation in mice lacking the T-cell leukaemia oncoprotein tal-1/SCL. *Nature (London)* **373,** 432–434.

Sieff, C. A., and Nathan, D. G. (1993). The anatomy and physiology of hematopoiesis. *In* "Hematology of Infancy and Childhood" (D. G. Nathan and F. A. Nathan eds.), Vol. 1, pp.156–215. W. B. Saunders, Philadelphia.

Smith, P. B., Flajnik, M. F., and Turpen, J. B. (1989). Experimental analysis of ventral blood island hematopoiesis in *Xenopus* embryonic chimeras. *Dev. Biol.* **131,** 302–312.

Spangrude, G. J., Heimfeld, S., and Weissman, I. L. (1988). Purification and characterization of mouse hematopoietic stem cells. *Science* **241,** 58–62.

Stainier, D. Y., Weinstein, B. M., Detrich, H. W., III. Zon, L. I., and Fishman, M. C. (1995). Cloche, an early acting zebrafish gene, is required by both the endothelial and hematopoietic lineages. *Development* **121,** 3141–3150.

Tam, P. P., and Behringer, R. R. (1997). Mouse gastrulation: The formation of a mammalian body plan, *Mech. Dev.* **68,** 3–25

Tan, J. C., Nocka, K., Ray, P., Traktman, P., and Besmer, P. (1990). The dominant W42 spotting phenotype results from a missense mutation in the c-Kit receptor kinase. *Science* **247,** 209–212.

Tavian, M., Coulombel, L., Luton, D., Clemente, H. S., Dieterlen-Liévre, F., and Peault, B. (1996). Aorta-associated CD34$^+$ hematopoietic cells in the early human embryo. *Blood* **87,** 67–72.

Tavian, M., Hallais, M. F., and Peault, B. (1999). Emergence of intraembryonic hematopoietic precursors in the pre-liver human embryo. *Development* **126,** 793–803.

Tavian, M., Robin, C., Coulombel, L., and Peault, B. (2001). The human embryo, but not its yolk sac, generates lympho-myeloid stem cells: Mapping multipotent hematopoietic cell fate in intraembryonic mesoderm. *Immunity* **15,** 487–495.

Terstappen, L. W., Huang, S., Safford, M., Lansdorp, P. M., and Loken, M. R. (1991). Sequential generations of hematopoietic colonies derived from single nonlineage-committed CD34$^+$CD38$^-$ progenitor cells. *Blood* **77,** 1218–1227.

Thompson, M. A., Ransom, D. G., Pratt, S. J., MacLennan, H., Kieran, M. W., Detrich, H. W., III, Vail, B., Huber, T. L., Paw, B., Brownlie, A. J., Oates, A. C., Fritz, A., Gates, M. A., Amores, A., Bahary, N., Talbot, W. S., Her, H., Beier, D. R., Postlethwait, J. H., and Zon, L. I. (1998). The cloche and spadetail genes differentially affect hematopoiesis and vasculogenesis. *Dev. Biol.* **197,** 248–269.

Turpen, J. B., Kelley, C. M., Mead, P. E., and Zon, L. I. (1997). Bipotential primitive-definitive hematopoietic progenitors in the vertebrate embryo. *Immunity* **7,** 325–334.

Vandenbunder, B., Pardanaud, L., Jaffredo, T., Mirabel, M. A., and Stehelin, D. (1989). Complementary patterns of expression of c-*ets 1*, c-*myb* and c-*myc* in the blood-forming system of the chick embryo. *Development* **107,** 265–274.

Wang, Q., Stacy, T., Binder, M., Marin-Padilla, M., Sharpe, A. H., and Speck, N. A. (1996). Disruption of the Cbfa2 gene causes necrosis and hemorrhaging in the central nervous system and blocks definitive hematopoiesis. *Proc. Natl. Acad. Sci. USA* **93,** 3444–3449.

Warren, A. J., Colledge, W. H., Carlton, M. B., Evans, M. J., Smith, A. J., and Rabbitts, T. H. (1994). The oncogenic cysteine-rich LIM domain protein rbtn2 is essential for erythroid development. *Cell* **78,** 45–57.

Weiss, M. J., Keller, G., and Orkin, S. H. (1994). Novel insights into erythroid development revealed through in vitro differentiation of GATA-1 embryonic stem cells. *Genes Dev.* **8,** 1184–1197.

Willett, C. E., Cortes, A., Zuasti, A., and Zapata, A. G. (1999). Early hematopoiesis and developing lymphoid organs in the zebrafish. *Dev. Dyn.* **214,** 323–336.

Yamada, Y., Warren, A. J., Dobson, C., Forster, A., Pannell, R., and Rabbitts, T. H. (1998). The T cell leukemia LIM protein Lmo2 is necessary for adult mouse hematopoiesis. *Proc. Natl. Acad. Sci. USA* **95,** 3890–3895.

Yoder, M. C. (2001). Introduction: Spatial origin of murine hematopoietic stem cells. *Blood* **98,** 3–5.

Yoder, M. C., and Hiatt, K. (1997). Engraftment of embryonic hematopoietic cells in conditioned newborn recipients. *Blood* **89,** 2176–2183.

Yoder, M. C., Hiatt, K., Dutt, P., Mukherjee, P., Bodine, D. M., and Orlic, D. (1997a). Characterization of definitive lymphohematopoietic stem cells in the day 9 murine yolk sac. *Immunity* **7,** 335–344.

Yoder, M. C., Hiatt, K., and Mukherjee, P. (1997b). *In vivo* repopulating hematopoietic stem cells are present in the murine yolk sac at day 9.0 postcoitus. *Proc. Natl. Acad. Sci. USA* **94,** 6776–6780.

4

Patterning the Sea Urchin Embryo: Gene Regulatory Networks, Signaling Pathways, and Cellular Interactions

Lynne M. Angerer and Robert C. Angerer*
Department of Biology
University of Rochester
Rochester, New York 14627

We discuss steps in the specification of major tissue territories of the sea urchin embryo that occur between fertilization and hatching blastula stage and the cellular interactions required to coordinate morphogenetic processes that begin after hatching. We review evidence that has led to new ideas about how this embryo is initially patterned: (1) Specification of most of the tissue territories is not direct, but proceeds gradually by progressive subdivision of broad, maternally specified domains that depend on opposing gradients in the ratios of animalizing transcription factors (ATFs) and vegetalizing (β-catenin) transcription factors; (2) the range of maternal nuclear β-catenin extends further than previously proposed, that is, into the animal hemisphere, where it programs many cells to adopt early aboral ectoderm characteristics; (3) cells at the extreme animal pole constitute a unique ectoderm region, lacking nuclear β-catenin; (4) the pluripotential mesendoderm is created by the combined outputs of ATFs and nuclear β-catenin, which initially overlap in the macromeres, and by an undefined early micromere signal; (5) later micromere signals, which activate

*To whom correspondence should be addressed. E-mail: rang@mail.rochester.edu

Current Topics in Developmental Biology, Vol. 53

159

Notch and Wnt pathways, subdivide mesendoderm into secondary mesenchyme and endoderm; and (6) oral ectoderm specification requires reprogramming early aboral ectoderm at about the hatching blastula stage. Morphogenetic processes that follow initial fate specification depend critically on continued interactions among cells in different territories. As illustrations, we discuss the regulation of (1) the ectoderm/endoderm boundary, (2) mesenchyme positioning and skeletal growth, (3) ciliated band formation, and (4) several suppressive interactions operating late in embryogenesis to limit the fates of multipotent cells. © 2003, Elsevier Science (USA).

I. Introduction

Patterning of embryos involves the specification in a three-dimensional array of different cell types whose developmental fates are established through a combination of inheritance of maternal determinants and signaling among cells or tissues. Frequently, these early patterning events occur during a limited number of cell divisions and in the absence of growth. The sea urchin embryo affords a classic model of such a developmental program. During cleavage the egg cytoplasm is subdivided through a relatively invariant series of cell divisions to generate the radially symmetric hatching blastula, which contains ~350 cells arranged in an epithelium, a single cell layer thick. Within this blastula, six initial major tissue territories are specified but none is yet morphologically distinguishable. After hatching, the cells of the embryo undergo only several more rounds of cell division, during which morphogenesis is executed. Mesenchyme cells ingress and endoderm invaginates, skeletal spicules form and, after 3 days, a bilaterally symmetric 2000-cell pluteus larva emerges, which has about 15 cell types. Sea urchin morphogenesis has been reviewed in detail (Ettensohn and Ingersoll, 1992) and is summarized briefly in Fig. 1.

The sea urchin embryo is a highly regulative developmental system. Classic experiments using blastomere isolation and recombination showed that only the micromeres, the vegetal-most 4 cells of the 16-cell embryo, are determined (Hörstadius, 1939, 1973). In isolation, or when transplanted to ectopic positions, micromere progeny differentiate along their normal path to form skeletogenic mesenchyme. In contrast, the fates of other blastomeres normally are dependent on the organizing activity of the micromeres. Their progeny are gradually specified during development and do not commit to specific fates until after morphogenesis begins. Macromere and mesomere lineages retain broader developmental capacities until the mesenchyme blastula/gastrula stage, 1 to 2 days after fertilization, when embryos contain between 400 and 800 cells. Thus, presumptive ectoderm can be induced to form endoderm and mesoderm cell types (Henry *et al.,* 1989; Khaner and Wilt, 1990; Ransick and Davidson, 1993; Sherwood and McClay, 2001; Wikramanayake and Klein, 1997), presumptive endoderm and secondary mesenchyme can exchange fates on perturbation of specific signaling pathways (Sherwood and McClay, 1999), and some secondary mesenchyme cells can replace primary mesenchyme cells surgically removed from the embryo (Ettensohn and McClay, 1988) or photochemically ablated (Ettensohn, 1990a). A major focus of this chapter is on the mechanisms that underlie the progressive restriction of developmental capacities of all but the most vegetal blastomeres of the embryo.

A. Developmental Axes

1. Animal–Vegetal Axis

Echinoderm eggs, like those of animals in most phyla, are endowed with a maternally determined animal–vegetal (A–V) axis. This polarity is revealed by extrusion of the polar bodies at the animal pole, by a small indentation in the clear extracellular "egg jelly" at the animal pole, and, in eggs of several species, by a concentration of pigment in a subequatorial band orthogonal to the A–V axis. Except for these features, the isolecithal egg lacks visible evidence of its inherent developmental polarity. However, bisection of unfertilized eggs through the equator or separation of embryos into vegetal and animal halves at the eight-cell stage demonstrates polarization of developmental potential along the A–V axis: Animal halves give rise only to poorly differentiated epithelial spheres, whereas blastomeres derived from vegetal halves undergo extensive regulation of fate and elaborate surprisingly normal structures and tissues, including ectoderm, which normally is derived mostly from animal blastomeres (Hörstadius, 1973; Maruyama *et al.,* 1985). Thus, maternal information sufficient to direct the specification of most tissues is restricted to the vegetal hemisphere, whereas normal differentiation of animal blastomeres to produce different ectoderm cell types requires inductive interactions from vegetal cells. The vegetal pole of the sea urchin embryo is functionally analogous (and perhaps homologous) to the organizer of *Xenopus* and zebrafish embryos and the node of avian and mammalian embryos. It is the site that initiates invagination at gastrulation, where vegetal blastomeres form an inductive center essential for specification of mesoderm and endoderm as well as for differentiation of ectoderm (Davidson *et al.,* 1998; Ettensohn and Sweet, 2000; McClay, 2000).

Fate mapping of early embryonic blastomeres shows that regions of the egg cytoplasm that give rise to mesoderm, endoderm, and ectoderm in the normal embryo are arrayed along the axis in that order from vegetal to animal (Fig. 2; see color insert). The first obvious morphological evidence of A–V asymmetry is seen at the fourth cleavage, when an asymmetric oblique division of the four vegetal blastomeres gives rise to four vegetal micromeres and four macromeres above them, while the four animal blastomeres divide meridionally to form a single tier of eight mesomeres. We have proposed that the 16-cell stage is a critical step in early patterning, as it partitions the embryo into three transcriptional regulatory domains, as discussed in detail below.

Figure 1 Normal development of the sea urchin pluteus larva (*Lytechinus variegatus* at 23°C). Time after fertilization is shown in parentheses: (A) fertilized egg (10 min); (B) 2-cell stage (1 h); (C) 4-cell stage (1.5 h); (D) 16-cell stage (3 h); (E) 64-cell stage (4.5 h); (F) prehatching blastula (6.5 h); (G) hatched blastula (8.5 h); (H) mesenchyme blastula (11 h); (I) early gastrula (14 h); (J) midgastrula (16 h); (K) late gastrula (18 h), plane of focus at the level of the archenteron; (L) same embryo, plane of focus at the level of the primary mesenchyme cell ring; (M) prism

stage (21 h); (N) early pluteus larva (24 h); (O) late pluteus larva (36 h). Reprinted from Ettensohn, C. A., and Ingersoll, E. P. (1992), by courtesy of Marcel Dekker, Inc. Scale bar in (A) = 25 μm. For the other commonly used species, *Strongylocentrotus purpuratus,* cultured at 15°C, the developmental times are longer, as follows: 2-cell stage, 1.5 h; 4-cell stage, 2.5 h; 16-cell stage, 5.5 h; 64-cell stage, 8 h; prehatching blastula (12 h); hatched blastula, 18 h; mesenchyme blastula, 24 h; early gastrula, 30 h; midgastrula, 36 h; late gastrula, 44 h; prism, 60 h; early pluteus, 72 h; late pluteus larva, 84 h. Echinoderm isolecithal eggs (A) are endowed with a maternally determined animal–vegetal (A–V) axis. The animal pole is the site of extrusion of the polar bodies and is marked by a small indentation in the clear extracellular "egg jelly" at the animal pole and, in eggs of several species, by a subequatorial pigment band (not shown); the animal pole is at the top in (D–M). About 1.5 h after fertilization the embryo begins a series of 7 or 8 cleavage divisions (B–E), which, at ∼12 h, have produced a hollow early blastula of about 150 cells, a single cell layer thick (F). The first obvious morphological evidence of A–V asymmetry occurs at the third cleavage (D), when an asymmetric oblique division of the four vegetal blastomeres gives rise to four vegetal micromeres and four macromeres above them (D, arrow), while the four animal blastomeres divide meridionally to form a single tier of eight mesomeres. Vegetal cells form a thickened columnar epithelium, the vegetal plate (G and H), which is the site of invagination at gastrulation (I, arrow). Cell division slows markedly at the end of cleavage and the ∼350-cell embryo secretes a zygotically produced hatching enzyme and hatches from the fertilization envelope as a ciliated, free-swimming "hatched blastula" (G). Morphogenesis begins when primary mesenchyme cells (PMCs) undergo an epithelial–mesenchymal transition and ingress into the blastocoel (H, arrow), producing the mesenchyme blastula. The PMCs leave behind their sister lineage, the small micromeres, which remain in the center of the vegetal plate, surrounded by concentric tori of cells that will give rise to various secondary mesenchyme (inner torus) and endoderm (outer torus) cell types. Gastrulation begins with an invagination in the center of the vegetal plate (I) that produces an archenteron extending about one-third the diameter of the blastocoel (J). During this period the PMCs migrate through the blastocoel and are guided by positional information produced by the ectoderm to form a ring of cells in the vegetal hemisphere orthogonal to the A–V axis and near the level of the future endoderm–ectoderm border (L). Within the ring are two bilaterally symmetric clusters of cells on the ventral (oral) side of the embryo (K and L, arrowheads). Simultaneously, the archenteron extends to fuse with the ectoderm wall (K), which then opens to form the stomodeum. During this process the tip of the archenteron is guided by the exploratory activity of filopodia that are extended by secondary mesenchyme cells (SMCs) at its tip (K, arrowheads). These adhere selectively to the ectodermal target and their contraction serves to guide the invaginating archenteron, and in some species appears to provide part of the force required to complete archenteron extension. Toward the end of gastrulation, subpopulations of SMCs begin to differentiate and emigrate from the tip of the archenteron. These include migratory pigment cells, which disperse over the basal surface of the aboral ectoderm and then embed in it, blastocoelar cells of unknown function, daughters of the small micromeres, and a few additional cells from other SMC lineages that form small coelomic pouches on either side of the esophagus (M, arrows) and about 20 cells that differentiate as striated muscle bands and wrap around the esophagus. When gastrulation is complete, the endoderm differentiates as a tripartite gut separated into distinct fore-, mid-, and hindgut by sphincters formed from primitive myoepithelial cells. PMCs secrete calcite spicules (L), which are extended during the next day to form the skeleton (L–O). Differentiation along the oral–aboral (O–A) axis becomes obvious during gastrulation, primarily in ectodermal tissues that form three separate, histologically distinct territories. The aboral cone of cells (M–O) differentiates as a squamous epithelium consisting of a single aboral ectoderm cell type. The oral region around the stomodeum also forms squamous cells, but with a distinct pattern of gene expression at late stages, and also includes some serotonergic neuronal cells above the mouth. At the border between oral and aboral epithelia, a ciliary band differentiates (M–O, arrows), consisting of three or four rows of cuboidal cells that include precursors of dopaminergic neurons. Extension of the skeletal spicules and expansion of the aboral ectoderm, constricted by the ciliary band, produce the characteristic pyramidal shape of the pluteus larva (O).

2. Oral–Aboral Axis

Sea urchin embryos are bilaterally symmetric deuterostomes. The plane that separates left and right sides of the embryo is defined by the A–V axis and an axis orthogonal to it, now commonly referred to as the oral–aboral (O–A) axis. O–A polarity is not determined maternally by prelocalized determinants, as shown by the classic experiments of producing twins from eggs bisected through the A–V axis or from separated two-cell blastomeres and quadruplets from four-cell blastomeres (Hörstadius, 1973). However, developmental asymmetry is established along this axis after fertilization as early as the first cleavage in some species (Cameron et al., 1989) and as late as the third cleavage (and perhaps later) in others (Kominami, 1988), as shown by lineage tracing studies. This asymmetry must be labile because four-cell blastomeres can regulate to produce normal embryos. The mechanism that provisionally establishes this axis at early stages is unknown. O–A orientation is unrelated to the sperm entry point at fertilization (Cameron et al., 1989) and bears a variable relationship to the initial cleavage planes in some species (Kominami, 1988). The earliest biochemical marker of this axis during cleavage and early blastula stages is an asymmetric gradient of redox potential, higher on the future oral side (Coffman and Davidson, 2001; Czihak, 1963). As discussed below, the earliest demonstrated asymmetry in gene expression along this axis is not evident until about the hatching blastula stage. Careful fate-mapping studies identified O–A asymmetry in the disposition of mesenchyme precursors in the vegetal plate at the mesenchyme blastula stage (Ruffins and Ettensohn, 1996). However, morphological O–A polarity is not visible until the early gastrula stage, when clusters of primary mesenchyme cells form at bilaterally symmetric, "ventrolateral" positions and begin to secrete the triradiate spicule rudiments (see Fig. 1).

The stability of the O–A axis depends on correct development of cells along the A–V axis. Nearly every experimental perturbation that interferes with normal A–V development results in loss of O–A polarity. Different manipulations may convert the ectoderm largely to either an oral or aboral fate (e.g., Angerer et al., 2000, 2001; Logan et al., 1999; Wikramanayake et al., 1995; Wikramanayake and Klein, 1997), completely block its differentiation, or apparently reorient the axis to align with the A–V axis (Hardin et al., 1992). This linkage between O–A and A–V patterning undoubtedly is related to the fact that both depend heavily on vegetal signaling initiated by the canonical Wnt pathway. As discussed in detail below, the activity of this pathway plays a key role in the establishment of broad gene regulatory domains that are the first steps in patterning the embryo.

B. The Big Picture

We discuss a model in which four broad regions arrayed along the A–V axis are first specified by maternal mechanisms. Two of these regions, the animal pole and early aboral ectoderm-like domains, reside in the animal tier of the eight-cell embryo (and in their daughters, the mesomeres). The other two correspond to the macromeres and micromeres (Fig. 3A, left; see color insert). Each of these regions is distinguished by a unique ratio of animalizing transcription factors (ATFs) and nuclear β-catenin (Fig. 3B, fourth cleavage). We emphasize the maternal processes that establish these ratios, the zygotic contributions that reinforce them, and the molecular pathways that implement specification of the major tissue territories of the embryo. The steps in this process that are currently supported by experimental evidence are schematized in Fig. 3.

First, at least two maternal mechanisms operate to endow blastomeres with different gene regulatory potentials as they separate along the A–V axis between the third and fifth cleavages. Early blastomeres have an inherent capacity to accumulate β-catenin in their nuclei (a process now termed "nuclearization") in a vegetal-to-animal concentration gradient (Logan *et al.,* 1999). This gradient is elaborated in a temporal and spatial wave during the first seven or eight cleavage divisions (Fig. 4G–K; see color insert). Nuclear β-catenin serves both as the determinant of micromere fate (Logan *et al.,* 1999) as well as the origin of the vegetal signaling mechanism (VSM; reviewed in Angerer and Angerer, 2000) that is required for normal specification of other blastomeres (Fig. 3B). This β-catenin asymmetry is superimposed on a maternal background of gene regulatory activity stored in the form of mRNAs and/or proteins that initially are uniformly distributed throughout the egg and embryo. At least some of these activities serve to antagonize the VSM and to bias blastomeres toward an early preectoderm fate (Angerer and Angerer, 2000; Kenny *et al.,* 2002). The molecular asymmetry of β-catenin and its downstream effectors, coupled with a maternally regulated asymmetry in the early cleavage pattern, partitions these animalizing and vegetalizing factors into differently sized blastomeres (Figs. 3B and 4). Thus, mesomeres, macromeres, and micromeres acquire distinct maternally sponsored gene regulatory potentials as soon as they are separated.

Second, depending on the ratio of their concentrations, β-catenin/TCF-Lef (T cell factor-lymphoid enhancer factor) and the ATFs activate, directly and indirectly, different target genes in different blastomeres along the A–V axis. Expression of the ATFs is zygotically upregulated in animal blastomeres (Kenny *et al.,* 1999; Wei *et al.,* 1999b), probably as the result of auto- and cross-regulatory interactions among different members of that cohort. β-Catenin/TCF-Lef and at least some of the ATFs antagonize each other's functions, further modulating the ratios of these vegetalizing and animalizing factors in different blastomeres along the A–V axis during cleavage (Kenny *et al.,* 1999; Logan *et al.,* 1999). A high initial ratio of β-catenin/ATFs is required for determination of micromere lineage fates by the fourth cleavage, whereas the lower ratios required for mesendoderm patterning are achieved much more gradually, between the fifth and ninth cleavages (Fig. 3B).

Third, the different combinations of activating and repressing transcription factors present in early blastomeres and zygotic factors whose synthesis they activate regulate region-specific production of signaling molecules that comprise the VSM (Sherwood and McClay, 1999; A. Wikramanayake, personal communication; Sweet *et al.*, 2002). Other known proximal effectors of β-catenin function as transcription regulators (Li *et al.*, 1999; Howard *et al.*, 2001; K. Akasaka, personal communication; summarized by Davidson *et al.*, 2002a,b). In addition to signaling that induces vegetal differentiation, a major function of the nuclear β-catenin-dependent gene regulatory pathway is to suppress ATF function in the vegetal blastomeres (Howard *et al.*, 2001) (Fig. 3B, eighth cleavage). These cells then are competent to send and receive various signals that are required for induction of endoderm (A. Wikramanayake, personal communication, 2001) and secondary mesenchyme (Logan *et al.*, 1999; McClay *et al.*, 2000; Sweet *et al.*, 1999, 2002). Nuclear β-catenin-dependent information also is required for patterning of ectoderm along the O–A axis (Wikramanayake and Klein, 1997), again by stimulating production of at least one repressor required for transfating preaboral ectoderm to oral fate (Angerer *et al.*, 2001).

Fourth, final fates are not assigned directly to most early blastomeres along the A–V axis. A reproducible cleavage pattern is superimposed on the maternal molecular asymmetries, which divides the egg cytoplasm among blastomeres of reasonably reproducible sizes and geometric arrangement. The potency and interactions of these blastomeres have been tested by experimental isolations and rearrangements and it is common to describe the regulation of fate specification in terms of these interactions. However, although blastomere fates can be mapped, final fates are assigned directly only to the small (coelomic mesenchyme; CM) and large (skeletogenic mesenchyme; SM) micromere daughters, and the only early cleavage plane that corresponds to a known signaling interface is that between large micromeres and macromeres and their progeny. Thus, although initial cell-autonomous biases are imposed on other blastomeres by their differential inheritance of maternal factors (Fig. 3A, left), borders between major territories (secondary mesenchyme, endoderm, and ectoderm) are not conditionally specified until nearly the end of cleavage (Fig. 3A, right). These territories continue to be further subdivided even later. We discuss two examples of this process—the creation and subsequent partitioning of mesendoderm into endoderm and secondary mesenchyme and of a broad early aboral ectoderm domain into oral and aboral epithelia.

Although the major territories of the pluteus are specified by the hatching blastula stage, for the most part these initial specifications are conditional. An extensive series of both positive and negative signals continues to be sent through gastrulation to refine borders, to maintain positional information, to create new tissues at the interface of different early territories or subregions of territories, and to guide most cells that retain broader developmental capacities to their final specific fates. We discuss our current understanding of signals that help to establish the ectoderm–endoderm boundary, that guide skeletogenic primary mesenchyme cells (PMCs) to their correct positions, that generate the ciliated band between ectoderm territories, and that restrict the developmental potency of some secondary mesenchyme cells and those within the archenteron.

II. Cell Fate Specification during the Premorphogenetic Phase of Sea Urchin Development

Subdivisions of the egg into an array of developmental domains begins with the maternally regulated partitioning of transcription factor activities to establish the four regulatory domains mentioned above and depicted in Fig. 3A. The animal pole domain (AnPD; purple in Fig. 3) is defined by high ATF and very low, or no, nuclear β-catenin activities; the preoral ectoderm (pre-AoeD; light blue in Fig. 3) is a broad region containing high levels of ATFs and low nuclear β-catenin; the macromere domain (macD; light blue with gold cross-hatching in Fig. 3) is a maternal mesendoderm domain that also contains high levels of ATFs but also significant amounts of nuclear β-catenin; and the vegetal domains [primary mesenchyme (PM) and coelomic mesenchyme (CM)] are characterized by low ATF and high levels of nuclear β-catenin (Fig. 3B). Transfer from maternal to zygotic control occurs gradually during cleavage and early blastula stages, leading at about the hatching blastula stage to the last major specification transitions—from preoral ectoderm to facial epithelium (FE) and aboral ectoderm (AOE) and from mesendoderm to secondary mesenchyme and endoderm (Fig. 3A).

A. Maternally Regulated Asymmetries

1. The Nonvegetal Domain: A Gene Regulatory Region Defined by Animalizing Transcription Factors Present in Mesomere and Macromere Lineages

a. VEB Gene Regulators.   The activity of the ATFs was revealed by the expression patterns of several genes that they regulate, termed very early blastula (VEB) genes. The two VEB genes extensively studied encode metalloproteases—the hatching enzyme [*HE* (Lepage *et al.*, 1992b); *SpHE* (Reynolds *et al.*, 1992)] and one related to bone morphogenetic protein 1 (BMP-1) and tolloid [*SpAN* (Reynolds *et al.*, 1992); *BP10* (Lepage *et al.*, 1992a)]. VEB genes are activated cell-autonomously by the eight-cell stage (Reynolds *et al.*, 1992; Ghiglione *et al.*, 1993). They remain active in macromeres and mesomeres and their progeny, but are shut off in the micromeres of 16-cell embryos. As cleavage continues, VEB mRNAs continue to accumulate in a contiguous region that gradually retracts from the vegetal pole (Reynolds *et al.*, 1992). *In vivo* promoter–reporter transgene assays defined compact regulatory regions of $\sim$300 bp in both *SpHE* and *SpAN* that are sufficient for mediating spatially correct transcription (Kozlowski *et al.,* 1996; Wei *et al.,* 1995). Similarly, elements sufficient for amplitude control and spatial regulation were identified relatively close to the *HE* transcription start site (Ghiglione *et al.,* 1997). In *SpHE* and *SpAN,* each regulatory region contains many different closely packed *cis*-acting elements, each of which confers positive regulatory activity. Three different subregions and several individual *cis* elements of the *SpHE* promoter were shown to independently mediate correct spatial expression from a basal promoter (Wei *et al.,* 1997, 1999a).

These and subsequent observations discussed below led us to propose that a major early transcriptional territory of the early embryo is a "nonvegetal" domain that includes a diverse set of positively acting transcription regulatory activities. On the basis of subsequent tests of their developmental effects, we termed these animalizing transcription factors, or ATFs (Angerer and Angerer, 2000). Several of these factors have now been cloned: a member of the Ets family (SpEts4; Wei *et al.,* 1999b), several members of the Sox family (SpSoxB1, SpSoxB2; Kenny *et al.,* 1999) and a CCAAT-binding factor (SpCBF [Li *et al.,* 1993; Li *et al.,* 2002]). Although for technical reasons the VEB genes were selected initially, in part on the basis of their strictly zygotic expression, the same pattern of mRNA accumulation has been observed for several other genes whose transcripts are also represented in the unfertilized egg (e.g., SpEts4, SpSoxB1, and SpSoxB2). In these cases, *in situ* hybridization shows an mRNA distribution that modulates during cleavage from uniform in eggs to the VEB pattern, as maternal transcripts are replaced by new zygotic synthesis (Kenny *et al.,* 1999; Wei *et al.,* 1999a).

   b. ATF asymmetry along the A–V axis.   We suggest that, downstream of the maternal mechanisms that supply the initial activities, a zygotic increase in ATF levels is achieved through a self-sustaining regulatory network in which individual ATF genes autoactivate and/or cross-regulate each other. Consistent with this idea is the fact that ATF mRNAs that have been analyzed accumulate in a VEB-like, nonvegetal pattern in the embryo (Wei *et al.,* 1999a; Kenny *et al.,* 1999). Currently, the only functional evidence to support this hypothesis is that misexpression of a dominant-negative version of the SpSoxB1 ATF (i.e., including the DNA-binding domain only) leads to reduction in SpSoxB1 protein levels (A. Kenny, unpublished observations). The first ATF activities are derived from low amounts of stored maternal protein and/or translation of maternal mRNA during the first few hours of development. For example, SpSoxB1 protein accumulates uniformly in nuclei through the eight-cell stage. However, a striking difference in SpSoxB1 nuclear concentrations along the A–V axis accompanies the asymmetric fourth cleavage: As shown in Fig. 4A, SpSoxB1 protein is present at much (15- to 20-fold) higher concentration in mesomere and macromere nuclei than in micromere nuclei (Kenny *et al.,* 1999). Asymmetric cleavage of vegetal eight-cell blastomeres provides one mechanism for this unequal partitioning. When cells are in mitosis, SpSoxB1 signals are undetectable on chromosomes and elevated in the cytoplasm (our unpublished observations). This shuttling between nucleus and cytoplasm provides a general mechanism by which micromeres inherit four- to fivefold less SpSoxB1 and of any other transcription factors that similarly partition with the cytoplasmic volume. Furthermore, micromeres also will acquire correspondingly lower amounts of uniformly distributed ATF mRNAs and thus will translate a smaller amount of the corresponding factors. In addition to asymmetric cleavage, other processes not yet elucidated also appear to reduce ATF concentrations an additional four- to fivefold in micromeres (Kenny *et al.*, 1999). If ATFs are autoregulatory or form a cross-regulatory network, then this initial asymmetry could be rapidly amplified after the 16-cell stage.

These events at the fourth cleavage establish an initial border of the nonvegetal domain corresponding to the interface between micromeres and macromeres. However, the border of the nonvegetal region originally observed at the VEB stage is closer to the equator (Fig. 4B) and, at the mesenchyme blastula stage, SpSoxB1 is excluded from most of the vegetal plate (Fig. 4C, double arrow), which forms from derivatives of veg_2 (Kenny *et al.*, 1999) (see Fig. 2 for description of veg_1 and veg_2 cells). This retreat of ATFs to nuclei of presumptive ectoderm correlates precisely with the advancing nuclear β-catenin wave (Fig. 4G–K) and is a response to this vegetalizing mechanism (see the next section). When β-catenin nuclear function is blocked by injecting mRNA encoding C-cadherin into one-cell zygotes, SpSoxB1 accumulates uniformly in all nuclei (Howard *et al.*, 2001). This indicates that the initial 16-cell stage asymmetry can neither be maintained nor increased in the absence of the nuclear β-catenin-dependent signaling mechanisms.

The extent to which the ATFs can independently support development is best represented by embryos in which all vegetalizing function is blocked by injection of C-cadherin mRNA and that consist of undifferentiated preectoderm. The fact that cadherin-treated embryos markedly overaccumulate SpSoxB1 (and probably other ATFs) (Howard *et al.*, 2001) suggests that the ectoderm of these embryos may not be representative of that of any phase of normal development. Cadherin-injected embryos are similar, but not identical, to animal-half embryos, which form dauerblastulae consisting of an epithelium that exhibits some polarity (Wikramanayake *et al.*, 1995, 1998; Logan *et al.*, 1999; Howard *et al.*, 2001). Below, we present arguments that this polarity aligns with the A–V axis and results from an inherent maternally sponsored low level of nuclear β-catenin in animal blastomeres, which supports differentiation of some animal blastomere progeny to an epithelium with aboral characteristics.

2. Maternal Vegetal Domain

a. Nuclear β-Catenin.   The vegetal counterpart of the ATFs is the single factor, β-catenin, (Logan *et al.*, 1999) acting as a transcriptional coactivator of TCF-Lef (Clevers and van de Wetering, 1997; Huang *et al.*, 2000; Vonica *et al.*, 2000). Two differences in the regulation of β-catenin and the ATFs are critical.

First, in contrast to the initial uniform distribution of the ATF system, the maternal mechanism that regulates nuclear entry of β-catenin is polarized, being most active initially in the micromeres and their progeny. Second, accumulation of (detectable) nuclear β-catenin is delayed for 5–6 h until the late 16- to 32-cell stage and thereafter progresses slowly across the blastomeres of the mesendoderm domain (MED), derived from macromere daughters, during the subsequent 3–4 cleavage divisions (Logan *et al.*, 1999) (Fig. 4G–K).

The accumulation of β-catenin in vegetal nuclei is thought to be initiated cell-autonomously because it occurs in approximately the correct fraction of cells of embryos that have been continuously dissociated into individual blastomeres beginning at the two-cell stage (Logan *et al.*, 1999). Several interesting questions about the temporal and spatial regulation of this process remain to be explored. First, nuclearization of β-catenin must be linked ultimately to the activation of signal transduction pathways about 5 h earlier at fertilization. The molecular nature of this linkage is an attractive subject for future study. Second, the cell-autonomous mechanism that endows different blastomeres with different capacities to accumulate nuclear β-catenin is not yet understood. An interesting observation is that β-catenin may be preferentially tethered to the plasma membrane of animal blastomeres in a complex involving BEP4 (butanol-extractable protein 4) (Romancino *et al.*, 2001). BEP4 mRNA is localized in eggs (Fig. 5A; see color insert) and animal blastomeres via specific proteins that bind to its 3′ untranslated region (UTR) (Montana *et al.*, 1997) and BEP4 protein is associated with the nonvegetal cortex (Costa *et al.*, 1997; Romancino *et al.*, 1998; Romancino and di Carlo, 1999). Treatment of embryos with an Fab fragment directed against BEP4 leads to nuclearization of β-catenin in more animal blastomeres, as does treatment with LiCl, which upregulates nuclear β-catenin levels. In both cases, the embryos are strongly vegetalized (Romancino *et al.*, 2001). These workers suggest that the antibody against BEP4 releases some β-catenin from the cell surface, which leads to an increased concentration in nuclei. Thus, one of the mechanisms regulating nuclear accumulation of β-catenin appears to involve cytoplasmic tethering in the animal hemisphere.

b. Asymmetric Cleavage.   Entry of β-catenin into vegetal nuclei elevates their β-catenin/ATF ratio. The highest ratio is achieved early on in the micromeres by two simultaneous processes: the entry of large amounts of β-catenin into micromere nuclei and the reduction in ATF levels by asymmetric cleavage and other undefined mechanisms. Previous work has shown that micromeres have special properties. They either inherit or construct a cytoplasmic domain that is distinct from that of their nonvegetal counterparts. For example, their maternal RNA sequence complexity is 25% lower than that of macromeres plus mesomeres (Ernst *et al.*, 1980; Rodgers and Gross, 1978), which indicates that they have the capacity either to selectively degrade or to specifically exclude a large number of different RNA molecules, perhaps including ATF mRNAs. They might also be able to specifically repress the translation of ATF mRNAs. A potential mechanism for sorting molecules would be their interactions with the cortical cytoskeleton, which retracts from the vegetal pole as micromeres form (Schroeder, 1980). This cortical retraction is strikingly displayed in *Arbacia,* where the pigment granules associated with the cortex are withdrawn from the micromeres (Fig. 5B).

c. ATF downregulation. Both high levels of nuclear β-catenin and the absence (or low levels) of at least some ATFs are essential for micromeres to function as an organizing center and to differentiate as skeletogenic mesenchyme. For example, when micromeres of normal embryos are replaced with those taken from embryos expressing C-cadherin, PMCs do not form, vegetal structures are not induced, and different ectodermal cell types do not differentiate (Logan *et al.,* 1999). The same phenotype results from injection of SpSoxB1 or SpSoxB2 mRNA at the one-cell stage (Kenny *et al.,* 2002). These effects show that at least some ATFs can completely antagonize the function of nuclear β-catenin as a vegetal determinant in micromeres. Because they are present at equal concentrations in animal and vegetal blastomeres at the eight-cell stage, they must be eliminated in order for micromeres to assume their maternally determined fate. As discussed below, more gradual elimination of at least some of the ATFs is also required for the differentiation of vegetal tissues derived from the mesendoderm domain.

3. Maternal Ectoderm Domains

Maternal mechanisms establish an initial asymmetry within the animal hemisphere of the sea urchin embryo. Animal-half embryos derived from eggs (Horstadius, 1939) or eight-cell embryos always show morphological polarization, with one side thickened and bearing elongated cilia and the other side thinning to a squamous epithelium—the classic dauerblastula phenotype (Hörstadius, 1973). Embryos of some species also express markers characteristic of aboral ectoderm (Wikramanayake and Klein, 1997). We propose that embryoids derived from animal halves are polarized along the A–V axis. This idea is based on the fact that all procedures that interfere with vegetal signaling in intact embryos (i.e., similar to separating animal hemispheres) also block development of O–A polarity, radializing embryos about the A–V axis (Emily-Fenouil *et al.,* 1998; Logan *et al.,* 1999; Vonica *et al.,* 2000; Howard *et al.,* 2001; Angerer *et al.,* 2000, 2001). As discussed further below, this interpretation of the phenotype differs from previous suggestions that polarity in animal-half embryos lies along the O–A axis (Wikramanayake and Klein, 1997). These alternatives could be distinguished by a lineage-tracing experiment that determines the frequency with which individual four-cell blastomeres, marked with a lineage tracer, make a similar contribution to both thick (AnPD) and thin epithelia (pre-AoeD) in animal halves: Our interpretation predicts that all blastomeres contribute to both, whereas if animal-half embryos were polarized along the O–A axis, some blastomeres would contribute to only thick or thin regions.

We further suggest that maternally sponsored patterning of ectoderm, specifically the production of the thin squamous ectoderm, requires nuclear β-catenin. This possibility has not seemed likely because β-catenin has not been detected in animal blastomere nuclei by immunostaining (Logan *et al.*, 1999). However, embryos in which β-catenin nuclear function has been blocked by injection of cadherin mRNA contain essentially only cuboidal epithelial cells that are most similar to those derived from the animal pole region of the embryo (Howard *et al.*, 2001; Li *et al.*, 1999; Logan *et al.*, 1999). Unlike animal-half embryos, they do not show signs of morphological or molecular patterning; in particular, they lack a thinner epithelial region that expresses aboral markers. These observations show that the information required for initial ectoderm polarity must be inherited from the egg and that it is cadherin-sensitive. Another observation supporting this interpretation is that animal blastomeres, the mesomeres, have the capacity to respond to LiCl, which stimulates nuclearization of β-catenin, and then to differentiate into vegetal cell types. The fact that macromeres require nuclear β-catenin to respond to micromere signals (Logan *et al.*, 1999) and produce vegetal tissues in the normal embryo suggests that the mesomeres probably do as well. We imagine that LiCl treatment enhances an inherent, low, but undetectable, level of nuclear β-catenin in animal blastomeres, driven by processes that were programmed in the egg, as has been suggested (Davidson *et al.*, 1998).

We therefore refer to the two morphologically distinct domains of animal-half embryos or dauerblastulae as the animal pole domain (AnPD) and the preaboral ectoderm domain (pre-AoeD) (Fig. 3A). It is important to note that whereas the capacity to form these distinct gene expression domains is maternally specified, molecular evidence of this division is not apparent in the distribution of zygotic transcripts until toward the end of cleavage (see below) and morphological differences appear much later. We believe that these domains are not abnormal properties of animal-half embryos but represent actual intermediates in the progressive specification of ectoderm tissue territories in the normal embryo because patterns of gene expression in the normal intact early blastula reveal the regulatory activities in the AnP and pre-Aoe domains, as discussed below.

a. AnPD. In normal embryos the AnPD gives rise to the thickened apical plate of the mesenchyme blastula, which elaborates a sensory structure that includes a tuft of long cilia. Although this region has not been fate mapped at high resolution, at least to a first approximation it corresponds to the oral hood region of the ciliated band. Several lines of evidence indicate that these animal pole cells constitute a distinct regulatory subregion of the presumptive ectoderm, which emerges toward the end of cleavage. In normal embryos this region is marked at the hatching/early mesenchyme blastula stage by accumulation of mRNA encoding the Nk2.1 factor (Peterson, personal communication). It also is revealed by the lack of accumulation or downregulation of several marker mRNAs that are found in the pre-AoeD [see below; *Spec1, Spec2a, SpEGFI, SpEGFII*, and *SpARS* (Fig. 6A; see color insert)]. One of these is *SpSoxB2*, which disappears from the AnPD during early blastula stages, about 150 cells (Fig. 6E). Cells at the animal pole also are the most resistant to experimentally enhanced vegetal signaling produced either by misexpression throughout the embryo of components of the canonical Wnt pathway, such as stabilized β-catenin (Wikramanayake *et al.*, 1998), TCF-Lef (Huang *et al.*, 2000; Vonica *et al.*, 2000), dominant-negative glycogen synthase kinase 3β (GSK3β) (Emily-Fenouil *et al.*, 1998); or of effectors of that pathway, such as SpKrl (Howard *et al.*, 2001) (Fig. 6F, left); or by treatments that upregulate β-catenin nuclearization, such as LiCl (Hörstadius, 1973) or anti-BEP4 antibodies (Romancino *et al.*, 2001). They also are refractory to the effects of misexpression of BMP-2/4, which converts all other ectoderm to an aboral-like phenotype (Fig. 6F, right) (Angerer *et al.*, 2000). In embryos treated with NiCl$_2$, which expands the oral ectoderm region, the animal pole region also assumes a distinct morphology (Hardin *et al.*, 1992).

b. Pre-AoeD. The pre-AoeD is fated to give rise to both oral and aboral ectoderm in normal embryos (Cameron *et al.*, 1989). This domain initially probably also includes most, or all, of the region that will become endoderm, on the basis of the following observations. Early distributions of mRNAs that are aboral ectoderm-specific at late stages include a significant portion of the vegetal hemisphere (Fig. 6A), which subsequently is respecified to endoderm by the vegetal signaling mechanism. In addition, a variety of experimental treatments lead endoderm to transfate to ectoderm (see Section III,A). However, in this section we consider the portion of the pre AoeD restricted to the animal hemisphere (Fig. 3A, light blue). This region is defined as preaboral because patterns of gene expression suggest that initially all cells within it are biased toward the aboral ectoderm fate. Genes that are activated early in ectoderm differentiation typically are first expressed throughout the presumptive ectoderm (except sometimes for a few cells at the animal pole) (Fig. 6A and B), and only during the mesenchyme blastula and gastrula stages do their transcripts become restricted to the single squamous epithelial cell type of aboral ectoderm (Fig. 6C and D). Examples include *Spec1* (Hardin *et al.*, 1988), *SpARS* (Yang *et al.*, 1993), and *SpEGFII* (Grimwade *et al.*, 1991), for each of which the broad expression pattern is evident at the hatching/early mesenchyme blastula stages, $\sim$350 cells. On the basis of mRNA *in situ* hybridization patterns, transcription of all of these genes must be downregulated in presumptive oral ectoderm at, or shortly after, the mesenchyme blastula stage. However, the exact timing cannot be precisely defined by these assays because mRNA half-lives are either unknown or relatively long (Cabrera *et al.*, 1984). For example, in Fig. 6A and B, transcripts are present at uniform concentrations throughout the pre-AoeD of early mesenchyme blastulae, but subsequent differential expression in oral and aboral ectoderm (Fig. 6C) leads to their being restricted to aboral ectoderm by the pluteus stage (Fig. 6D). Only in the case of aboral ectoderm-specific genes that are not activated until after the hatching blastula stage, such as *SpHox7* (formerly called *SpHbox1*; Angerer *et al.*, 1989), is expression restricted to the presumptive aboral region, when it is first detectable by *in situ* hybridization at the early mesenchyme blastula stage. No genes have been identified that are specifically expressed in presumptive oral ectoderm during cleavage or early blastula stages, nor has any mRNA been observed to modulate from an early broad distribution in ectoderm to restricted expression in the oral territory. These observations also are consistent with the view that the first bias of most animal blastomeres, excluding the AnPD, is toward the aboral ectoderm fate.

Whereas maternal mechanisms initiate expression of aboral ectoderm-specific genes in the pre-AoeD, communication among cells of the embryo is required to achieve their full expression. For example, in *Strongylocentrotus purpuratus,* although *Spec1* transcripts have been detected in some cells of embryos completely dissociated into individual cells beginning at the two-cell stage (Stephens *et al.*, 1989), their concentration is significantly lower than that in control embryos (Hurley *et al.*, 1989). The intercellular interactions that are required to achieve quantitatively normal levels of *Spec1* transcripts must be among pre-AoeD blastomeres, that is, a community effect, because the concentration of this mRNA is high in animal-half embryoids. In contrast, in *Lytechinus,* signals from vegetal cells must be involved because animal halves fail to express the comparable aboral marker, *LpS1* (Wikramanayake *et al.*, 1995). We expect that the transition from pre-Aoe to aboral ectoderm (AOE) in both species requires several kinds of intercellular interactions.

c. Gene Regulatory Regions within the Animal Hemisphere.

There are two major possibilities for autonomous molecular mechanisms that construct the AnPD and pre-AoeD, which are not mutually exclusive. First, unidentified maternal determinants may specify cell fates in these domains. This seems unlikely for the AnPD because region-specific transcription has not been detected until about 15 h after fertilization. We favor the alternative that cells in these two domains share ATF regulatory activities but have different capacities to nuclearize β-catenin during the fifth to eighth cleavage stages. In this view, during late cleavage and blastula stages the vegetal border of the AnPD marks the functional range of nuclear β-catenin and/or its effectors. This interpretation is both parsimonious and consistent with several different experimental observations. As mentioned above, animal-half embryos of all species tested have a pathway(s) that can be stimulated by LiCl, whose activation results in differentiation of vegetal structures or aboral ectoderm (Livingston and Wilt, 1989; Wikramanayake and Klein, 1997). For example, lower levels of LiCl can cause animal halves of *Lytechinus pictus* embryos to express the aboral marker *LpS1* on their thinner sides, implying that levels of nuclear β-catenin higher than those provided by purely maternal mechanisms are required for activation of aboral ectoderm genes in the pre-AoeD in this species. Higher LiCl induces vegetal differentiation in all species. However, because LiCl may have effects other than stabilizing β-catenin (Ransick *et al.*, 2002) and the lack of aboral ectoderm differentiation in cadherin-treated embryos may result from interference with the vegetal signaling mechanism, it is important to establish whether cadherin blocks morphological polarization in animal halves and whether polarity can be rescued by an activating form of TCF-Lef. Second, micromeres can induce ectopic axes when they are transplanted to the animal pole of 16-cell embryos (Hörstadius, 1935; Ransick and Davidson, 1993) or are combined with animal-half embryos at the same stage (Hörstadius, 1973). The simplest interpretation of this is that the mesomeres of 16-cell embryos have intact the same pathways as vegetal blastomeres, which allows them to receive micromere signals. As noted above, reception of micromere signals by overlying vegetal blastomeres of normal embryos requires β-catenin function, but this requirement has not been tested for mesomeres. Third, the older literature provides a provocative report (Hörstadius, 1936) that an$_2$ blastomeres (see Fig. 2 for description of an$_1$ and an$_2$ cells) extracted from embryos strongly vegetalized with LiCl could function exactly like micromeres. These cells in combination with normal animal halves produced a relatively normal pluteus. One interpretation of the molecular mechanism underlying this result is that LiCl elevated β-catenin levels in the an$_2$ cell nuclei, overcame initial levels of ATF activity, and activated the signaling mechanism normally restricted to micromeres. Interestingly, Sweet *et al.* (2002) have shown that misexpression of one downstream effector of β-catenin function, LvDelta, in mesomeres similarly endows them with an organizing capacity like that of micromeres.

4. Maternally Specified Mesendoderm Domain

Mesendoderm is defined here as a pluripotent region that can give rise to either endoderm or diverse secondary mesenchyme cell types, depending on the signals it receives. (This definition excludes the primary mesenchyme and small micromere lineages, which are separated and determined at the fifth cleavage.) Veg$_2$ progeny that separate at the sixth cleavage produce most of the mesendoderm derivatives, whereas veg$_1$ progeny yield both ectoderm and endoderm.

We previously proposed that MED specification begins at the 16-cell stage, when the large macromeres inherit a unique ratio of ATFs and nuclear β-catenin (Angerer and Angerer, 2000). The proposal that macromeres represent a unique transcription domain is substantiated by the observation that a gene encoding the transcription factor SpKrox1 is activated and functions specifically in macromeres and their progeny (Wang *et al.*, 1996). At hatching, expression of *SpKrox1* marks the vegetal plate, including presumptive secondary mesenchyme and endoderm, and it is downregulated as the mesendoderm invaginates. This expression pattern suggested an early role for SpKrox1 in mesendoderm specification. Consistent with such a role, SpKrox functions downstream of β-catenin and tests of the effects of expressing in embryos a dominant-negative variant of SpKrox1 (the SpKrox1

DNA-binding domain fused to the Engrailed repression domain) indicate that it functions upstream of several genes involved in mesendoderm differentiation, including SpWnt8 (Davidson *et al.*, 2002b).

At the 32-cell stage, β-catenin accumulates in macromere daughter nuclei and thereafter its nuclearization progresses slowly across the blastomeres of the mesendoderm during the subsequent three or four cleavage divisions. This wave of nuclear β-catenin accumulation is thought to be initiated by cell autonomous mechanisms because embryos dissociated into single cells beginning at the two-cell stage have approximately the correct fraction of cells with detectable β-catenin in their nuclei (Logan *et al.*, 1999). Whether these mechanisms are sufficient for the full amplitude of the β-catenin wave or, alternatively, whether this requires augmentation via receptor–ligand-mediated communication among cells of the embryo is not yet clear.

B. Zygotic Patterning Mechanisms

1. Gene Regulators Activated Downstream of β-Catenin

Rapid progress has been made in elucidating the zygotic gene regulatory cascade that emanates from nuclear β-catenin and is responsible for specification and differentiation of mesendoderm tissues [e.g., see Davidson *et al.* (2002a,b) and the Davidson lab web site (http://www.its.caltech.edu/~mirsky/endomes.htm), which frequently updates network components and structure]. We focus here on several genes positioned close to β-catenin in the hierarchy, and for which developmental roles are relatively well established. These include both transcription factors whose zygotic expression is activated by the β-catenin wave and signaling ligands that mediate major inductive interactions.

Transcription factors positioned proximal to β-catenin in the regulatory hierarchy include both activators and repressors of gene expression. As discussed above, *SpKrox* mRNA accumulates specifically in macromeres at the time when β-catenin is first detectable in their nuclei, suggesting that this gene is a direct target of β-catenin and its transcriptional binding partner, TCF-Lef. *SpKrox* encodes a homeodomain-containing transcription activator, the downstream targets (not necessarily direct) of which include SpWnt8 (see below) and SpOtx (Davidson *et al.*, 2002b). SpOtx is an important ubiquitous early activator that also is present as a maternal protein (Li *et al.*, 1997). These, and probably other factors, both spatially restricted and uniformly distributed, activate the program required for mesendoderm specification and, ultimately, differentiation of endoderm and secondary mesenchyme tissues.

Interestingly, two other genes that are proximal to β-catenin encode repressors of activities that interfere with vegetal differentiation. The first of these, *SpKrl* ("Krüppel-like") is expressed transiently in both the micromere lineages and the

MED (Howard *et al.,* 2001). *SpKrl* encodes a protein related in its DNA-binding domain (but not elsewhere) to *Drosophila* Krüppel. SpKrl expression, activated by β-catenin/TCF-Lef, mediates the repression of SpSoxB1, and probably of other ATF levels. *SpKrl* mRNA accumulation begins at the 16-cell stage and subsequently its transcription closely follows the wave of β-catenin nuclearization (Fig. 4D–K) *SpKrl* is likely to be a direct target of β-catenin/TCF-Lef because upregulation of β-catenin activity by LiCl causes increased *SpKrl* mRNA accumulation in the absence of protein synthesis; that is, *SpKrl* is an immediate early target of β-catenin function. SpKrl functions as a repressor in sea urchin embryos because injection of either *SpKrl* mRNA or of a message encoding a chimeric protein consisting of the SpKrl DNA-binding domain linked to the *Drosophila* Engrailed repression domain (Fig. 6F, left) has a similar phenotypic effect. In both cases the domain of SpSoxB1 protein accumulation is restricted dramatically toward the animal pole and ectoderm fails to differentiate. Injection of mRNA encoding stabilized β-catenin produces the same phenotype, which is consistent with SpKrl's mediating the repressive function of nuclear β-catenin. Conversely, when synthesis of SpKrl protein is blocked by morpholino knockdown, both the concentration of SpSoxB1 in nuclei and its domain of expression increase and the differentiation of endoderm and of at least some secondary mesenchyme cell types is blocked (Howard *et al.,* 2001). Because loss of SpKrl causes upregulation of SpSoxB1 and suppresses vegetal differentiation, SpKrl may mediate most of the repressive functions of β-catenin.

These results suggest that an important β-catenin-dependent function that is mediated by SpKrl is to clear animalizing factors such as SpSoxB1 from the MED (the region marked by the double arrow in Fig. 4C). Clearing the ATFs is required because, as mentioned above, they can strongly antagonize the β-catenin initiated vegetalizing mechanism. For example, injection of mRNA encoding SpSoxB1 produces a phenotype indistinguishable from that which results from misexpressing cadherin, that is, uniform epithelial balls of animal polelike ectoderm (Kenny *et al.,* 2002). These embryoids are completely devoid of vegetal tissues, and even the maternally determined PMCs fail to differentiate, suggesting that ectopic expression of sufficiently high ATF levels can directly antagonize β-catenin nuclear function. How this Sox factor does this in sea urchin embryos has not yet been determined. Competition between Sox proteins and TCF-Lef for binding to the same *cis*-acting elements is one possibility, because these factors share consensus DNA-binding sites and, in fact, SpSoxB1 and TCF-Lef can bind *in vitro* to the same *cis* element in the *SpKrl* promoter (Newman and Angerer, unpublished observations). In addition, competition through protein–protein interactions has been documented in *Xenopus* embryos: Several Sox factors, including one closely related to SpSoxB1, can bind directly to β-catenin and block its interaction with TCF-Lef (Zorn *et al.,* 1999). Our observation that an SoxB1 protein mutated to eliminate DNA-binding activity nevertheless produces a dominant-negative phenotype implies that SoxB1 function involves protein–protein interactions, although it does not appear to bind directly to β-catenin (A. Levine, unpublished observations, 2001).

The second repressor critical for vegetal specification is pmar1, a paired-class homeodomain protein (Oliveri *et al.*, 2002). *pmar1* expression is strictly zygotic and the mRNA accumulates transiently and exclusively in the micromere lineages, beginning at the time that β-catenin is first detectable in micromere nuclei. The apparent homolog of *pmar1* in *Hemicentrotus pulcherrimus, micro1*, has a similar expression pattern, although traces of this or a closely related mRNA were also reported in mesomeres (Kitamura *et al.*, 2002). *pmar1* expression responds to experimental manipulation of β-catenin levels; on the basis of the early time of its activation, it is probable, but not yet demonstrated, that regulation by β-catenin/TCF-Lef is direct. Injection of *pmar1* mRNA at the one-cell stage produces a remarkable phenotype in which the more animal cells appear to be converted to the skeletogenic (primary) mesenchyme fate. The converted cells adopt a mesenchymal morphology and express molecular markers characteristic of both the signaling function of micromeres (i.e., the Delta ligand; see below) and the skeletogenic differentiation program. The fact that a pmar1 DNA-binding domain–Engrailed repression domain fusion produces the same phenotype indicates that pmar1 functions as a repressor. Because most cells of the embryo can be converted to a micromere-like fate by pmar1 action, Oliveri *et al.* (2002) propose that it represses a broadly distributed repressor(s) of micromere fate (thus, it is named "*p*aired-class *m*icromere *a*ntirepressor"). The hypothesized ubiquitous repressor(s) have the unexpected function of preventing most of the embryo from assuming the micromere fate. Interestingly, this is a demonstrated effect of overexpression of SpSoxB1. However, this factor is unlikely to be sufficient to execute the ubiquitous repressor function because its loss via morpholino-mediated knockdown does not increase the number of PMCs (our unpublished results, 2001). The functional properties of the converted cells have been demonstrated by showing that mesomeres that misexpress *pmar1* share the ability of micromeres to induce ectopic guts (McClay and Oliveri, personal communication). These observations and other tests of regulatory interactions place *pmar1* downstream of β-catenin and upstream of genes involved in micromere specification and differentiation (Oliveri *et al.*, 2002). The global repressor target(s) of pmar1 remain to be identified.

2. Signaling Events and Molecules Involved in Mesendoderm Patterning

Specification of the MED in *Strongylocentrotus purpuratus* requires an early signal(s) from micromeres and their progeny between the fourth and sixth cleavage divisions. Removal of the micromeres at the 16-cell stage virtually eliminates subsequent expression of the general mesendoderm marker, Endo16, and gastrulation is blocked or greatly delayed (Ransick and Davidson, 1995). As micromere progeny are removed at successively later stages, Endo16 expression increases, but complete and timely gastrulation is still compromised. The molecular identity of this early signal has not been established, but it provides regulatory information that is superimposed on maternally initiated cell-autonomous β-catenin and the ATFs (Fig. 3A). The early signal is dependent on nuclear β-catenin, because micromeres overexpressing C-cadherin do not send it (Logan *et al.*, 1999). Similarly, macromere progeny overexpressing cadherin cannot receive it (Logan *et al.*, 1999). However, the early micromere signal is unlikely to be sufficient for the complete induction of endoderm and mesenchyme because, as discussed below, additional, and probably distinct, signals emanating from cells of the micromere lineages (Delta and Wnt8) are required for these presumptive tissues to continue their developmental programs.

Although the micromere removal experiments suggest that signaling through the 60-cell stage is sufficient to initiate early steps in mesendoderm specification, there is emerging evidence that micromere progeny continue to send signals through the blastula stage that promote vegetal fates. Thus, micromere progeny retain the capacity to induce mesenchyme and endoderm differentiation in animal halves of *Hemicentrotus pulcherrimus* embryos through the blastula stage (Minokawa and Amemiya, 1999). In an extension of this study, micromeres were separated from macromeres plus mesomeres, these two cell populations were cultured separately, and then they were recombined at different times. These two embryo regions retained the ability to interact up to the hatching blastula stage, and the recombinant embryos showed partial rescue of the delayed gastrulation and endoderm differentiation defects that are characteristic of micromereless embryos (Ishizuka *et al.*, 2001). Currently, the later signaling mechanism that is best characterized is activation of the Notch pathway in *Lytechinus variegatus* by the LvDelta ligand (Sherwood and McClay, 1999; Sweet *et al.*, 2002).

Specification of secondary mesenchyme is spatially and temporally tightly correlated with a striking internalization of the Notch receptor, both in normal embryos (Sherwood and McClay, 1997) and in those in which the sizes of the emerging SMC and endoderm territories are experimentally altered (Sherwood and McClay, 1999). In normal embryos of *Lytechinus variegatus,* Notch protein accumulates on all apical cell surfaces of the embryo during cleavage. At the eighth or ninth cleavage (mesenchyme blastula stage), cells at the vegetal pole, which correspond precisely to the SMC precursors in the fate map (Ruffins and Ettensohn, 1996), suddenly downregulate Notch levels and internalize the protein in vesicles, producing a Notch "hole" (Fig. 7A). Slightly later, Notch is upregulated on the apical surfaces of the adjacent cells that will form endoderm (Sherwood and McClay, 1997). Sweet *et al.* (1999) showed that one signal to Notch originates in micromere progeny: When micromeres were removed, several types of secondary mesenchyme cells, which were identified both morphologically and by a panel of specific antibodies, failed to differentiate. These embryos failed to internalize Notch but instead showed uniform vegetal expression of apical Notch, consistent

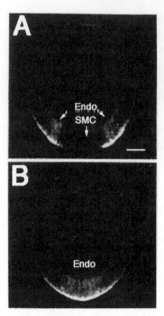

Figure 7 Micromeres are required for Notch activation in presumptive secondary mesenchyme. (A) Normal *Lytechinus variegatus* mesenchyme blastula stained for Notch. Within the mesendoderm, Notch is present at high concentrations on the apical surface of presumptive endoderm (Endo) but not presumptive secondary mesenchyme (SMC). Loss of Notch at the surface reflects its internalization, which is an indication of Notch signaling activity. (B) Embryos of the same age in which micromeres were removed at the 16-cell stage, showing that Notch internalization did not occur. These embryos have expanded endoderm and lack secondary mesenchyme. (See text for details.) From Sweet, H. C., Hodor, P. G., and Ettensohn, C. A. (1999). *Development* **126,** 5255–5265, with permission from Company of Biologists Ltd. Scale bar in (A) = 25 μm.

with conversion of presumptive SMCs to endoderm fate (Fig. 7B). Conversely, micromeres could induce animal blastomeres to form several different SMC types, although in lower numbers than in normal embryos.

The time at which micromere lineages signal through Notch has been determined by surgical manipulations of embryos. Heterochronic transplants of early micromeres to progressively older hosts (McClay *et al.,* 2000) showed that the signal could be received after transplantation of micromeres as late as the eighth to tenth cleavage. By photoablation of micromeres at progressively later stages, Sweet and Ettensohn (personal communication) have demonstrated that this cannot be received until between the seventh and tenth cleavages in *Lytechinus variegatus*. This window matches the observed timing of Notch internalization at about the ninth cleavage (Sherwood and McClay, 1997). Furthermore, lineage tracing of individual cells in the vegetal plate of mesenchyme blastulae indicates that SMC and endoderm lineages are completely separated by this time (Ruffins and Ettensohn, 1996). These results argue that the Notch pathway mediates a late signal sent from micromere progeny to macromere progeny at about the early mesenchyme blastula stage.

A *Delta* homolog, *LvDelta,* has been identified, whose mRNA accumulates at the time and site expected for the Notch ligand. *LvDelta* mRNA appears first in the large micromere lineages, beginning at the 60-cell stage, that is, about 1 cell division before Notch is internalized in the overlying cells. It is subsequently downregulated in the micromeres when they ingress into the blastocoel and then accumulates in vegetal macromere derivatives between the mesenchyme blastula and early gastrula stages (Sweet *et al.,* 2002).

The effects of experimental alterations of Notch and Delta activities firmly support a central role for this pathway in SMC specification. Injection at the one-cell stage of increasing doses of mRNA encoding a dominant-negative Notch variant (dnNotch) reduces the domain of SMCs and expands endoderm and the region of strong apical Notch staining correspondingly. Morpholino-mediated knockdown of LvDelta expression produces a similar phenotype, with greatly reduced numbers of several SMC types (pigment and blastocoelar cells and muscle fibers; Sweet *et al.,* 2002). Conversely, injection of mRNA encoding a constitutively active Notch receptor progressively channels mesendoderm differentiation toward secondary mesenchyme fates (Sherwood and McClay, 1999), just as does overexpression of LvDelta (Sweet *et al.,* 2002). The sites of Notch and LvDelta function have been determined with chimeric embryos, by selectively inhibiting their functions in some blastomeres. Expression of dnNotch blocks the ability of micromereless embryos to respond to transplanted normal micromeres, but micromeres expressing dnNotch are competent to induce SMCs when transplanted to normal hosts (McClay *et al.,* 2000). That LvDelta provides the signal from micromeres that is required for differentiation of pigment cells and some blastocoelar cells was shown by producing chimeric embryos in which LvDelta function is blocked in micromeres by morpholino knockdown. Similar experiments show that LvDelta produced at later times by macromere progeny is required for specification of additional SMC types (Sweet and Ettensohn, 2002).

The second signaling pathway involved in mesendoderm patterning is the Wnt pathway (Wikramanayake, personal communication; Davidson *et al.,* 2002b). SpWnt8 mRNA begins to accumulate in micromeres of the 16-cell embryo and then expression follows the wave of entry of β-catenin into nuclei (Fig. 4M). Thus, Wnt8 is positioned to affect specification of all mesendoderm, and introduction of an mRNA encoding a dominant-negative SpWnt8 variant in intact embryos does result in loss of both endoderm and secondary mesenchyme (Wikramanayake, personal communication). Although Wnt8 is necessary for both SMC and endoderm differentiation, it is only sufficient to produce ectopic endoderm: Injection of Wnt8 mRNA into one-cell embryos causes production of multiple ectopic guts, but not overproduction of mesenchyme cell types. Wnt8 can also induce ectopic guts, but not mesenchyme, when expressed in isolated animal-half embryos. Thus, at least ectopically in the animal hemisphere region,

Wnt8 can induce endoderm in the apparent absence of Delta–Notch function. Further tests of Wnt8 and Notch pathways are required to understand their actions and interactions determining, for example, whether ectopic expression of activated Notch in addition to Wnt8 in animal-half embryos would convert ectopic endoderm to mesoderm. In summary, the available data are consistent with a model in which Wnt8 signaling is required for mesendoderm specification, and continued Wnt8 signaling is required to activate the network of genes required for endoderm differentiation (Davidson *et al.,* 2002a,b), while Delta–Notch functions to divert mesendoderm to SMC fate.

3. Respecification of Part of Preaboral Ectoderm to Form the Facial Epithelial Domain

In Section II,A,3 we presented arguments that the first patterning of ectoderm is the separation of a small animal pole domain from a larger region with aboral ectoderm characteristics (pre-AoeD). A key regulatory factor required for specification of the pre-AoeD is SpOtx, which is ubiquitously expressed when aboral ectoderm-specific genes are first activated (Li *et al.,* 1997). Misexpression of an SpOtx–Engrailed chimera that represses Otx target genes blocks aboral ectoderm differentiation and allows expression of a marker expressed in differentiated oral ectoderm (Fig. 8A and B; see color insert). Conversely, embryos that lack aboral ectoderm as a result of cadherin mRNA injection can be partially rescued by coinjection of *SpOtx* mRNA (Li *et al.,* 1999). Because the genes whose transcription is rescued include at least one, *actin CyIIIa,* that is not directly regulated by SpOtx, it follows that this factor works upstream in the aboral ectoderm regulatory hierarchy; that is, that many gene(s) expressed in the pre-AoeD are driven either directly or indirectly by SpOtx.

The realization that a portion of the pre-AoeD is subsequently respecified to an oral, facial epithelial (FE) fate emerged from studies of the transcription factor SpGsc (Angerer *et al.,* 2001). SpGsc is a transcription repressor that accumulates at about the hatching blastula stage in nuclei of presumptive FE cells, where it antagonizes SpOtx function. Thus, misexpression of SpGsc converts all ectoderm to the FE fate, whereas mRNA encoding an activating variant (the viral VP16 activation domain fused to the SpGsc DNA-binding domain) drives all ectoderm to the aboral fate. Loss of SpGsc function by morpholino-mediated translational interference likewise allows expression of an aboral marker in all cells that express *SpOtx* at this stage (Fig. 8C and D). One mechanism for the SpGsc/SpOtx antagonism, that is, direct competition for binding to the same *cis* elements, was demonstrated by binding studies *in vitro* and by the ability of SpGsc to downregulate an SpOtx-driven transgene construct *in vivo.* Interestingly, the time when the first downregulation of aboral ectoderm gene transcription is evident in the emerging oral region corresponds to the time that SpGsc appears. This competition between SpGsc and SpOtx supports the concept of an initial preaboral ectoderm, SpOtx-dependent

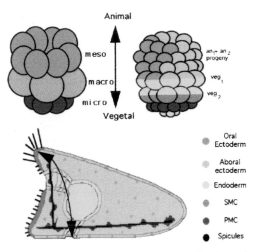

Figure 4.2 Fate map of the sea urchin embryo. Lineages of major regions of the sea urchin embryo are illustrated at the 16-cell, 60-cell, and pluteus stages. At the 16-cell stage, A–V polarity is demonstrated by unequal blastomere sizes: 4 vegetal micromeres, 4 macromeres, and 8 mesomeres. At the 60-cell stage, there are 8 vegetal micromere granddaughters that will give rise to the skeletogenic PMC lineage and 4 vegetal small micromere daughters that contribute to the coelomic rudiments. The macromeres have produced two tiers, each of 8 blastomeres, called veg_1 (more animal tier) and veg_2 (more vegetal tier). The mesomeres have divided to give 2 tiers of 8 cells each at the 32-cell stage, termed an_1 and an_2; these divide again to produce 32 an_1 + an_2 progeny in the 60-cell embryo. Major tissues of the pluteus larva include oral and aboral ectodermal epithelia separated by the ciliary band; endoderm differentiated as fore-, mid- and hindgut separated by myoepithelial sphincters; secondary mesenchyme derivatives including pigment cells, blastocoelar cells, esophagaeal muscle fibers, and coelomic rudiments (pink); and a supporting skeleton synthesized by the primary mesenchyme cells. From Angerer, L. M. and Angerer, R. C. (2000) Animal–vegetal axis patterning mechanisms in the early sea urchin embryo. *Dev. Biol.* **218**, 1–12.

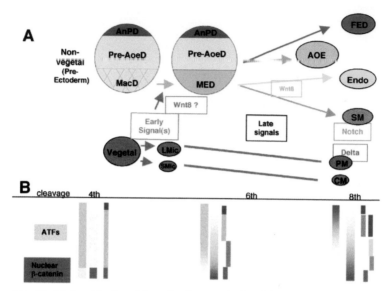

Figure 4.3 Model: Specification of the major tissue territories of the sea urchin embryo is initiated by the creation of a series of broader gene regulatory domains that are subsequently remodeled largely through the activity of the canonical Wnt signaling pathway. (**A**) Maternal mechanisms initially divide the embryo into nonvegetal and vegetal gene regulatory domains that specify preectoderm and skeletogenic/coelomic mesenchyme developmental states, respectively. Within the nonvegetal domain are three gene regulatory subdomains, defined by their unique ratios of animalizing transcription factors (ATFs) to nuclear β-catenin (**B**). These are the animal pole domain (AnPD; purple), the preoral ectoderm domain (pre-AoeD; light blue), and the macromere domain (MacD; light blue with yellow cross-hatching). Signal(s) from micromeres that are likely to include SpWnt8 are sent between 16- and 32-cell stages and are required to convert the macromere domain to mesendoderm (MED; orange). At about the eighth or ninth cleavage, an additional signal(s), the Delta homolog, is sent to Notch on overlying cells and this signaling is required to convert them from a general early MED fate to secondary mesenchyme (pink). In addition, SpWnt8 (see Fig. 4) is secreted from both micromeres and their progeny (red) and mesendoderm cells (orange) and is required for their (yellow) differentiation. In the animal hemisphere a portion of the pre-AoeD is re-specified to differentiate as facial epithelium (FED; dark blue), in part, by the appearance around hatching blastula stage of repressors, such as SpGsc and activated p3a2. (**B**) Corresponding changes with developmental time in the distributions of ATFs (blue, thick box), represented by SpSoxB1, and nuclear β-catenin (red, thick box). In this model, low, and as yet undetectable, levels of nuclear β-catenin are proposed to extend into the Pre-AoeD (light pink). Distinct gene regulatory domains are represented by thin lines color coded as in panel A. (See text for further discussion.)

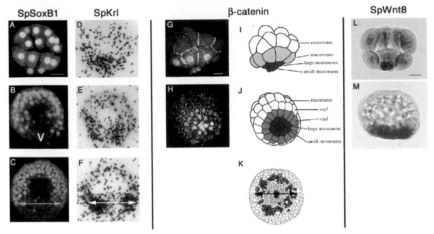

Figure 4.4 **Vegetal specification mechanisms**: (A–C) The concentration of the animalizing transcription factor, SpSoxB1, is strongly reduced in vegetal cells beginning at the fourth or fifth cleavage. The region lacking SpSoxB1 expands with the advancing vegetal wave of nuclear β-catenin (G–K) and the expression of one of its target genes, encoding the SpKrl repressor (D–F). The images in (A), (B), and (C) are reconstructed partial stacks obtained by confocal microscopy. SpSoxB1 protein distribution determined with SpSoxB1 antiserum at the 16 to 32-cell stage (A), ~200-cell, early blastula (B), early mesenchyme blastula (C). In (A) nuclei were also stained with DAPI and the two signals merged. Lower FITC signals relative to DAPI signals are indicated by pseudocolor progressing from yellow to red. These data demonstrate that the SpSoxB1 to DNA ratio is highest in the macromeres, intermediate in the mesomeres and lowest in the micromeres. Reprinted from Kenny, A. P., Kozlowski, D. J., Olcksyn, D. W., Angerer, L. M., and Angerer, R. C. (1999) *Development* **126**, 5473-5483, with permission of Company of Biologists Ltd. (D–F) *SpKrl* mRNA is concentrated in vegetal precursors of endoderm and mesenchyme. ^{33}P-labeled antisense RNA probe for *SpKrl* mRNA was hybridized to sections of 60-cell embryos (D), very early blastula (9 h, ~180 cells) (E), and early mesenchyme blastula (18 h, ~300 cells) (F). Inverted dark field images are shown. Reprinted from Howard, E. W., Newman, L. A., Oleksyn, D. W., Angerer, L. M., and Angerer, R. C. (2000). *Development* **128**, 365–375, with permission of Company of Biologists Ltd. Embryos in A––F are shown vegetal pole down [marked by *V* in (B)]. (G–K) Pattern of nuclear β-catenin accumulation at fifth cleavage, seventh cleavage and mesenchyme blastula stages. At fifth (G and I) and seventh (H and J) cleavages, β-catenin is found in nuclei of both the large and small micromeres and the macromeres and their progeny, but is absent from nuclei of mesomeres. Drawing of a vegetal pole view of a mesenchyme blastula embryo (K); the progeny of the large micromeres and veg$_2$ cells have downregulated the levels of nuclear β-catenin while the small micromere progeny (center of vegetal pole region) retain nuclear β-catenin and a subset of veg$_1$ cells that form a ring three to five cells wide has acquired this protein in their nuclei. Before invagination of the archenteron, the ring of veg$_1$, nuclear β-catenin-positive cells is approximately two cells wide. Reprinted from Logan, C. Y., Miller, J. R., Ferkowicz, M. J., and McClay, D. R. (1999). *Development* **126**, 345-357, with permission of Company of Biologists Ltd. Double arrows in (C), (F), and (K) indicate similar regions in the different embryos, illustrating that retraction of ATF signals correlates with the expanding domain of *SpKrl* expression and nuclear β-catenin accumulation. (L, M) Whole mount *in situ* hybridization detecting *SpWnt8* mRNA at the 16-cell (L) and early blastula stages (M); embryos are oriented vegetal pole down. As is the case for *SpKrl*, expression of *SpWnt8* depends on nuclear β-catenin. (Images provided by Drs. A. H. Wikramanayake and W. H. Klein.) Scale bars = 25 μm. Scale bars in (A), (G) and (L) apply to (A–F), (G and H) and (L and M), respectively.

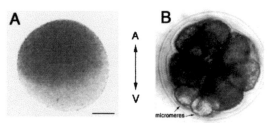

Figure 4.5   Early cortical domains: (**A**) BEP4 protein is localized to the animal hemisphere of the egg as shown by immunostaining. Reprinted from Romancino, D. P., and Di Carlo, M. (1999). *Mech. Dev.* **87**, 3–9, with permission. BEP4 protein may suppress nuclearization of β-catenin in animal blastomeres. (See text for details.) (**B**) Pigment granules are excluded from the micromeres of 16-cell embryos of *Arbacia punctulata*, illustrating that the embryo has the capacity to fractionate egg components along the A–V axis at the 16-cell stage. Both BEP4 and pigment granules are associated with the cytoskeleton in the cortex. Scale bar in (**A**) = 25 μm.

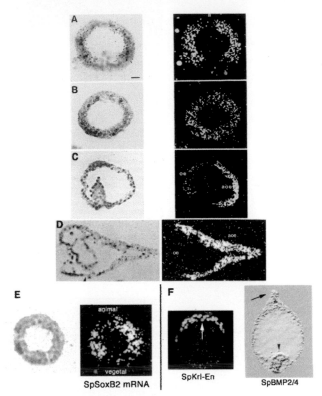

SpSoxB2 mRNA | SpKrl-En | SpBMP2/4

Figure 4.6 Ectoderm territories: Expression of the aboral ectoderm-specific marker gene *SpARS*, which encodes arylsulfatase. (**A** and **B**) At the early mesenchyme blastula stage, *SpARS* is expressed in the pre-AoeD, but not in cells at the animal (a) or vegetal (v) poles. The section in (**A**) passes through the A–V axis, whereas that in (**B**) is orthogonal to it and passes through the equator of the embryo. This pattern reflects the preaboral ectoderm domain (pre-AoeD). (**C**) At the gastrula stage, *SpARS* mRNA concentration is reduced in the presumptive oral ectoderm (oc or facial epithelium), but increased in the presumptive aboral ectoderm (aoe). (**D**) At the pluteus stage, *SpARS* transcripts are restricted to the aboral ectoderm. From Yang, Q., Kingsley, P. D., Kozlowski, D. J., Angerer, R. C., and Angerer, L. M. (1993). *Dev. Growth Differ*. **35**, 139–151, reproduced with permssion. This pattern is typical of a number of other aboral ectoderm-specific genes and reveals the existence of the animal pole and preaboral ectoderm domains at the blastula stage and the subsequent repression of aboral ectoderm markers in the facial epithelial domain. (See text for details.) (**E**) *SpSoxB2* transcripts are depleted at the animal poles of very early blastulas, indicating that the animal pole region is defined at least by seventh or eighth cleavage stages (L. Angerer, unpublished observation). (**F**) **Left**: Immunostaining of *SpSoxB1* in embryos injected with mRNA encoding SpKrl-engrailed (SpKrl-En), which, like SpKrl, down regulates SpSoxB1 expression in all but the animal pole domain (white arrow). Reprinted from Howard, E. W., Newman, L. A., Oleksyn, D. W., Angerer, L. M., and Angerer, R. C. (2000). *Development* **128**, 365–375, with permission of Company of Biologists Ltd. **Right**: Image of an embryo injected with SpBMP2/4 mRNA, in which most of the ectoderm region expands at the expense of the endoderm (arrowhead) and forms a squamous epithelium. The ectoderm in the animal pole domain responds differently to this treatment. Both misexpression experiments support the existence of the animal pole domain. Reprinted from Angerer, L. M., Oleksyn, D., Dale, L. and Angerer, R. C. (2000). *Development* **127**, 1105–1114, with permission of Company of Biologists Ltd. Scale bar in (A) = 10 μm.

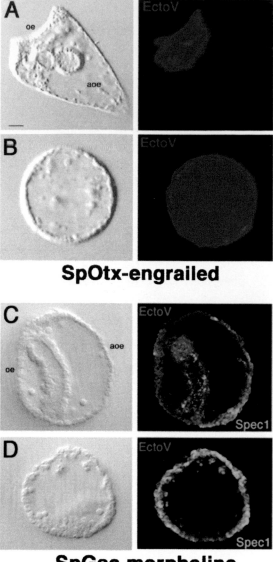

SpOtx-engrailed

SpGsc morpholino

Figure 4.8 SpOtx and SpGsc are required for aboral and oral ectoderm differentiation, respectively. (**A, B**) Embryos injected with an mRNA encoding a fusion protein containing the DNA-binding domain of SpOtx linked to the dominant repression domain from *Drosophila* engrailed, express a marker of differentiated oral ectoderm (oe) [Ecto V, red; (**A**), pluteus] throughout the entire ectoderm (**B**). Reprinted from Li, X., Wikramanayake, A. H., Klein, W. H. (1999). *Dev. Biol.* **212**, 425–439, with permission. (**C** and **D**) In contrast, embryos depleted of SpGsc by injection of a mor-pholino express a marker of differentated aboral ectoderm (aoe) [Spec1, green; (**C**), late gastrula] throughout the ectoderm. SpGsc, a transcription repressor, which is expressed specifically in the oral ectoderm, competes with SpOtx, a ubiquitous transcription activator required for aboral ectoderm dif-ferentiation, for binding to the same *cis*-regulatory elements. (See text for details.) Reprinted from Angerer, L. M., Oleksyn, D. W., Levine, A., Li, X., Klein, W. H., and Angerer, R. C. (2001). *Development* **28**, 4393–4404, with permission of Company of Biologists Ltd. Scale bar in (A) = 25 μm.

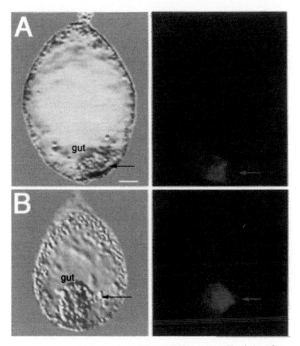

Figure 4.9 **The ectoderm–endoderm boundary shifts vegetally in embryos injected with SpBMP2/4 mRNA**. Immunostaining with Endo1, which marks mid- and hindgut regions in the fully differentiated pluteus larva [(**A**), arrow], or with EctoV, which labels foregut [(**B**), arrow], shows that even though the size of the gut is severely reduced, it is nevertheless correctly patterned. Reprinted from Angerer, L. M., Oleksyn, D., Dale, L., and Angerer, R. C. (2000), *Development* **127**, 1105 1114, with permission of Company of Biologists Ltd. Scale bar in (A) = 25 μm.

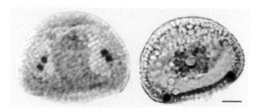

Figure 4.10 **Whole mount *in situ* hybridization shows that the gene encoding Orthopedia, a homeobox-containing transcription factor, is expressed bilaterally in two pairs of cells in the gastrula (A)**, and that these cells are located in the ectoderm (**B**) at sites overlying the positions of the ventrolateral PMC clusters. Because misexpression of Otp alters the sites of skeletogenesis, this protein may function in that signaling because misexpression of Otp alters the sites of skeletogenesis, this protein may promote the production of signals sent from ectoderm to PMCs.

state that subsequently is suppressed in the presumptive FED by SpGsc. It also fits nicely with older observations that O–A polarity is not stable until the beginning of morphogenesis—again, shortly after SpGsc accumulates in the facial epithelium. What limits SpGsc expression to the FED is not known and will require analysis of its *cis*-regulatory system.

SpOtx and SpGsc are undoubtedly critical factors in establishing and/or maintaining O–A polarity, but they are not the only regulators of ectodermal patterning along this axis. For example, expression of *actin CyIIIa*, one of the canonical aboral ectoderm-specific genes that is apparently not subject to direct SpOtx/SpGsc regulation, is repressed in the FED by a different factor, p3a2 (Coffman and Davidson, 2001, and references cited therein). Thus, *actin CyIIIa* transcription is regulated spatially by a combination of ubiquitous activators coupled with territory-specific repressors (Kirchhamer and Davidson, 1996). Interestingly, p3a2 activity, rather than distribution, is spatially regulated along the O–A axis and it is sensitive to the O–A gradient of redox potential (Coffman and Davidson, 2001) that is evident between eight-cell and mesenchyme blastula stages (Czihak, 1963). Thus, effective repression of *actin CyIIIa* transcription by p3a2 may be directly linked to the first known biochemical asymmetry along the O–A axis. However, it is not yet clear when the level of activity of p3a2 becomes sufficient to effectively repress *actin CyIIIa* transcription in the FED. At early stages (up to about the hatching blastula stage), the binding of p3a2 to DNA is antagonized by a second abundant factor, p3a1 (Höög *et al.,* 1991; Zeller *et al.,* 1995), and the p3a1 level is higher or equivalent to that of p3a2 until early blastula stages. In addition, the fact that redox asymmetry must be maintained throughout early blastula stages for differences in p3a2 activity to be observed (Coffman and Davidson, 2001) suggests that this factor becomes an effective repressor of *actin CyIIIa* transcription at about the same time as when SpGsc accumulates.

The production or activation of repressors such as SpKrl, pmar1, SpGsc, and p3a2 in specific regions emerges as a major recurring spatial regulatory mechanism in the sea urchin embryo, as has been discussed in some detail (Davidson, 2001). This theme is repeated at later stages for genes expressed specifically in endoderm. For example, the regulatory region of *Endo16* includes several modules that mediate repression in ectoderm and mesenchymal tissues (Yuh *et al.,* 2001; Yuh and Davidson, 1996). As is the case for the SpGsc and SpKrl repressors, the activity of repressors of *Endo16* is regulated by a LiCl-sensitive signaling pathway (Yuh and Davidson, 1996), probably the canonical Wnt pathway. Thus, a prominent feature of the gene regulatory networks in this system is the initial activation of many genes in broad regions, followed by the emergence of localized repressors that limit the developmental potentials of cells by antagonizing the function of those broadly distributed transcription activators.

A second gene encoding a transcription factor that is expressed early in ectoderm development in a localized manner is *PlHbox12*. Its transcription begins between the 4- and 60-cell stages and is restricted to one side of the nonvegetal domain (di Bernardo *et al.,* 1995). Interestingly, *PlHbox12* expression is blocked in dissociated embryos, making it the earliest known gene whose activation depends on intercellular communication, contrasting with activation of the VEB genes, which is entirely cell-autonomous. Whether *PlHbox12* expression aligns with the O–A axis is not known.

The model of ectoderm patterning discussed here differs from a previous proposal that ectoderm initially is specified as oral ectoderm. That interpretation was based on several experiments using EctoV, an extracellular epitope expressed in differentiated oral ectoderm (facial epithelium and ciliated band) of larvae (Coffman and McClay, 1990). EctoV is expressed when embryos are perturbed by a variety of different experimental manipulations, many of which cause limited differentiation. It is strongly expressed in animal-half embryos separated at the eight-cell stage (Wikramanayake *et al.,* 1995) and in 3-day embryos in which the function of SpOtx is blocked (Li *et al.,* 1999) or when β-catenin vegetal signaling is blocked by diverse agents [e.g., cadherin (Wikramanayake *et al.,* 1998), SpSoxB1 or SpSoxB2 (Kenny *et al.,* 2002), dnLef (Vonica *et al.,* 2000), or constitutively active GSK3β (Emily-Fenouil *et al.,* 1998)]. These are all treatments that suppress aboral ectoderm differentiation. However, these nuclear β-catenin-deficient, EctoV-positive embryos do not express SpGsc, which, as discussed above, is necessary for specification of the FED and is expressed long before EctoV accumulates in differentiated oral ectoderm of normal embryos (Angerer *et al.,* 2001). These results led us to propose that all preectoderm cells have the capacity to eventually produce EctoV, but they do it only when aboral ectoderm differentiation is prevented. In other words, EctoV accumulation is not always indicative of differentiated facial epithelial ectoderm, but inhibiting its accumulation is a property of differentiating aboral ectoderm.

Patterning of ectoderm also requires cell–cell interactions within that tissue. Our laboratory showed that BMP2/4 signaling has a strong effect on ectoderm differentiation (Angerer *et al.,* 2000). BMP2/4 mRNA appears about the same time as *SpGsc* message and its misexpression results in the conversion of all ectoderm, except animal pole cells, to the aboral phenotype (Fig. 6F, right). Conversely, loss of BMP2/4 function via injection of *Xenopus noggin* mRNA promotes differentiation of oral cell types. BMP2/4 function in promoting aboral ectoderm differentiation appears to be mediated by preferential expression in that tissue of Smad5, a downstream transcription effector in this pathway. Aboral ectoderm differentiation is strongly suppressed when Smad5 function is inhibited by dominant-negative interference with a nonphosphorylatable Smad5 mutant (our unpublished observations). It is interesting to note that the regulation of BMP2/4 differs from that of SpGsc because it does not depend on β-catenin (Angerer *et al.,* 2001). Therefore, BMP2/4 signaling is part of a parallel pathway that is required to maintain aboral ectoderm differentiation whereas SpGsc, localized in the presumptive FED, is required to suppress it.

Ectoderm patterning during the premorphogenetic phase of development creates three subdomains—the AnPD, the FED, and the AOE. Subsequent patterning in the ectoderm occurs during the morphogenetic phase, when other cell types become detectable either morphologically or by marker expression. In particular, interactions between the FED and AOE induce the ciliated band, as discussed below.

III. Interactions between Conditionally Specified Tissue Territories in Defining Borders and Coordinating Morphogenesis

Although major tissue territories are conditionally specified before morphogenesis begins, in most cases the precise borders between territories are not yet fixed. In addition, subdivisions of these territories (e.g., ciliated band, neurons, gut regions, muscle, and some different SMC types) are not defined until during and after gastrulation and most territories retain remarkable regulative capacities during this time. In the early specifications discussed above, the general patterning mechanism involves vegetal signaling activating genes that modify the differentiation that otherwise would be supported by maternally entrained nonvegetal ATFs; that is, the signaling is predominantly unidirectional. In the subsequent events discussed below, progression to the determined state involves reciprocal signaling between tissues that have already been provisionally specified. In most cases, the molecular mechanisms that underlie these interactions remain to be elucidated. We discuss a subset of these events that includes interactions between (1) endoderm and ectoderm to regulate the border between them, (2) ectoderm and PMCs required for patterning the PMC syncytium, and (3) facial epithelium and aboral ectoderm to induce the ciliated band that separates these territories. Finally, we close our discussion by briefly reviewing the role of negative intercellular interactions in patterning the sea urchin embryo, focusing on interactions that prevent transfating of SMCs to PMCs, and on the patterning of endoderm.

A. Endoderm–Ectoderm Border

Lineage-tracing experiments have revealed that veg_1 blastomeres contribute to endoderm as well as ectoderm (Logan and McClay, 1997, 1998; Ransick and Davidson, 1998). Their progeny do not involute until the late gastrula stage. The relative contributions of individual blastomeres to ectoderm and endoderm are variable, both within and among *Lytechinus variegatus* embryos (Logan and McClay, 1997), and occasionally endoderm cells descend even from an_2 blastomeres (Sherwood and McClay, 2001). These observations lead to the conclusion that the endoderm–ectoderm (E–E) border is not fixed by lineage, but is negotiated by cell–cell interactions that probably occur during the mesenchyme blastula and gastrula stages.

BMP signaling has been implicated in establishment of the endoderm–ectoderm boundary (Angerer *et al.*, 2000). The *BMP2/4* homolog is expressed transiently in presumptive ectoderm, with its mRNA level peaking at the mesenchyme blastula stage and diminishing as gastrulation begins. Misexpression of BMP2/4 by injection of the mRNA at the one-cell stage shifts the E–E boundary toward the vegetal pole in a dose-dependent manner, greatly reducing the size of the archenteron and decreasing the number of pigment cell SMC derivatives. In contrast, interference with BMP2/4 signaling by injection of mRNA encoding *Xenopus* Noggin (which binds and inactivates BMP2/4) changes ectoderm from large squamous to compact cuboidal cells and a greater fraction of the embryo is devoted to endoderm. Interestingly, as the size of the archenteron is progressively reduced by increasing BMP2/4 levels, it nevertheless retains molecular patterning along the A–V axis and expresses both foregut and mid/hindgut markers (Fig. 9; see color insert). That is, BMP2/4 does not simply truncate the gut, eliminating the more animal derivatives. On the basis of this observation, it was proposed that one function of BMP2/4 is to help define the limit of the E–E border. As discussed above, misexpressed BMP2/4 converts most of the cells of the embryo to an aboral ectoderm phenotype, and our model predicts that it would stabilize the transition of the pre-AoeD to definitive AOE. In the normal embryo, BMP2/4 is produced at the same time that β-catenin enters veg_1 blastomeres. If at that time it promotes AOE determination, as we have proposed, then we suspect that these cells would then become refractory to endodermalizing signals.

It is expected that some mechanism operates in presumptive endoderm to counter BMP2/4 signaling and Notch has been proposed to play such a role (Sherwood and McClay, 2001). Notch protein accumulates to highest levels in mesenchyme blastulae on the apical membranes of presumptive endoderm (Fig. 7A). The position of the ectoderm–endoderm boundary can be shifted in the animal or vegetal direction by misexpressing, throughout the embryo, either constitutively active or dominant-negative forms of this receptor, respectively. Overexpression of constitutively active Notch in veg_1 blastomeres on the vegetal side of the border leads to changes in the fate of cells derived from overlying an_2 progeny. Corresponding changes are seen in the range of detectable nuclear β-catenin, raising the interesting possibility that, in normal embryos, Notch functions to promote or stabilize the nuclearization of β-catenin in veg_1 progeny during the eighth to ninth cleavage (mesenchyme blastula stage; Sherwood and McClay, 2001).

B. Ectoderm Signaling and Skeletogenesis

The ectoderm provides a guide map to position PMCs migrating within the blastocoel, first to the vegetal hemisphere and then to form a ring located near the endoderm–ectoderm (E–E) boundary perpendicular to the A–V axis, with two bilateral clusters on the future oral side of the embryo. There the PMCs fuse into a syncytium and begin to secrete calcite spicules. The biology of PMCs and the regulation of skeletogenesis have been reviewed in detail (Ettensohn *et al.*, 1997).

Evidence for A–V polarity of PMC positional cues is that PMCs transplanted to abnormal positions in the animal hemisphere will migrate vegetally to form a normal ring (Malinda and Ettensohn, 1994). During their migratory phase between the mesenchyme blastula and early gastrula stages, the PMCs apparently monitor their positions by extending long filopodia (termed "cytonemes" in other systems; Ramirez-Weber and Kornberg, 1999) that explore the basal surfaces of ectodermal cells and the extracellular matrix (Malinda *et al.*, 1995; Miller *et al.*, 1995). Migration vegetally can be abrogated if a ring of ectoderm cells around the equator is photoablated, presumably interfering with the ability of the PMCs to sample their environment (Malinda and Ettensohn, 1994). An extracellular matrix component that may attract the PMCs is ECM3, which is associated with fibers that are organized in a band within the blastocoel at the position of the PMC ring (Hodor *et al.*, 2000). The position of the PMC ring is linked to the E–E boundary because both can be caused to shift together by various experimental perturbations. For example, misexpression of SpBMP2/4 (Angerer *et al.*, 2000) or Smad5 (our unpublished observations, 2001) causes both to shift vegetally, whereas inhibition of SpBMP2/4 signaling by *noggin* misexpression has the opposite effect. Extreme examples are provided by embryos strongly vegetalized by treatments that expand the domain of nuclear β-catenin, in which PMCs and spicules are found associated with the small ectodermal cap at the animal pole (Emily-Fenouil *et al.*, 1998; Wikramanayake *et al.*, 1998).

The ectoderm underlying the ventrolateral PMC clusters is thickened and may provide signals for their formation and for the initiation of spiculogenesis. Many experimental observations support this view. For example, treatments that radialize the ectoderm, that is, that abolish oral–aboral polarity, also cause the PMC distribution to be radialized, typically with multiple small clusters distributed around the ring (Hardin *et al.*, 1992; Angerer *et al.*, 2000). This effect on PMC patterning is non-cell-autonomous because PMCs from experimentally altered embryos form normal bilateral clusters on the oral side when transplanted into normal hosts, whereas PMCs from normal embryos transplanted into treated host embryos do not (Hardin *et al.*, 1992; Tan *et al.*, 1998; di Bernardo *et al.*, 1999). Similar conclusions were obtained through heterochronic transplant experiments, which showed that only when ectoderm differentiated sufficiently to express guidance cues were the PMCs correctly organized (Ettensohn, 1990a). Finally, skeleton size is determined by signals sent from the ectoderm and not by the number of PMCs (Ettensohn, 1990b), because addition of as many as twice the normal number of PMCs does not affect skeletal morphology. Ectoderm continues to send signals as the larval arms grow and the spicules elongate. When overlying ectoderm is surgically ablated from the arms, spicule growth rates are markedly reduced (Ettensohn and Malinda,

1993). These signals presumably serve to coordinate ectoderm cell division and differentiation with skeletogenesis.

Perhaps the most intriguing protein shown to be involved in interactions between ectoderm and PMCs is orthopedia (Otp), a homeodomain-containing protein, presumably a transcription factor (di Bernardo *et al.*, 1999). During the early gastrula stage, when PMC clusters form on the oral side, *otp* is transcribed in only two pairs of ectoderm cells immediately adjacent to these clusters (Fig. 10; see color insert). Cells expressing *otp* mRNA are associated with ectopic sites of spiculogenesis in embryos in which the A–V and O–A axes have been perturbed with LiCl and $NiCl_2$, respectively. As development proceeds and the skeleton is elaborated, *otp* transcripts also accumulate in ectoderm cells near the tips of the growing spicules. Additional evidence that Otp target gene products may help regulate spiculogenesis is the observation that embryos injected with *otp* mRNA at the one-cell stage make an abnormal number of radially deposited spicules. Whether Otp functions upstream in the sequence of ectoderm–PMC interactions and is involved in positioning PMC clusters, or, alternatively, whether it functions downstream in reciprocal cell–cell interactions that regulate spicule growth is not yet clear. Loss-of-function assays, such as morpholino-mediated translation interference, will be required to resolve this question.

Another protein required for regulation of spiculogenesis is Pl-nectin, which is found in the apical ectodermal extracellular matrix (Zito *et al.*, 2000). Severe and specific defects in spicule growth and patterning are observed in embryos treated with Fab fragments of a monoclonal antibody against Pl-nectin, but no defects result when the same antibody is injected into the blastocoel. Although antibodies against other apical extracellular matrix components [hyalin (Adelson and Humphreys, 1988) and fibropellin (Burke *et al.*, 1991)] that interfere with ectoderm–matrix interactions also cause a block in gastrulation, no defects in gut formation are observed in embryos treated with anti-Pl-nectin. These results suggest that Pl-nectin might be required to make ectoderm competent to send the appropriate signals for PMC differentiation. The fact that it is also present in the endoderm apical matrix raises the possibility that its role in patterning is permissive. Finally, high-resolution microscopy has revealed that PMC filopodia can penetrate the ectoderm layer and reach the apical ECM (McClay, personal communication). Consequently, Pl-nectin's role in spiculogenesis could be mediated through a direct interaction with the PMCs.

C. Formation of the Ciliated Band

At the border between facial and aboral epithelia, a ciliated band differentiates, consisting of three or four rows of cuboidal cells that include neuronal precursors. This band is contiguous with the ciliated band in the oral hood (acron), in which arise the serotonergic neurons (Burke, 1978). The oral hood derives at least approximately from the AnPD discussed above and it is autonomously specified because ciliated band and serotonergic neurons differentiate in animal-half embryos isolated at the eight-cell stage, that is, before vegetal signaling events (Wikramanayake and Klein, 1997). Lineage-tracing studies in *Strongylocentrotus purpuratus* indicate that 8 animal blastomeres of the 16-cell embryo make variable contributions to the ciliated band and that different individual vegetal blastomeres also give rise to mesendoderm derivatives as well as to both oral (facial epithelium) and aboral ectoderm cell types (Cameron *et al.*, 1990). When borders between 16-cell blastomeres approximate the position of the ciliated band, the contribution in different embryos of the progeny of adjacent blastomeres is variable (Cameron *et al.*, 1993): Sometimes all ciliated band cells derive from one or the other of the adjacent blastomeres and sometimes they come from both lineages. Such analyses demonstrate that the ciliated band is specified by cell–cell interactions, not by lineage. These interactions probably do not begin until after aboral and facial epithelia are specified, that is, at about the time that the embryo hatches. A distinct program of gene expression in presumptive ciliated band cells has not been detected until the late gastrula stage, when *SpEGFII* (Grimwade *et al.*, 1991) and *Spec3* (which encodes a structural protein of cilia) (Eldon *et al.*, 1987) are downregulated and upregulated, respectively. Thus, ciliated band specification is likely to occur well downstream of respecification of some of the pre-AoeD to FE fate at about the hatching blastula stage. It is likely that differentiated signaling functions of the FE and the AOE may be required as part of this specification.

Several other observations support the idea that ciliated band formation requires differentiated facial and aboral epithelial tissues. Many experimental manipulations that prevent differentiation of either facial epithelium or aboral ectoderm produce embryos lacking a ciliated band. Examples include misexpression of SpGsc (Angerer *et al.*, 2001), SpOtx (Mao *et al.*, 1996), SpBMP2/4 or Noggin (Angerer *et al.*, 2000), SpSoxB1 (Kenny *et al.*, 2002), and cadherin (Logan *et al.*, 1999). In other situations the relative positions of these two tissues can be altered, which causes a corresponding shift in the position of the band. For example, treatment of embryos with $NiCl_2$ completely radializes the embryo and produces a ring of aboral ectoderm perpendicular to the A–V axis surrounding the anus (Hardin *et al.*, 1992). In that case the ciliated band forms around the embryo immediately adjacent to the aboral ectoderm ring.

D. The Role of Negative Signals in Restricting Cell Fates

Many regions of the sea urchin embryo maintain relatively broad developmental capacities even during the postmorphogenetic phase of development. This state of conditional specification means that cells in particular regions have started to express tissue-specific genes, but require signals or other environmental cues to restrict developmental potentials and to maintain and complete those programs. Just as experiments involving recombination of early blastomeres may inadvertently imply that there is signaling at their interfaces, so have similar studies created the impression that repressive interactions operate primarily between early blastomeres. Many studies have shown that different tiers of blastomeres up to the 60-cell stage can, in isolation, give rise to more cell types than those predicted from fate maps. For example, the fact that the veg$_2$ tier isolated from the 60-cell embryo can eventually give rise to all cell types of the embryo, including ectoderm and primary mesenchyme (Khaner and Wilt, 1991), does not indicate when repressive signals are sent in the normal embryo that limit fates to mesendoderm. Similarly, the difference in differentiation of intact animal halves of eight-cell embryos versus pairs of mesomeres (Henry *et al.,* 1989) may not be the result of negative lateral interactions among these early blastomeres, but instead may reflect a community effect that comes into play at later times. It is now clear that the sea urchin embryo has an extensive repertoire of negative regulatory mechanisms that persist late into embryogenesis and serve to restrict cell fates. Experimental elimination of these repressive interactions leads to some remarkable late changes of fate. We consider two of the best characterized of these—the signals sent by PMCs to SMCs to prevent them from differentiating as skeletogenic mesenchyme and those that restrict fates of different endodermal subregions.

1. The Secondary Mesenchyme Cell–Primary Mesenchyme Cell Decision

Some SMCs retain the capacity through gastrula stages to transfate to PMCs, but are prevented from doing so by negative signals from the PMCs (Ettensohn and McClay, 1988). Removal of PMCs from the mesenchyme blastula does not prevent embryos from elaborating normal skeletal spicules and normal numbers of functional PMCs. Ettensohn (1990b) has shown that SMCs can "count" PMCs because the number of SMCs that switch fates is directly related to the number of original PMCs removed. Experiments in which PMCs are removed at different times or in which heterochronic chimeras of PMCs and PMC-depleted embryos of different ages are created show that these signals must be sent and received during several hours when the SMCs emigrate from the tip of the archenteron (Ettensohn *et al.,* 1997; Ettensohn and Ruffins, 1993). After that time, the SMCs are no longer competent to receive repressive signals, although these continue to be sent by the PMCs. Thus, in normal embryos SMCs must receive repressive signals from PMCs, which prevent their adopting the PMC fate. Counts of the different SMC types in PMC-depleted embryos suggest that presumptive pigment cells provide the major reservoir of potential converting PMCs (Ettensohn and Ruffins, 1993), which may relate to the fact that pigment cells derive primarily from the most vegetal SMC precursors.

The molecules and signaling pathways responsible for communication between PMCs and SMCs remain to be identified. The idea that signaling molecules freely diffuse through the blastocoel is not appealing. Instead, the fact that PMCs and

SMCs explore their environment with long, thin filopodia (Malinda *et al.*, 1995; Miller *et al.*, 1995) suggests that the signals could be exchanged between these cells in a more intimate fashion.

2. Patterning of the Endoderm

Although fate-mapping experiments show that precursors giving rise to different regions of the gut are present at specific positions in the vegetal plate of the early mesenchyme blastula (Logan and McClay, 1997; Ransick and Davidson, 1998), commitment of these cells to specific fates does not occur until morphogenesis is well underway. For example, when *Lytechinus variegatus* embryos are dissociated into single cells at progressively later stages, significant expression of endoderm-specific markers (LvN1.2 and Endo1) occurs only when dissociation is done after early gastrula stages (Chen and Wessel, 1996). McClay and Logan (1996) showed that specification of different regions of the gut remains conditional at least until the end of gastrulation. Thus, parts can be regenerated after surgical removal and interchanged segments of the gut differentiate according to their new positions and will adopt the fate of their neighbors when transplanted to host embryo archentera. Heterochronic transplants between donor endoderm fragments of varying ages and host archentera of a constant age demonstrate that competence of the donor tissue to regulate to a new endoderm cell type characteristic of the host persists until after archenteron extension is complete. Even more surprising is the finding that when the entire archenteron is removed even late in gastrulation, a new complete tripartite gut is regenerated, along with SMC derivatives that include esophageal muscle cells and coelomic sacs. In these experiments, the mesenchyme and endoderm are thought to regenerate primarily from veg_1, some progeny of which contribute to these tissues in normal embryos. However, it would not be surprising if some animal cell progeny also contributed to the regenerated gut because fate mapping of an_2 blastomeres shows that they can occasionally contribute to endoderm in normal embryos (Sherwood and McClay, 2001). The finding that cells giving rise to endoderm have strong regulative capacities means that the size of endoderm subregions must be controlled in part by negative signals operating even in the late gastrula, which restrict the developmental potencies of neighboring cells in these regions. If ectoderm can also transfate to endoderm in these "gutless wonders" (McClay and Logan, 1996), then it is likely that in the normal embryo endoderm signals to overlying ectoderm not to express its latent endoderm potency.

These experiments suggest that the region containing the vegetal plate/ archenteron and perhaps flanking cells can be described as a classic "field" in which cell fates are assigned by their position within the field. Missing parts of the field can be replaced, even in the extreme case in which only its perimeter remains. A fascinating question that remains to be examined is whether the molecular mechanisms responsible for endoderm regeneration are the same as those that initially are required for its patterning.

IV. Outlook

Extraordinary progress has been made in identifying the major pathways that control cell fate specification along the A–V and O–A axes of the sea urchin embryo. The development of methods to perturb gene function by antisense morpholino or dominant-negative interference, the ability to micromanipulate the embryo, and the relative ease with which gene regulatory mechanisms can be elucidated have made this a powerful system with which to analyze developmental regulatory networks. Differential screens based on knocking out the function of β-catenin, Notch, ATFs, or other regulators are being used to establish epistatic relationships among downstream effectors that include transcription factors and signaling pathways. Because the sea urchin embryo system is so amenable to this kind of approach, detailed understanding of the molecular mechanisms that regulate its patterning can be expected soon.

Acknowledgments

We thank Drs. Charles Ettensohn, David McClay, Athula Wikramanayake, Marta Di Carlo, William Klein, and Giovanni Spinelli for allowing us to reproduce their work here. We have benefited greatly in formulating the ideas presented here from stimulating discussions with these colleagues, with Eric Davidson and members his laboratory, and with recent members of our group, including David Kozlowski, Zheng Wei, Alan Kenny, and Eric Howard. This work was supported by a grant to R.C.A. from the NIH (GM25553).

References

Adelson, D. L., and Humphreys, T. (1988). Sea urchin morphogenesis and cell-hyalin adhesion are perturbed by a monoclonal antibody specific for hyalin. *Development* **104,** 391–402.

Angerer, L. M., and Angerer, R. C. (2000). Animal–vegetal axis patterning mechanisms in the early sea urchin embryo. *Dev. Biol.* **218,** 1–12.

Angerer, L. M., Dolecki, G. J., Gagnon, M. L., Lum, R., Wang, G., Yang, Q., Humphreys, T., and Angerer, R. C. (1989). Progressively restricted expression of a homeo box gene within the aboral ectoderm of developing sea urchin embryos. *Genes Dev.* **3,** 370–383.

Angerer, L. M., Oleksyn, D. W., Logan, C. Y., McClay, D. R., Dale, L., and Angerer, R. C. (2000). A BMP pathway regulates cell fate allocation along the sea urchin animal–vegetal embryonic axis. *Development* **127,** 1105–1114.

Angerer, L. M., Oleksyn, D. W., Levine, A. L., Li, X., Klein, W. H., and Angerer, R. C. (2001). Sea urchin goosecoid links fate specification along the animal–vegetal and oral–aboral embryonic axes. *Development* **128,** 4393–4404.

Burke, R. D. (1978). The structure of the nervous system of the pluteus larva of *Strongylocentrotus purpuratus*. *Cell Tissue Res.* **191,** 233–247.

Burke, R. D., Myers, R. L., Sexton, T. L., and Jackson, C. (1991). Cell movements during the initial phase of gastrulation in the sea urchin embryo. *Dev. Biol.* **146,** 542–557.

Cabrera, C. V., Lee, J. J., Ellison, J. W., Britten, R. J., and Davidson, E. H. (1984). Regulation of cytoplasmic mRNA prevalence in sea urchin embryos: Rates of appearance and turnover for specific sequences. *J. Mol. Biol.* **174,** 85–111.

Cameron, R. A., Fraser, S. E., Britten, R. J., and Davidson, E. H. (1989). The oral–aboral axis of a sea urchin embryo is specified by first cleavage. *Development* **106,** 641–647.

Cameron, R. A., Fraser, S. E., Britten, R. J., and Davidson, E. H. (1990). Segregation of oral from aboral ectoderm precursors is completed at fifth cleavage in the embryogenesis of *Strongylocentrotus purpuratus. Dev. Biol.* **137,** 77–85.

Cameron, R. A., Britten, R. J., and Davidson, E. H. (1993). The embryonic ciliated band of the sea urchin, *Strongylocentrotus purpuratus,* derives from both oral and aboral ectoderm. *Dev. Biol.* **160,** 369–376.

Chen, S. W., and Wessel, G. M. (1996). Endoderm differentiation in vitro identifies a transitional period for endoderm ontogeny in the sea urchin embryo. *Dev. Biol.* **175,** 57–65.

Clevers, H., and van de Wetering, M. (1997). TCF/LEF factors earn their wings. *Trends Genet.* **13,** 485–489.

Coffman, J. A., and Davidson, E. H. (2001). Oral–aboral axis specification in the sea urchin embryo. *Dev. Biol.* **230,** 18–28.

Coffman, J. A., and McClay, D. R. (1990). A hyaline layer protein that becomes localized to the oral ectoderm and foregut of sea urchin embryos. *Dev. Biol.* **140,** 93–104.

Costa, C., Rinaldi, A. M., Romancino, D. P., Cavalcante, C., Vizzini, A., and Di Carlo, M. (1997). Centrifugation does not alter spatial distribution of "BEP4" mRNA in *Paracentrotus lividus* eggs. *FEBS Lett.* **410,** 499–501.

Czihak, G. (1963). Entwicklungsphysiologische Untersuchungen an Echiniden (Verteilung und Bedeutung der Cytochromoxydase). *Wilhelm Roux Arch. Entwicklungsmech. Org.* **154,** 272–292.

Davidson, E. H. (2001). "Genomic Regulatory systems: Development and Evolution." Academic Press, San Diego, CA.

Davidson, E. H., Cameron, R. A., and Ransick, A. (1998). Specification of cell fate in the sea urchin embryo: Summary and some proposed mechanisms. *Development* **125,** 3269–3290.

Davidson, E. H., Rast, J. P., Oliveri, P., Ransick, A., Calestani, C., Yuh, C., *et al.* (2002a). A genomic regulatory network for development. *Science* **295,** 1669–1678.

Davidson, E. H., Rast, J. P., Oliveri, P., Ransick, A., Calestani, C., Yuh, C., *et al* (2002b). A provisional regulatory gene network for specification of endomesoderm in the sea urchin embryo. *Dev. Biol.* **246,** 162–190.

di Bernardo, M., Russo, R., Oliveri, P., Melfi, R., and Spinelli, G. (1995). Homeobox-containing gene transiently expressed in a spatially restricted pattern in the early sea urchin embryo. *Proc. Natl. Acad. Sci. USA* **92,** 8180–8184.

di Bernardo, M., Castagnetti, S., Bellomonte, D., Oliveri, P., Melfi, R., Palla, F., and Spinelli, G. (1999). Spatially restricted expression of PlOtp, a *Paracentrotus lividus Orthopedia*-related homeobox gene, is correlated with oral ectodermal patterning and skeletal morphogenesis in late-cleavage sea urchin embryos. *Development* **126,** 2171–2179.

Eldon, E. D., Angerer, L. M., Angerer, R. C., and Klein, W. H. (1987). Spec3: Embryonic expression of a sea urchin gene whose product is involved in ectodermal ciliogenesis. *Genes Dev.* **1,** 1280–1292.

Emily-Fenouil, F., Ghiglione, C., Lhomond, G., Lepage, T., and Gache, C. (1998). GSK3β/shaggy mediates patterning along the animal–vegetal axis of the sea urchin embryo. *Development* **125,** 2489–2498.

Ernst, S. G., Hough-Evans, B. R., Britten, R. J., and Davidson, E. H. (1980). Limited complexity of the RNA in micromeres of sixteen-cell sea urchin embryos. *Dev. Biol.* **79,** 119–127.

Ettensohn, C. A. (1990a). Cell interactions in the sea urchin embryo studied by fluorescence photoablation. *Science* **248,** 1115–1118.

Ettensohn, C. A. (1990b). The regulation of primary mesenchyme cell patterning. *Dev. Biol.* **140,** 261–271.

Ettensohn, C. A., and Ingersoll, E. P. (1992). "Morphogenesis of the sea urchin embryo." *In* Morphogenesis (E. F. Rossomando and S. Alexander, eds.), Marcel Dekker, Inc., New York.

Ettensohn, C. A., and Malinda, K. M. (1993). Size regulation and morphogenesis: A cellular analysis of skeletogenesis in the sea urchin embryo. *Development* **119,** 155–167.

Ettensohn, C. A., and McClay, D. R. (1988). Cell lineage conversion in the sea urchin embryo. *Dev. Biol.* **125,** 396–409.

Ettensohn, C. A., and Ruffins, S. W. (1993). Mesodermal cell interactions in the sea urchin embryo: Properties of skeletogenic secondary mesenchyme cells. *Development* **117,** 1275–1285.

Ettensohn, C. A., and Sweet, H. C. (2000). Patterning the early sea urchin embryo. *Curr. Top. Dev. Biol.* **50,** 1–44.

Ettensohn, C. A., Guss, K. A., Hodor, P. G., and Malinda, K. M. (1997). The morphogenesis of the skeletal system of the sea urchin embryo. *In:* "Reproductive Biology of Invertebrates" (J. R. Collier, ed.), Vol. VII, pp. 225–265. Oxford & IBH, Oxford.

Ghiglione, C., Lhomond, G., Lepage, T., and Gache, C. (1993). Cell-autonomous expression and position-dependent repression by Li^+ of two zygotic genes during sea urchin early development. *EMBO J.* **12,** 87–96.

Ghiglione, C., Emily-Fenouil, F., Lhomond, G., and Gache, C. (1997). Organization of the proximal promoter of the hatching-enzyme gene, the earliest zygotic gene expressed in the sea urchin embryo. *Eur. J. Biochem.* **250,** 502–513.

Grimwade, J. E., Gagnon, M. L., Yang, Q., Angerer, R. C., and Angerer, L. M. (1991). Expression of two mRNAs encoding EGF-related proteins identifies subregions of sea urchin embryonic ectoderm. *Dev. Biol.* **143,** 44–57.

Hardin, J., Coffman, J. A., Black, S. D., and McClay, D. R. (1992). Commitment along the dorsoventral axis of the sea urchin embryo is altered in response to $NiCl_2$. *Development* **116,** 671–685.

Hardin, P. E., Angerer, L. M., Hardin, S. H., Angerer, R. C., and Klein, W. H. (1988). *Spec2* genes of *Strongylocentrotus purpuratus:* Structure and differential expression in embryonic aboral ectoderm cells. *J. Mol. Biol.* **202,** 417–431.

Henry, J. J., Amemiya, S., Wray, G. A., and Raff, R. A. (1989). Early inductive interactions are involved in restricting cell fates of mesomeres in sea urchin embryos. *Dev. Biol.* **136,** 140–153.

Hodor, P. G., Illies, M. R., Broadley, S., and Ettensohn, C. A. (2000). Cell–substrate interactions during sea urchin gastrulation: Migrating primary mesenchyme cells interact with and align extracellular matrix fibers that contain ECM3, a molecule with NG2-like and multiple calcium-binding domains. *Dev. Biol.* **222,** 181–194.

Höög, C., Calzone, F. J., Cutting, A. E., Britten, R. J., and Davidson, E. H. (1991). Gene regulatory factors of the sea urchin embryo. II. Two dissimilar proteins, P3A1 and P3A2, bind to the same target sites that are required for early territorial gene expression. *Development* **112,** 351–364.

Hörstadius, S. (1935). Über die determination in verlaufe der eiachse bei seeigeln. *Pubbl. Stn. Zool. NapoliII* **14,** 251.

Hörstadius, S. (1936). *Wilhelm Roux Arch. Entwicklungsmech. Org.* **138,** 197.

Hörstadius, S. (1939). The mechanics of sea urchin development, studied by operative methods. *Biol. Rev. Cambr. Philos. Soc.* **14,** 132–179.

Hörstadius, S. (1973). "Experimental Embryology of Echinoderms." Clarendon Press, Oxford.

Howard, E. W., Newman, L. A., Oleksyn, D. W., Angerer, R. C., and Angerer, L. M. (2001). *SpKrl:* A direct target of β-catenin regulation required for endoderm differentiation in sea urchin embryos. *Development* **128,** 365–375.

Huang, L., Li, X., El-Hodiri, H. M., Dayal, S., Wikramanayake, A. H., and Klein, W. H. (2000). Involvement of Tcf/Lef in establishing cell types along the animal–vegetal axis of sea urchins. *Dev. Genes. Evol.* **210,** 73–81.

Hurley, D. L., Angerer, L. M., and Angerer, R. C. (1989). Altered expression of spatially regulated embryonic genes in the progeny of separated sea urchin blastomeres. *Development* **106,** 567–579.

Ishizuka, Y., Minokawa, T., and Amemiya, S. (2001). Micromere descendants at the blastula stage are involved in normal archenteron formation in sea urchin embryos. *Dev. Genes. Evol.* **211,** 83–88.

Kenny, A. P., Kozlowski, D. J., Oleksyn, D. W., Angerer, L. M., and Angerer, R. C. (1999). SpSoxB1, a maternally encoded transcription factor asymmetrically distributed among early sea urchin blastomeres. *Development* **126,** 5473–5483.

Kenny, A. P., Oleksyn, D. W., Angerer, R. C., and Angerer, L. M. (2002). Sox transcription factors are required for gastrulation of the sea urchin embryo. In preparation.

Khaner, O., and Wilt, F. (1990). The influence of cell interactions and tissue mass on differentiation of sea urchin mesomeres. *Development* **109,** 625–634.

Khaner, O., and Wilt, F. (1991). Interactions of different vegetal cells with mesomeres during early stages of sea urchin development. *Development* **112,** 881–890.

Kirchhamer, C. V., and Davidson, E. H. (1996). Spatial and temporal information processing in the sea urchin embryo: Modular and intramodular organization of the *CyIIIa* gene cis-regulatory system. *Development* **122,** 333–348.

Kitamura, K., Nishimura, Y., Kubotera, N., Higuchi, Y., and Yamaguchi, M. (2002). Transient activation of the micro1 homeobox gene family in the sea urchin (*Hemicentrotus pulcherrimus*) micromere. *Dev. Genes Evol.* **212,** 1–10.

Kominami, T. (1988). Determination of dorso–ventral axis in early embryos of the sea urchin, *Hemicentrotus pulcherrimus*. *Dev. Biol.* **127,** 187–196.

Kozlowski, D. J., Gagnon, M. L., Marchant, J. K., Reynolds, S. D., Angerer, L. M., and Angerer, R. C. (1996). Characterization of a *SpAN* promoter sufficient to mediate correct spatial regulation along the animal–vegetal axis of the sea urchin embryo. *Dev. Biol.* **176,** 95–107.

Lepage, T., Ghiglione, C., and Gache, C. (1992a). Spatial and temporal expression pattern during sea urchin embryogenesis of a gene coding for a protease homologous to the human protein BMP-1 and to the product of the *Drosophila* dorsal-ventral patterning gene tolloid. *Development* **114,** 147–163.

Lepage, T., Sardet, C., and Gache, C. (1992b). Spatial expression of the hatching enzyme gene in the sea urchin embryo. *Dev. Biol.* **150,** 23–32.

Li, X., Chuang, C. K., Mao, C. A., Angerer, L. M., and Klein, W. H. (1997). Two Otx proteins generated from multiple transcripts of a single gene in *Strongylocentrotus purpuratus*. *Dev. Biol.* **187,** 253–266.

Li, X., Bhattacharya, C., Dayal, S., Maity, S., and Klein, W. H. (2002). Ectoderm gene activation in sea urchin embryos mediated by the CCAAT-binding factor. *Differentiation* **70,** 109–119.

Li, X., Wikramanayake, A. H., and Klein, W. H. (1999). Requirement of SpOtx in cell fate decisions in the sea urchin embryo and possible role as a mediator of β-catenin signaling. *Dev. Biol.* **212,** 425–439.

Li, Z., Kalasapudi, S. R., and Childs, G. (1993). Isolation and characterization of cDNAs encoding the sea urchin (Strongylocentrotus purpuratus) homologue of the CCAAT binding protein NF-Y A subunit. *Nucleic Acids Res.* **21,** 4639.

Livingston, B. T., and Wilt, F. H. (1989). Lithium evokes expression of vegetal-specific molecules in the animal blastomeres of sea urchin embryos. *Proc. Natl. Acad. Sci. USA* **86,** 3669–3673.

Logan, C. Y., and McClay, D. R. (1997). The allocation of early blastomeres to the ectoderm and endoderm is variable in the sea urchin embryo. *Development* **124,** 2213–2223.

Logan, C. Y., and McClay, D. R. (1998). The lineages that give rise to the endoderm and the mesoderm in the sea urchin embryo. *In:* "Cell Fate and Lineage Determination." Academic Press, San Diego, CA.

Logan, C. Y., Miller, J. R., Ferkowicz, M. J., and McClay, D. R. (1999). Nuclear β-catenin is required to specify vegetal cell fates in the sea urchin embryo. *Development* **126,** 345–357.

Malinda, K. M., and Ettensohn, C. A. (1994). Primary mesenchyme cell migration in the sea urchin embryo: Distribution of directional cues. *Dev. Biol.* **164,** 562–578.

Malinda, K. M., Fisher, G. W., and Ettensohn, C. A. (1995). Four-dimensional microscopic analysis of the filopodial behavior of primary mesenchyme cells during gastrulation in the sea urchin embryo. *Dev. Biol.* **172,** 552–566.

Mao, C. A., Wikramanayake, A. H., Gan, L., Chuang, C. K., Summers, R. G., and Klein, W. H. (1996). Altering cell fates in sea urchin embryos by overexpressing SpOtx, an orthodenticle-related protein. *Development* **122,** 1489–1498.

Maruyama, Y. K., Nakaseko, Y., and Yagi, S. (1985). Localization of cytoplasmic determinants responsible for primary mesenchyme formation and gastrulation in the unfertilized egg of the sea urchin *Hemicentrotus pulcherrimus*. *J. Exp. Zool.* **236,** 155–163.

McClay, D. R. (2000). Specification of endoderm and mesoderm in the sea urchin. *Zygote* **8,** S41.

McClay, D. R., and Logan, C. Y. (1996). Regulative capacity of the archenteron during gastrulation in the sea urchin. *Development* **122,** 607–616.

McClay, D. R., Peterson, R. E., Range, R. C., Winter-Vann, A. M., and Ferkowicz, M. J. (2000). A micromere induction signal is activated by β-catenin and acts through Notch to initiate specification of secondary mesenchyme cells in the sea urchin embryo. *Development* **127,** 5113–5122.

Miller, J., Fraser, S. E., and McClay, D. (1995). Dynamics of thin filopodia during sea urchin gastrulation. *Development* **121,** 2501–2511.

Minokawa, T., and Amemiya, S. (1999). Timing of the potential of micromere descendants in echinoid embryos to induce endoderm differentiation of mesomere descendants. *Dev. Growth Differ.* **41,** 535–547.

Montana, G., Bonura, A., Romancino, D. P., Sbisa, E., and Di Carlo, M. (1997). A 54-kDa protein specifically associates the 3′ untranslated region of three maternal mRNAs with the cytoskeleton of the animal part of the *Paracentrotus lividus* egg. *Eur. J. Biochem.* **247,** 183–189.

Oliveri, P., Carrick, D. M., and Davidson, E. H. (2002). A regulatory gene network that directs micromere specification in the sea urchin embryo. *Dev. Biol.* **246,** 209–228.

Ramirez-Weber, F. A., and Kornberg, T. B. (1999). Cytonemes: Cellular processes that project to the principal signaling center in *Drosophila* imaginal discs. *Cell* **97,** 599–607.

Ransick, A., and Davidson, E. H. (1993). A complete second gut induced by transplanted micromeres in the sea urchin embryo. *Science* **259,** 1134–1138.

Ransick, A., and Davidson, E. H. (1995). Micromeres are required for normal vegetal plate specification in sea urchin embryos. *Development* **121,** 3215–3222.

Ransick, A., and Davidson, E. H. (1998). Late specification of Veg1 lineages to endodermal fate in the sea urchin embryo. *Dev. Biol.* **195,** 38–48.

Ransick, A., Rast, J. P., Miokawa, T., Calestani, C., and Davidson, E. H. (2002). New early zygotic regulators expressed in endomesoderm of sea urchin embryos discovered by differential array hybridization. *Dev. Biol.* **246,** 132–147.

Reynolds, S. D., Angerer, L. M., Palis, J., Nasir, A., and Angerer, R. C. (1992). Early mRNAs, spatially restricted along the animal–vegetal axis of sea urchin embryos, include one encoding a protein related to tolloid and BMP-1. *Development* **114,** 769–786.

Rodgers, W. H., and Gross, P. R. (1978). Inhomogeneous distribution of egg RNA sequences in the early embryo. *Cell* **14,** 279–288.

Romancino, D. P., and di Carlo, M. (1999). Asymmetrical localization and segregation of *Paracentrotus lividus* bep4 maternal protein. *Mech. Dev.* **87,** 3–9.

Romancino, D. P., Montana, G., and di Carlo, M. (1998). Involvement of the cytoskeleton in localization of *Paracentrotus lividus* maternal BEP mRNAs and proteins. *Exp. Cell. Res.* **238,** 101–109.

Romancino, D. P., Montana, G., Dalmazio, S., and di Carlo, M. (2001). Bep4 protein is involved in patterning along the animal–vegetal axis in the *Paracentrotus lividus* embryo. *Dev. Biol.* **234,** 107–119.

Ruffins, S. W., and Ettensohn, C. A. (1996). A fate map of the vegetal plate of the sea urchin (*Lytechinus variegatus*) mesenchyme blastula. *Development* **122,** 253–263.

Schroeder, T. E. (1980). Expressions of the prefertilization polar axis in sea urchin eggs. *Dev. Biol.* **79,** 428–443.

Sherwood, D. R., and McClay, D. R. (1997). Identification and localization of a sea urchin Notch homologue: Insights into vegetal plate regionalization and Notch receptor regulation. *Development* **124,** 3363–3374.

Sherwood, D. R., and McClay, D. R. (1999). LvNotch signaling mediates secondary mesenchyme specification in the sea urchin embryo. *Development* **126,** 1703–1713.

Sherwood, D. R., and McClay, D. R. (2001). LvNotch signaling plays a dual role in regulating the position of the ectoderm–endoderm boundary in the sea urchin embryo. *Development* **128,** 2221–2232.

Stephens, L., Kitajima, T., and Wilt, F. (1989). Autonomous expression of tissue-specific genes in dissociated sea urchin embryos. *Development* **107,** 299–307.

Sweet, H. C., Hodor, P. G., and Ettensohn, C. A. (1999). The role of micromere signaling in Notch activation and mesoderm specification during sea urchin embryogenesis. *Development* **126,** 5255–5265.

Sweet, H. C., Gehring, M., and Ettensohn, C. A. (2002). LvDelta is a mesoderm-inducing signal in the sea urchin embryo and can endow blastomeres with organizer-like properties. *Development* **129,** 1945–1955.

Tan, H., Ransick, A., Wu, H., Dobias, S., Liu, Y. H., and Maxson, R. (1998). Disruption of primary mesenchyme cell patterning by misregulated ectodermal expression of *SpMsx* in sea urchin embryos. *Dev. Biol.* **201,** 230–246.

Vonica, A., Weng, W., Gumbiner, B. M., and Venuti, J. M. (2000). TCF is the nuclear effector of the β-catenin signal that patterns the sea urchin animal–vegetal axis. *Dev. Biol.* **217,** 230–243.

Wang, W., Wikramanayake, A. H., Gonzalez-Rimbau, M., Vlahou, A., Flytzanis, C. N., and Klein, W. H. (1996). Very early and transient vegetal-plate expression of *SpKrox1*, a Krüppel/Krox gene from *Stronglyocentrotus purpuratus*. *Mech. Dev.* **60,** 185–195.

Wei, Z., Angerer, L. M., Gagnon, M. L., and Angerer, R. C. (1995). Characterization of the *SpHE* promoter that is spatially regulated along the animal–vegetal axis of the sea urchin embryo. *Dev. Biol.* **171,** 195–211.

Wei, Z., Angerer, L. M., and Angerer, R. C. (1997). Multiple positive cis elements regulate the asymmetric expression of the *SpHE* gene along the sea urchin embryo animal vegetal axis. *Dev. Biol.* **187,** 71–78.

Wei, Z., Angerer, L. M., and Angerer, R. C. (1999a). Spatially regulated SpEts4 transcription factor activity along the sea urchin embryo animal–vegetal axis. *Development* **126,** 1729–1737.

Wei, Z., Angerer, R. C., and Angerer, L. M. (1999b). Identification of a new sea urchin ets protein, SpEts4, by yeast one-hybrid screening with the hatching enzyme promoter. *Mol. Cell. Biol.* **19,** 1271–1278.

Wikramanayake, A. H., and Klein, W. H. (1997). Multiple signaling events specify ectoderm and pattern the oral–aboral axis in the sea urchin embryo. *Development* **124,** 13–20.

Wikramanayake, A. H., Brandhorst, B. P., and Klein, W. H. (1995). Autonomous and non-autonomous differentiation of ectoderm in different sea urchin species. *Development* **121,** 1497–1505.

Wikramanayake, A. H., Huang, L., and Klein, W. H. (1998). β-Catenin is essential for patterning the maternally specified animal–vegetal axis in the sea urchin embryo. *Proc. Natl. Acad. Sci. USA* **95,** 9343–9348.

Yang, Q., Kingsley, P. D., Kozlowski, D. J., Angerer, R. C., and Angerer, L. M. (1993). Immunochemical analysis of arylsulfatase accumulation in sea urchin embryos. *Dev. Growth Differ.* **35,** 139–151.

Yuh, C. H., and Davidson, E. H. (1996). Modular cis-regulatory organization of *Endo16*, a gut-specific gene of the sea urchin embryo. *Development* **122,** 1069–1082.

Yuh, C. H., Bolouri, H., and Davidson, E. H. (2001). Cis-regulatory logic in the *endo16* gene: Switching from a specification to a differentiation mode of control. *Development* **128,** 617–629.

Zeller, R. W., Britten, R. J., and Davidson, E. H. (1995). Developmental utilization of SpP3A1 and SpP3A2: Two proteins which recognize the same DNA target site in several sea urchin gene regulatory regions. *Dev. Biol.* **170,** 75–82.

Zito, F., Nakano, E., Sciarrino, S., and Matranga, V. (2000). Regulative specification of ectoderm in skeleton disrupted sea urchin embryos treated with monoclonal antibody to Pl-nectin. *Dev. Growth. Differ.* **42,** 499–506.

Zorn, A. M., Barish, G. D., Williams, B. O., Lavender, P., Klymkowsky, M. W., and Varmus, H. E. (1999). Regulation of Wnt signaling by Sox proteins: XSox17 α/β and XSox3 physically interact with β-catenin. *Mol. Cell* **4,** 487–498.

Index

PS, *see* Parasegment
Ptc, see Patched
PTHrP, *see* Parathyroid hormone-related peptide

R

Replication, asynchronous, genomic imprinting,
  129–130
Retinal pigment epithelium, Hedgehog
  signaling, 74
rhomboid, 62
RPE, *see* Retinal pigment epithelium

S

Sclerotome, Hedgehog signaling, 54–56
Sea urchin embryo
  morphogenesis
    ciliated band formation, 188–189
    ectoderm signaling, 186–188
    endoderm–ectoderm border, 185–186
    negative signals in cell fate, 189–191
    skeletogenesis, 186–188
  patterning
    animal–vegetal axis, 161
    A–V axis, maternal mechanisms, 164–167
    oral–aboral axis, 164
    overview, 160–161
  premorphogenetic phase, cell fate
    gene activators, 176–178
    maternal ectoderm domains, 171–175
    maternal mesendoderm domain, 175–176
    maternal nonvegetal domain, 169–171
    mesendoderm patterning, 178–182
    nonvegetal domain, 167–169
    pre-AoeD respecification, 182–185
Secondary mesenchymal cells, sea urchin,
  180–181, 190–191
Shh, see Sonic hedgehog
Skeletogenesis, sea urchin morphogenesis,
  186–188
SLOS, *see* Smith–Lemli–Opitz syndrome
SMCs, *see* Secondary mesenchymal cells
Smith–Lemli–Opitz syndrome, 78–79
SMOH, cancer, 81–82
Somatic stem cells, 16–17
Sonic hedgehog
  circulatory system, 14
  face and head, 15–16
  gut development, 20–21
  hair and feather morphogenesis, 23–26

Hedgehog signaling, 6–7
human congenital malformation, 76, 78
kidney and adrenal gland, 27
lateral asymmetry, 27–32
lungs and trachea, 43–45
mammary gland, 46
mature neurons, 13
myogenesis, 46–49
neural crest, 50
pancreas development, 50–52
pituitary gland, 53
proliferation and survival, 12–13
prostate gland, 54
sclerotome, 55–56
tongue and taste, 68
tooth, 69–70
vertebrate limb and fin, 32–43
visual system, 71
Sonic you, Hedgehog signaling, 74
SpGsc, 182–183
SpKrl, 176–178
Spleen, mouse hematopoiesis, 144
SSC, *see* Somatic stem cells
Survival, Hedgehog signaling, 12–13
syu, see Sonic you

T

Taste, Hedgehog signaling, 67–68
TCF-Lef, sea urchin cell fate, 177–178
Tongue, Hedgehog signaling, 67–68
Tooth, Hedgehog signaling, 68–70
Trachea, Hedgehog signaling, 43–45
Tumorigenesis, Hedgehog signaling, 82–83

V

VBI, *see* Ventral blood island
VEB, *see* Very early blastula genes
Ventral blood island, *Xenopus* hematopoiesis,
  148–149
Vertebrate fin, Hedgehog signaling, 32–43
Vertebrate limb, Hedgehog signaling, 32–43
Very early blastula genes, sea urchin cell fate,
  167–168
Visual system, Hedgehog signaling, 70–75

W

Wg, see Wingless
Wingless, Hedgehog signaling, 61–62, 64

Contents of Previous Volumes